MCP Mathematics

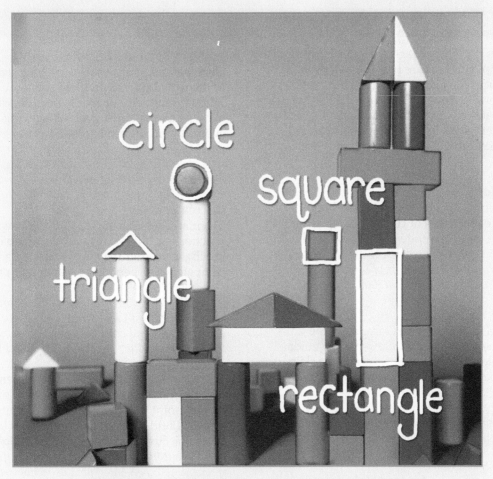

circle

square

triangle

rectangle

Richard Monnard • Royce Hargrove

TEACH THE COVER: What shapes can you find in the blocks?

ISBN 0-7652-6055-7
Printed in the United States of America
6 7 8 9 10 V0CR 14 13 12

Project Staff
Art & Design: Robert Dobaczewski, Kathleen Ellison, Senja Lauderdale, David Mager, Jim O'Shea, Angel Weyant
Editorial: Stephanie P. Cahill, Gina Dalessio, Phyllis Dunsay, Mary Ellen Gilbert, Dena Kennedy, Theresa McCarthy, Marilyn Sarch
Marketing: Doug Falk, Clare Harrison
Production/Manufacturing: Irene Belinsky, Lawrence Berkowitz, Louis Campos, Diane Fristachi, Pamela Gallo, Leslie Greenberg, Suellen Leavy, Ruth Leine, Karyn Mueller, Michele Uhl
Publishing Operations: Carolyn Coyle, Richetta Lobban

1-800-321-3106
www.pearsonlearning.com

MCP Mathematics

About the Program

This comprehensive math program will help students in Grades K to 6 develop a solid mathematics background. It is designed to encourage critical-thinking skills, active participation, and mastery of skills within the context of problem-solving situations. The program's developmental sequence introduces and extends skills taught in most mathematics curricula, such as number sense, operations, algebra, geometry, data collection and analysis, logic, and probability.

A Research-Based Approach

The program offers a strong pedagogical approach that is research based. First, students are provided a developed model that introduces the lesson concept. Then, students are given guided practice opportunities to get started. Next, abundant practice is provided for mastery. Finally, students can apply their skills to problem-solving and enrichment activities. An overview of this approach follows.

Direct Instruction

Each lesson begins with a developed model that demonstrates the algorithm or concept in a problem-solving situation and gets students actively involved in the situation. . . . *Students taught using direct instruction were found to perform better on tests of computation and year-end tests (Crawford and Snider 2000).* *

Guided Practice

A Getting Started section provides samples of the concept or skill being taught and allows you to guide and observe students as they begin to apply the skills learned. *When low-achieving students were taught using such direct instruction methods as . . . guided practice, . . . they showed improved mastery of basic skills, solved computation and word problems correctly in less time, and had higher self-ratings of academic motivation (Kame´enui, Carnine, Darch, and Stein 1986; Din 1998; Ginsburg-Block and Fantuzzo 1998).* *

Independent Practice

The Practice section can be used to develop and master basic skills by allowing students to independently practice the algorithms and to apply learning from the lesson or from previous lessons. *Research indicates that providing students with extended practice appears to serve two purposes: re-teaching of the skill for students who had not yet mastered it and relating of the previously learned skill to new skills, resulting in the formation of interrelationships among concepts that improved retention and yielded higher achievement test results (Hett 1989).* *

Higher-Order Thinking Skills

A collection of real-life word problems in the Problem Solving section provides application opportunities related to the lesson concepts. *Frequent practice with word problems is associated with higher-order skill development (Coy 2001). This finding is especially true when the word problems present familiar real life situations (Coy 2001).* *

Problem-Solving Strategies

Once per chapter, a special lesson introduces students to the techniques of problem solving using Polya's four-step model. The Apply activities in these lessons allow students to use problem-solving strategies in everyday situations. *Students who received instruction in problem-solving processes showed better performance on tests of skills, tasks, and problem solving, as well as on a measure of the transfer of learning (Durnin et al. 1997).* *

Calculator Usage

Calculator lessons in Grades 3 to 6 teach students the functions and the skills needed to use calculators intelligently after they've developed a foundation of competence in pencil-and-paper computation. *. . . Students who used calculators in mathematics instruction had more positive attitudes toward mathematics and a higher math self-concept. . . . A special curriculum developed around calculators has been shown to improve mathematics achievement (Hembree and Dessart 1986).* *

Frequent and Cumulative Assessment

Chapter Tests provide both students and teachers with a checkpoint that tests all the skills taught in a chapter. You will find Alternate Chapter Tests based on the same objectives in the Teacher's Edition. *Assessment has been found to be most effective when it is a frequent and well-integrated aspect of mathematics instruction (Brosnan and Hartog 1993).* Cumulative Assessments maintain skills that have been taught in all previous chapters. A standardized test format is used starting at the middle of the second-grade text. The Teacher's Edition also contains Alternate Cumulative Assessments. *. . . Frequent cumulative tests result in higher levels of achievement than do infrequent tests or tests related only to content since the last test (Dempster 1991).* *

Remedying Common Errors

The comprehensive Teacher's Edition provides abundant additional help for teachers, including a four-step lesson plan that walks the teacher through the lesson, pointing out common errors and providing intervention strategies. *Curricula with an error correction component were found to result in higher scores for computation, math concepts, and problem solving (Stefanich and Rokusek 1992; Crawford and Snider 2000).* *

*Research compiled by PRES Associates, Inc. (2004). Research Support for MCP Mathematics (unpublished).

Using the Student Edition

Use Each Page of a Lesson for Direct Instruction and Practice

Each two-page lesson focuses on one main objective. The pages begin with a developed model that students can actively work on as you walk them through it. This is followed by practice of the lesson's objective. You can begin the process of individual mastery by assigning practice exercises that students can work on independently.

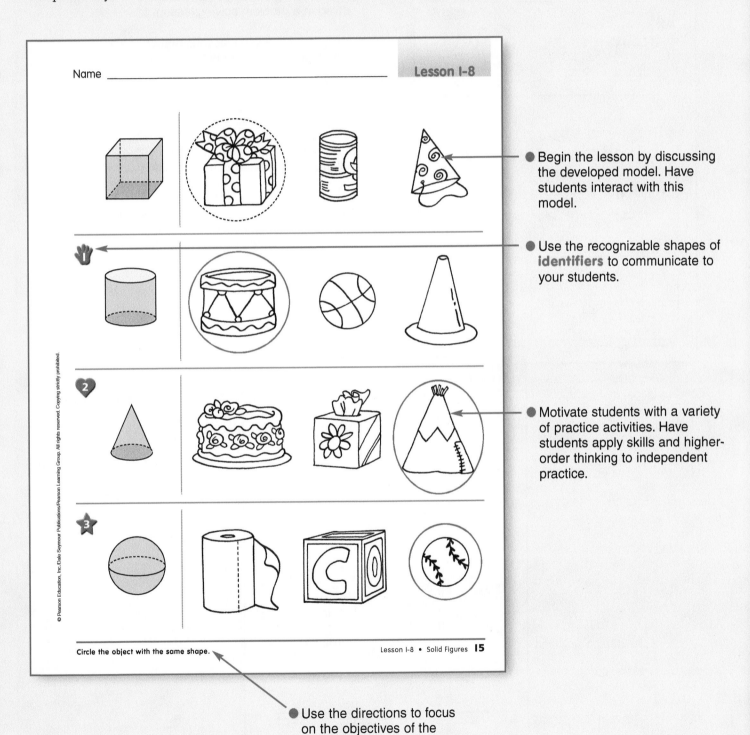

Name _____   Lesson 1-8

● Begin the lesson by discussing the developed model. Have students interact with this model.

● Use the recognizable shapes of **identifiers** to communicate to your students.

● Motivate students with a variety of practice activities. Have students apply skills and higher-order thinking to independent practice.

Circle the object with the same shape.   Lesson 1-8 • Solid Figures **15**

● Use the directions to focus on the objectives of the lesson page.

Using the Student Edition

Teach Problem Solving and Apply Skills

Problem-solving lessons draw on chapter concepts and skills to highlight key problem-solving strategies. The Chapter Challenge provides practice, applies new skills to known content, and encourages higher-order thinking.

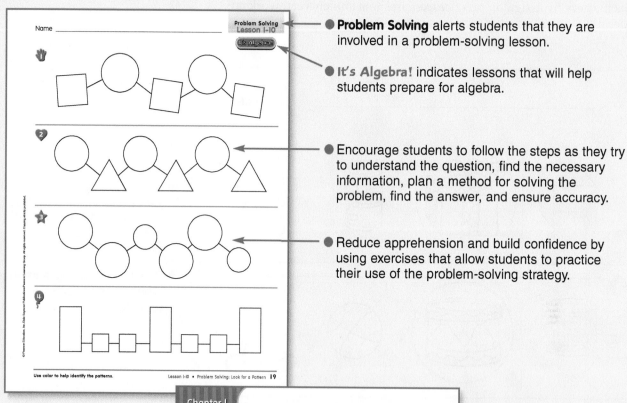

● **Problem Solving** alerts students that they are involved in a problem-solving lesson.

● **It's Algebra!** indicates lessons that will help students prepare for algebra.

● Encourage students to follow the steps as they try to understand the question, find the necessary information, plan a method for solving the problem, find the answer, and ensure accuracy.

● Reduce apprehension and build confidence by using exercises that allow students to practice their use of the problem-solving strategy.

● Extend chapter skills and make learning concepts fun with interesting **Challenge** activities.

Using the Teacher's Edition

Plan Ahead Using the Chapter Opener

Save time and ease lesson planning by using this overview of useful chapter information and activities.

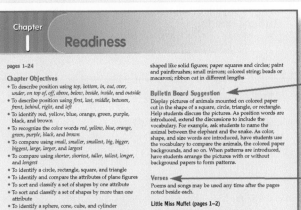

Chapter 1 — Readiness

pages 1–24

Chapter Objectives
- To describe position using *top, bottom, in, out, over, under, on top of, off, above, below, beside, inside,* and *outside*
- To describe position using *first, last, middle, between, front, behind, right,* and *left*
- To recognize *red, yellow, blue, orange, green, purple, black,* and *brown*
- To recognize the color words *red, yellow, blue, orange, green, purple, black,* and *brown*
- To compare using *small, smaller, smallest, big, bigger, biggest, large, larger,* and *largest*
- To compare using *shorter, shortest, taller, tallest, longer,* and *longest*
- To identify a circle, rectangle, square, and triangle
- To identify and compare the attributes of plane figures
- To sort and classify a set of shapes by one attribute
- To sort and classify a set of shapes by more than one attribute
- To identify a sphere, cone, cube, and cylinder
- To identify and compare the attributes of solid figures
- To match objects to outlines of their shapes
- To recognize symmetry in the environment
- To identify shapes with a line of symmetry
- To identify the matching parts of a symmetrical shape
- To identify patterns using color, size, or shape
- To complete a pattern
- To identify the object that comes next in a pattern

Materials
You will need the following:
*large cardboard box; *hand stamp or stickers; *yarn or string; *3 boxes with lids; *several color name cards for each color; *tape; *colored construction paper; *containers; *paper bag; *magazines; *blue and green blocks; *2 different-sized tissue boxes; *large and small colored blocks

Students will need the following:
Reproducible Masters RM1–RM4; container with a lid; paper clips; crayons; large and small blocks in different colors; sticks; blocks; magazines or catalogs; small objects of different colors; objects of various sizes; 12-in. ruler; yardstick; crayons in assorted lengths; picture books in assorted sizes; cardboard circles, squares, triangles, and rectangles; objects that are the same color; attribute blocks; 2 yarn loops; set of animal pictures; classroom objects shaped like plane shapes; large models of solid figures; modeling clay; building blocks shaped like spheres, cubes, cylinders, and cones; classroom objects

*Indicates teacher demonstration materials

T1a

shaped like solid figures; paper squares and circles; paint and paintbrushes; small mirrors; colored string; beads or macaroni; ribbon cut in different lengths

Bulletin Board Suggestion
Display pictures of animals mounted on colored paper cut in the shape of a square, circle, triangle, or rectangle. Help students discuss the pictures. As position words are introduced, extend the discussions to include the vocabulary. For example, ask students to name the animal between the elephant and the snake. As color, shape, and size words are introduced, have students use the vocabulary to compare the animals, the colored paper backgrounds, and so on. When patterns are introduced, have students arrange the pictures with or without background papers to form patterns.

Verses
Poems and songs may be used any time after the pages noted beside each.

Little Miss Muffet (pages 1–2)

Little Miss Muffet sat on her tuffet
Eating her curds and whey.
Along came a spider and sat down beside her
And frightened Miss Muffet away.

Hoot in the Chute (pages 1–2)

Is there a creature in our laundry chute?
Surely there is! I hear its hoot!
"Oh, no, you don't! Just look at this!
It can't be washed the way it is!
It's wrong side out! It's right side in!
Come on now! Please! You must begin
To turn your clothes all right side out!
Should you forget, you'll hear me shout!
Right side out! Right side out!
I'll not accept the wrong side out!
It's right side out! Not right side in!
That's outside out! Not outside in!"

Statue on the Curb (pages 3–4)

We're crossing the street
Watch out and beware!
The cars, trucks, and buses
Are all going somewhere!
We stop at the corner,
Look left and look right;
Then step off the curb
If nothing's in sight.
But if there's a car
Or truck down the way,
Pretend we are statues
And on the curb, stay!

• Use the **Bulletin Board Suggestion** to create a theme-related display that can be completed lesson by lesson.

• Motivate students and reinforce mathematics concepts by using instructional poems and songs.

• All **Materials** needed for chapter lessons are listed collectively. Demonstration items are marked with asterisks.

Little Spot's Surprise

Little Spot, the ladybug, had a cold. Everyone loved Little Spot and wanted to do something to help, but they didn't know what to do.

"She catches colds because she lives in that drafty old tree," said Mrs. Robin. "She should move over behind that big rock. It would keep the wind off of her. But she loves that big old drafty tree."

"She needs a warm little house," said Mr. Inchworm. "Why don't we build Little Spot a nice warm house to live in?"

"Yes," said Mr. Cutworm. "After all, Little Spot has always been nice to everyone. She wants all of us to be friends."

"I don't know anything about building houses," said Mrs. Robin.

"Well, I don't either," said Mr. Cutworm, "but if we all do what we do best, I think we'll get it done."

Ms. Spider, who hardly ever agreed with anyone, said, "I think he's right. And I would like to help. Little Spot is always nice to me, even when the rest of you aren't!"

When Little Spot's friends went to tell her the good news, she was curled up fast asleep under a leaf. So they just let her sleep and began their work.

Mr. Inchworm measured everything to fit. Mr. Cutworm busily cut the leaves and twigs exactly where Mr. Inchworm had measured them. Mrs. Robin gathered grass and straw. Ms. Spider quickly spun a web in and out of the leaves, twigs, grass, and straw. She fastened it all together as quickly as each new piece was put on.

The friends worked very quickly and very quietly. They didn't want to wake Little Spot. Soon, however, it began to grow dark. So several of Little Spot's firefly friends came over and lit the branch with light.

When the house was finished, Mrs. Robin gently woke Little Spot.

"Little Spot, dear, get up. We have a surprise for you."

"What?" asked Little Spot. "A surprise for me?"

"Yes," said Ms. Spider. "We decided to build you a nice warm house. We hope you like it."

"Oh, I love it!" said Little Spot. And it made her feel much, much better just to see how her friends had worked together. Little Spot knew her friends loved her as she loved them.

"Little Spot's Surprise," p. RM1

Activities
Use Reproducible Master RM1, "Little Spot's Surprise," to help students relate the story to the mathematical ideas presented.

• Help students realize the need to work together when a task is time-consuming, difficult, or impossible for one person to do alone. Hold up a long jump rope and tell students that you will hold it a distance off the floor for them to jump over it. As you try to hold the jump rope a few inches off the floor, students should notice that you need another person to hold the other end.

• Have students use building blocks or similar materials to build a house or other structure. Tell students that, as a class, they are to build the structure using the given materials. Divide the class into small groups. Assign each group a part of the construction to complete, such as the foundation, frame, roof, or landscaping.

T1b

• Introduce the chapter with a story that uses appealing characters and prompts students to think about mathematics in applied situations.

• Ease classroom management by using a facsimile of the accompanying **reproducible master** provided at point of use.

• Prompt students to prepare for the chapter ahead by using one or more of the story-related activities.

Using the Teacher's Edition

Instruct Using a Four-Step Lesson Plan

The Teacher's Edition is designed and organized with you in mind. Consistent four-step lesson plans on two pages will help you make efficient use of your planning time and will help you deliver effective instruction to meet diverse student needs.

- Ease classroom management with reduced student pages that provide point-of-use information and answers to exercises.

- Use **Getting Started** to review the **Vocabulary** and **Objectives** and to set a clear course for the lesson goal.

- Reduce class preparation time by gathering **Materials** early.

- Use the **Warm Up** exercises to help students brush up on skills at the beginning of each day's lesson.

- **Teach** gives practical suggestions for introducing the objective and developing the skill. You will find specific suggestions for a variety of effective presentation models. Most include ideas for the use of manipulatives.

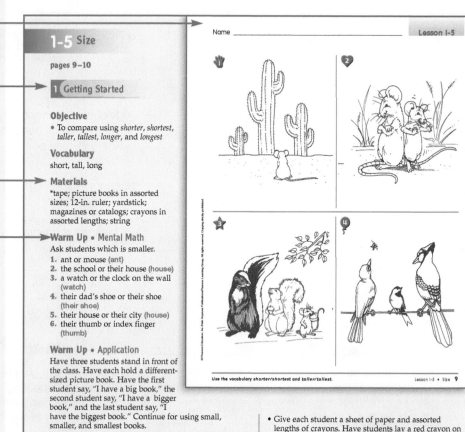

1-5 Size
pages 9–10

1 Getting Started

Objective
- To compare using *shorter, shortest, taller, tallest, longer,* and *longest*

Vocabulary
short, tall, long

Materials
*tape; picture books in assorted sizes; 12-in. ruler; yardstick; magazines or catalogs; crayons in assorted lengths; string

Warm Up • Mental Math
Ask students which is smaller.
1. ant or mouse (ant)
2. the school or their house (house)
3. a watch or the clock on the wall (watch)
4. their dad's shoe or their shoe (their shoe)
5. their house or their city (house)
6. their thumb or index finger (thumb)

Warm Up • Application
Have three students stand in front of the class. Have each hold a different-sized picture book. Have the first student say, "I have a big book," the second student say, "I have a bigger book," and the last student say, "I have the biggest book." Continue for using small, smaller, and smallest books.

2 Teach

Develop Skills and Concepts Have one student hold up a 12-in. ruler and another student hold up a yardstick. Ask students who has the taller stick and who has the shorter stick. Have another student join the group and hold up a pencil. Ask who now has the tallest stick and who has the shortest. Help students use position words along with size words to make more comparisons to describe the group.

- Have students cut out a picture of a tall person and a short person from a magazine or catalog. Have them show their pictures and tell which person is taller and which is shorter. Vary the activity by having students compare the size of animals in pictures.

- Give each student a sheet of paper and assorted lengths of crayons. Have students lay a red crayon on their papers and then lay a longer crayon next to it. Ask which crayon is shorter and which is longer. Tell students to lay another crayon between the first two and tell which is the longest and which is the shortest. Repeat for different crayons, having students name the color of each as they make length comparisons.

- Draw three lines of different lengths on the board, one under the other. Compare the lines by pointing to the shortest first and then each in turn as you say, *This line is long, but this line is longer and this line is the longest.* Then, reverse the procedure, beginning with the longest line, and say, *This line is short, but this line is shorter and this line is the shortest.* Help several students compare the lengths verbally. Then, have students lay out crayons or other objects of different lengths to make more comparisons.

T9

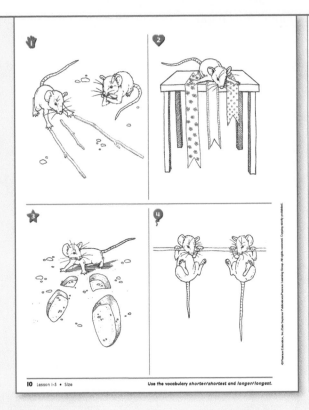

10 Lesson I-5 • Size

Use the vocabulary *shorter/shortest* and *longer/longest.*

For Mixed Abilities

Common Errors • Intervention
Some students may need additional practice with the vocabulary. Draw a line across the board for a starting point. Have students draw tall, short, and long sticks on the board to represent taller, longer, and shorter.

Enrichment • Application
1. Pair students. Have each pair draw a playground scene. Then, have pairs exchange pictures and tell three position facts about the new picture. (Possible answer: The girl is on the slide.)

2. Have students draw an object in the middle of a sheet of paper and then draw something shorter on the left and something taller on the right.

ESL/ELL STRATEGIES

Practice the comparative and superlative forms of words. Ask three students to line up. Describe the differences in their height using *tall, taller,* and *tallest* and then have students repeat. Repeat the activity with short pencils.

● Use the **Common Errors • Intervention** feature to explore a common error pattern and to provide remediation to struggling students. Collectively, all the Common Errors features in any chapter constitute a complete set of the common errors that students are likely to make when working in that chapter.

● **Enrichment** activities are a direct extension of the skills being taught. Some students may do these activities on their own while you work with those students who need more help.

● **ESL/ELL Strategies** are offered twice per chapter. These activities will help you provide insights into English vocabulary and increase comprehension of mathematical concepts. Specific techniques cited ensure that learning is taking place. The techniques also remove potential language barriers for English-language learners at beginning levels of proficiency.

3 Practice

Using page 9 Have students look at the mouse in Exercise 1. Ask what the mouse is looking at. (cacti) Ask if anyone knows where cacti are found. (desert) Tell students to circle the shortest cactus and tell if it is on the left or right. (left) Tell students to put an X on the tallest cactus and tell if it is on the left, on the right, or in the middle. (middle)

• Use the same strategy with Exercises 2 to 4, using the concepts of shorter, shortest, taller, and tallest.

Using page 10 Tell students to look over the pictures and find the one where a mouse is on top of something. (Exercise 2) Ask a student to use the words *top* or *bottom* and *right* or *left* in a sentence to tell about that picture. Then, have a student similarly tell about Exercise 3. (Possible answer: The mouse is behind the footprints.)

• Have students point to the mice in Exercise 1. Ask them if the sticks are the same length. (no) Ask if the middle stick is the longest. (no) Tell them to then mark an X on the stick that is the longest and to circle the stick that is the shortest.

• Similarly, discuss the other exercises on this page, using the concepts of shorter, shortest, longer, and longest.

4 Assess

Have students work in pairs. Put a line of tape on the floor. Give each pair a piece of string longer than, shorter than, and the same length as the tape. Tell students to find the piece of string that is the same length as the tape. Repeat to have them identify the piece that is longer and the piece that is shorter.

T10

● **Practice** offers practical suggestions for introducing the skill and effective presentation of the model for each page of the lesson, as well as guidelines to assist your students as they practice these pages independently.

● **Assess** provides you with a short question or activity that can be used to quickly evaluate if students have mastered the objective of the lesson.

Assessment

Assess Often With Chapter and Alternate Tests

A variety of assessments and information help you track your students' mastery of spatial sense, number sense, and problem solving.

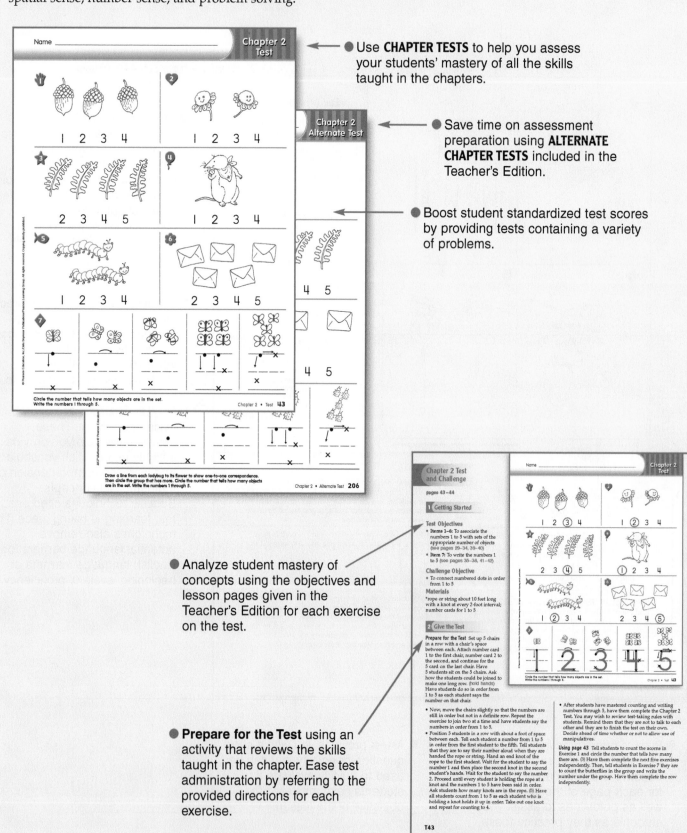

- Use **CHAPTER TESTS** to help you assess your students' mastery of all the skills taught in the chapters.

- Save time on assessment preparation using **ALTERNATE CHAPTER TESTS** included in the Teacher's Edition.

- Boost student standardized test scores by providing tests containing a variety of problems.

- Analyze student mastery of concepts using the objectives and lesson pages given in the Teacher's Edition for each exercise on the test.

- **Prepare for the Test** using an activity that reviews the skills taught in the chapter. Ease test administration by referring to the provided directions for each exercise.

Scope and Sequence

Levels	K	A	B	C	D	E	F
Readiness							
Using attributes of size, shape, and color	●	●	●				
Sorting and classifying	●	●					
Spatial relationships	●						
Numeration							
One-to-one correspondence	●						
Understanding numbers	●	●	●	●	●	●	●
Writing numbers	●	●					
Counting objects	●	●	●				
Sequencing numbers	●	●	●	●	●		
Numbers before and after	●	●	●	●	●		
Ordering numbers	●	●	●	●	●	●	●
Comparing numbers	●	●	●	●	●	●	●
Grouping numbers	●	●	●	●	●		
Ordinal numbers	●	●	●	●			
Number words		●	●	●		●	●
Expanded numbers				●	●	●	●
Place value	●	●	●	●	●	●	●
Skip-counting	●	●	●	●	●		
Roman numerals				●	●		●
Rounding numbers			●	●	●	●	●
Squares and square roots						●	
Primes and composites				●	●	●	●
Multiples			●	●	●	●	●
Least common multiples						●	●
Greatest common factors						●	●
Exponents						●	●
Addition							
Addition facts	●	●	●	●	●	●	●
Fact families		●	●	●	●	●	●
Missing addends		●	●	●	●	●	●
Adding money	●	●	●	●	●	●	●
Column addition		●	●	●	●	●	●
Two-digit addends		●	●	●	●	●	●
Multidigit addends			●	●	●	●	●
Addition with regrouping		●	●	●	●	●	●
Basic properties of addition			●	●	●	●	●
Estimating sums			●	●	●	●	●
Addition of fractions				●	●	●	●

Scope and Sequence

Levels	K	A	B	C	D	E	F
Addition (continued)							
Addition of mixed numbers				●	●	●	●
Addition of decimals				●	●	●	●
Addition of customary measures						●	●
Addition of integers						●	●
Subtraction							
Subtraction facts	●	●	●	●	●	●	●
Fact families		●	●	●	●	●	●
Missing subtrahends			●	●	●		
Subtracting money	●	●	●	●	●	●	●
Two-digit numbers		●	●	●	●	●	●
Multidigit numbers			●	●	●	●	●
Subtraction with regrouping		●	●	●	●	●	●
Zeros in the minuend			●	●	●	●	●
Basic properties of subtraction				●	●	●	●
Estimating differences			●	●	●	●	●
Subtraction of fractions				●	●	●	●
Subtraction of mixed numbers						●	●
Subtraction of decimals				●	●	●	●
Subtraction of customary measures						●	●
Subtraction of integers						●	●
Multiplication							
Multiplication facts			●	●	●	●	●
Fact families			●	●	●	●	●
Missing factors				●	●	●	●
Multiplying money				●	●	●	●
Multiplication by powers of ten				●	●	●	●
Multidigit factors				●	●	●	●
Multiplication with regrouping				●	●	●	●
Basic properties of multiplication			●	●	●	●	●
Estimating products				●	●	●	●
Multiples				●	●	●	●
Least common multiples						●	●
Multiplication of fractions						●	●
Factorization						●	●
Multiplication of mixed numbers						●	●
Multiplication of decimals					●	●	●
Exponents						●	●
Multiplication of integers						●	●

Scope and Sequence

Levels	K	A	B	C	D	E	F
Division							
Division facts			●	●	●	●	●
Fact families			●	●	●	●	●
Divisibility rules				●	●	●	●
Two-digit quotients				●	●	●	●
Remainders				●	●	●	●
Multidigit quotients					●	●	●
Zeros in quotients					●	●	●
Division by multiples of ten					●	●	●
Two-digit divisors					●	●	●
Properties of division					●	●	●
Averages					●	●	●
Greatest common factors						●	●
Division of fractions						●	●
Division of mixed numbers						●	●
Division of decimals						●	●
Division of integers							●
Money							
Counting pennies	●	●	●	●	●		
Counting nickels	●	●	●	●	●		
Counting dimes	●	●	●	●	●		
Counting quarters		●	●	●	●		
Counting half-dollars		●	●	●	●		
Counting dollar bills		●	●	●	●		
Writing dollar and cent signs		●	●	●	●	●	●
Matching money with prices	●	●	●				
Determining amount of change		●	●	●	●		
Determining sufficient amount		●	●				
Determining which coins to use		●	●				
Addition	●	●	●	●	●	●	●
Subtraction	●	●	●	●	●	●	●
Multiplication				●	●	●	●
Division				●	●	●	●
Estimating amounts of money			●	●	●	●	●
Rounding amounts of money			●	●	●	●	●
Buying from a table of prices		●	●	●	●	●	●
Fractions							
Understanding equal parts	●	●	●	●			
One-half	●	●	●	●	●		

Scope and Sequence

Levels	K	A	B	C	D	E	F
Fractions (continued)							
One-fourth	●	●	●	●	●		
One-third	●	●	●	●	●		
Identifying fractional parts of figures	●	●	●	●	●	●	●
Identifying fractional parts of sets		●	●	●	●	●	●
Finding unit fractions of numbers				●	●		
Equivalent fractions				●	●	●	●
Comparing and ordering fractions				●	●	●	●
Simplifying fractions				●	●	●	●
Mixed numbers				●	●	●	●
Addition of fractions				●	●	●	●
Subtraction of fractions				●	●	●	●
Addition of mixed numbers					●	●	●
Subtraction of mixed numbers						●	●
Multiplication of fractions						●	●
Multiplication of mixed numbers						●	●
Division of fractions						●	●
Division of mixed numbers						●	●
Renaming fractions as decimals				●	●	●	●
Renaming fractions as percents						●	●
Decimals							
Place value				●	●	●	●
Reading decimals				●	●	●	●
Writing decimals				●	●	●	●
Converting fractions to decimals					●	●	●
Writing parts of sets as decimals				●	●	●	●
Comparing decimals				●	●	●	●
Ordering decimals				●	●	●	●
Addition of decimals				●	●	●	●
Subtraction of decimals				●	●	●	●
Rounding decimals					●	●	●
Multiplication of decimals					●	●	●
Division of decimals						●	●
Renaming decimals as percents						●	●
Geometry							
Polygons	●	●	●	●	●	●	●
Sides and vertices of polygons		●	●	●	●	●	●
Faces, edges, and vertices		●	●	●	●	●	●
Solid figures	●	●	●	●	●	●	●

Scope and Sequence

Levels	K	A	B	C	D	E	F
Geometry (continued)							
Symmetry	●	●	●	●	●	●	●
Lines and line segments				●	●	●	●
Rays and angles				●	●	●	●
Measuring angles						●	●
Transformations			●	●	●	●	●
Congruency				●	●	●	●
Similar figures					●	●	●
Circles					●	●	●
Triangles				●	●	●	●
Quadrilaterals				●	●	●	●
Measurement							
Nonstandard units of measure	●	●					
Customary units of measure		●	●	●	●	●	●
Metric units of measure		●	●	●	●	●	●
Renaming customary measures				●	●	●	●
Renaming metric measures				●	●	●	●
Selecting appropriate units		●	●	●	●		
Estimating measures		●	●	●	●		
Perimeter by counting		●	●				
Perimeter by formula				●	●		●
Area of polygons by counting		●	●	●	●		
Area of polygons by formula					●	●	●
Volume by counting				●	●	●	
Volume by formula					●	●	●
Addition of measures						●	●
Subtraction of measures						●	●
Circumference of circles							●
Area of circles							●
Surface area of space figures							●
Estimating temperatures				●	●	●	●
Reading temperature scales		●	●	●	●	●	●
Time							
Ordering events	●						
Calendars	●	●	●	●	●		
Telling time to the hour	●	●	●	●	●		
Telling time to the half-hour		●	●	●	●		
Telling time to the five-minutes		●	●	●	●		
Telling time to the minute			●	●	●		

Scope and Sequence

Levels	K	A	B	C	D	E	F
Time (*continued*)							
Understanding A.M. and P.M.				●	●	●	●
Elapsed time			●	●	●	●	●
Graphing							
Tallies	●	●	●	●	●	●	
Bar graphs		●	●	●	●	●	●
Picture graphs	●	●	●	●	●		
Line graphs				●	●	●	●
Circle graphs						●	
Line plots			●			●	
Stem-and-leaf plots						●	●
Histograms							●
Ordered pairs				●	●	●	●
Statistics and Probability							
Understanding probability			●	●	●	●	●
Listing outcomes					●	●	●
Mean, median, and mode				●	●	●	●
Writing probabilities					●	●	●
Compound probability							●
Making predictions							●
Tree diagrams					●	●	●
Ratios and Percents							
Understanding ratios					●	●	●
Equal ratios						●	●
Proportions						●	●
Scale drawings						●	●
Ratios as percents						●	●
Percents as fractions						●	●
Fractions as percents						●	●
Finding the percents of numbers						●	●
Integers							
Understanding integers						●	●
Addition of integers						●	●
Subtraction of integers						●	●
Multiplication of integers							●
Division of integers							●
Graphing integers on coordinate planes							●

Scope and Sequence

Levels	K	A	B	C	D	E	F
Problem Solving							
Act it out	●	●	●	●	●	●	●
Choose a strategy	●	●	●	●	●		
Choose the correct operation	●	●	●	●	●		
Collect and use data		●	●	●	●	●	●
Determine missing or extra data		●	●	●	●		
Draw a picture or diagram	●	●	●	●	●	●	●
Identify a subgoal						●	●
Look for a pattern	●	●	●	●	●	●	●
Make a graph	●	●	●	●	●	●	●
Make a list		●	●	●	●	●	●
Make a model					●		
Make a table		●	●	●		●	●
Make a tally graph						●	
Restate the problem					●	●	●
Solve a simpler but related problem				●	●	●	●
Solve multistep problems				●	●	●	●
Try, check, and revise	●	●	●	●	●	●	●
Use a formula						●	●
Use a four-step plan				●	●	●	●
Use an exact answer or an estimate				●	●		
Use logical reasoning		●	●	●	●		●
Work backward				●	●	●	●
Write a number sentence		●	●	●	●		
Calculators							
Calculator codes				●	●	●	●
Equals key				●	●	●	●
Addition/subtraction keys				●	●	●	●
Multiplication/division keys				●	●	●	●
Clear key				●	●		
Calculators: Real-World Applications							
Averages						●	●
Formulas						●	●
Money					●	●	●
Percents						●	●
Repeating decimals						●	●
Statistics						●	●

Scope and Sequence

Levels	K	A	B	C	D	E	F
Algebra							
Patterns	●	●	●	●	●	●	●
Completing number sentences	●	●	●	●	●	●	●
Properties of numbers				●	●	●	●
Numerical expressions				●	●	●	●
Evaluating numerical expressions					●	●	●
Algebraic expressions						●	●
Evaluating algebraic expressions						●	●
Order of operations				●	●	●	●
Integers						●	●
Addition and subtraction of integers						●	●
Multiplication and division of integers							●
Ordered pairs			●	●	●	●	●
Function tables				●	●	●	●
Graphing a rule or an equation						●	●
Variables				●	●	●	●
Equations				●	●	●	●
Solve addition and subtraction equations						●	●
Solve multiplication and division equations						●	●
Model problem situations with equations						●	●
Solve inequalities							●
Formulas						●	●
Properties of equality						●	●

Contents

pages 1–24

Chapter Objectives

- To describe position using *top, bottom, in, out, over, under, on top of, off, above, below, beside, inside,* and *outside*
- To describe position using *first, last, middle, between, front, behind, right,* and *left*
- To identify red, yellow, blue, orange, green, purple, black, and brown
- To recognize the color words *red, yellow, blue, orange, green, purple, black,* and *brown*
- To compare using *small, smaller, smallest, big, bigger, biggest, large, larger,* and *largest*
- To compare using *shorter, shortest, taller, tallest, longer,* and *longest*
- To identify a circle, rectangle, square, and triangle
- To identify and compare the attributes of plane figures
- To sort and classify a set of shapes by one attribute
- To sort and classify a set of shapes by more than one attribute
- To identify a sphere, cone, cube, and cylinder
- To identify and compare the attributes of solid figures
- To match objects to outlines of their shapes
- To recognize symmetry in the environment
- To identify shapes with a line of symmetry
- To identify the matching parts of a symmetrical shape
- To identify patterns using color, size, or shape
- To complete a pattern
- To identify the object that comes next in a pattern

Materials

You will need the following:
*large cardboard box; *hand stamp or stickers; *yarn or string; *3 boxes with lids; *several color name cards for each color; *tape; *colored construction paper; *containers; *paper bag; *magazines; *blue and green blocks; *2 different-sized tissue boxes; *large and small colored blocks

Students will need the following:
Reproducible Masters RM1–RM4; container with a lid; paper clips; crayons; large and small blocks in different colors; sticks; blocks; magazines or catalogs; small objects of different colors; objects of various sizes; 12-in. ruler; yardstick; crayons in assorted lengths; picture books in assorted sizes; cardboard circles, squares, triangles, and rectangles; objects that are the same color; attribute blocks; 2 yarn loops; set of animal pictures; classroom objects shaped like plane shapes; large models of solid figures; modeling clay; building blocks shaped like spheres, cubes, cylinders, and cones; classroom objects

*Indicates teacher demonstration materials

shaped like solid figures; paper squares and circles; paint and paintbrushes; small mirrors; colored string; beads or macaroni; ribbon cut in different lengths

Bulletin Board Suggestion

Display pictures of animals mounted on colored paper cut in the shape of a square, circle, triangle, or rectangle. Help students discuss the pictures. As position words are introduced, extend the discussions to include the vocabulary. For example, ask students to name the animal between the elephant and the snake. As color, shape, and size words are introduced, have students use the vocabulary to compare the animals, the colored paper backgrounds, and so on. When patterns are introduced, have students arrange the pictures with or without background papers to form patterns.

Verses

Poems and songs may be used any time after the pages noted beside each.

Little Miss Muffet (pages 1–2)

Little Miss Muffet sat on her tuffet
Eating her curds and whey.
Along came a spider and sat down beside her
And frightened Miss Muffet away.

Hoot in the Chute (pages 1–2)

Is there a creature in our laundry chute?
Surely there is! I hear its hoot!
"Oh, no, you don't! Just look at this!
It can't be washed the way it is!
It's wrong side out! It's right side in!
Come on now! Please! You must begin
To turn your clothes all right side out!
Should you forget, you'll hear me shout!
Right side out! Right side out!
I'll not accept the wrong side out!
It's right side out! Not right side in!
That's outside out! Not outside in!"

Statue on the Curb (pages 3–4)

We're crossing the street
Watch out and beware!
The cars, trucks, and buses
Are all going somewhere!
We stop at the corner,
Look left and look right;
Then step off the curb
If nothing's in sight.
But if there's a car
Or truck down the way,
Pretend we are statues
And on the curb, stay!

Little Spot's Surprise

Little Spot, the ladybug, had a cold. Everyone loved Little Spot and wanted to do something to help, but they didn't know what to do.

"She catches colds because she lives in that drafty old tree," said Mrs. Robin. "She should move over behind that big rock. It would keep the wind off of her. But she loves that big old drafty tree."

"She needs a warm little house," said Mr. Inchworm. "Why don't we build Little Spot a nice warm house to live in?"

"Yes," said Mr. Cutworm. "After all, Little Spot has always been nice to everyone. She wants all of us to be friends."

"I don't know anything about building houses," said Mrs. Robin.

"Well, I don't either," said Mr. Cutworm, "but if we all do what we do best, I think we'll get it done."

Ms. Spider, who hardly ever agreed with anyone, said, "I think he's right. And I would like to help. Little Spot is always nice to me, even when the rest of you aren't!"

When Little Spot's friends went to tell her the good news, she was curled up fast asleep under a leaf. So they just let her sleep and began their work.

Mr. Inchworm measured everything to fit. Mr. Cutworm busily cut the leaves and twigs exactly where Mr. Inchworm had measured them. Mrs. Robin gathered grass and straw. Ms. Spider quickly spun a web in and out of the leaves, twigs, grass, and straw. She fastened it all together as quickly as each new piece was put on.

The friends worked very quickly and very quietly. They didn't want to wake Little Spot. Soon, however, it began to grow dark. So several of Little Spot's firefly friends came over and lit the branch with light.

When the house was finished, Mrs. Robin gently woke Little Spot.

"Little Spot, dear, get up. We have a surprise for you."

"What?" asked Little Spot. "A surprise for me?"

"Yes," said Ms. Spider. "We decided to build you a nice warm house. We hope you like it."

"Oh, I love it!" said Little Spot. And it made her feel much, much better just to see how her friends had worked together. Little Spot knew her friends loved her as she loved them.

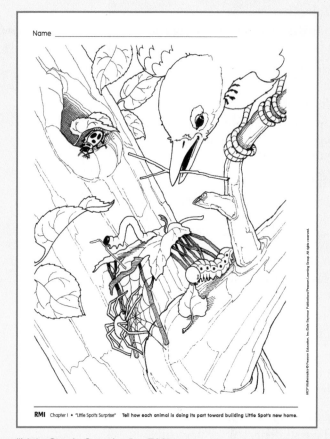

Name _____

RM1 Chapter I • "Little Spot's Surprise"   Tell how each animal is doing its part toward building Little Spot's new home.

"Little Spot's Surprise," p. RM1

Activities

Use Reproducible Master RM1, "Little Spot's Surprise," to help students relate the story to the mathematical ideas presented.

• Help students realize the need to work together when a task is time-consuming, difficult, or impossible for one person to do alone. Hold up a long jump rope and tell students that you will hold it a distance off the floor for them to jump over it. As you try to hold the jump rope a few inches off the floor, students should notice that you need another person to hold the other end.

• Have students use building blocks or similar materials to build a house or other structure. Tell students that, as a class, they are to build the structure using the given materials. Divide the class into small groups. Assign each group a part of the construction to complete, such as the foundation, frame, roof, or landscaping.

Lesson 1-1

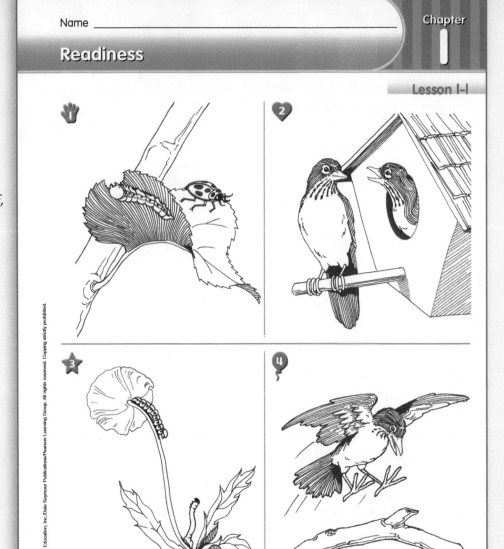

Use the vocabulary *top, bottom, in, out, over,* and *under.*

Lesson 1-1 • Position **1**

1 Getting Started

Objective
- To describe position using *top, bottom, in, out, over, under, on top of, off, above, below, beside, outside,* and *inside*

Vocabulary
top, bottom, in, out, over, under, on top of, off, above, below, beside, outside, inside

Materials
*large cardboard box; Reproducible Master RM2; container with a lid; paper clips; crayons; blocks; sticks

Warm Up • Mental Math
Have students tell the name of
1. a friend
2. a person in the room
3. someone older than they are
4. a baby they know
5. an adult or grown-up in their family
6. a person not in their family

Warm Up • Application
Have students tell the name of
1. their street
2. a book they like
3. the town they live in
4. a room in their home
5. a television show
6. a nursery rhyme

2 Teach

Develop Skills and Concepts Point to the top and bottom of a large cardboard box, identifying each in a complete sentence. Place an object in the box and have a student take it out of the box, each time describing the position. Continue this activity by placing the object under, on top of, over, above, below, inside, outside, and beside the box. Then, place the box on the floor. Repeat all the positions and have various students describe the situations.

- Read the traditional rhyme "Little Miss Muffet" on page T1a. Have students act out the rhyme. Rhymes such as the following may also be used to stress position words: "Jack and Jill," "Hickory Dickory Dock," "Little Jack Horner," "Jack Be Nimble," and "Diddle Diddle Dumpling."

- Give students copies of Reproducible Master RM2. Have students find the branch that has an X on it. Tell students to trace it. Ask them to find the other branches on the page and put an X on them also. Ask students to find the set of animals that has a circle around it and trace the circle. Continue until each set of animals has a circle around it. Use position words to discuss the location of the animals and the branches.

- Read "Hoot in the Chute" on page T1a. Show some clothing items and have students identify the outside and inside of each. Discuss clues that help us determine the outside and the inside. (seams, labels, color, pattern, and so on)

3 Practice

Using page 1 Demonstrate on the board how to mark an object with an X or a circle. Use position words whenever possible to describe the objects.

2 Lesson I-I • Position   Use the vocabulary *on top of, off, above, below, beside, outside,* and *inside.*

For Mixed Abilities

Common Errors • Intervention

Some students will need additional practice with the vocabulary. Use a large cardboard box for students to position themselves in and around it. Have students use sentences to say where they are.

Enrichment • Application

1. Have students practice turning items of clothing inside out and outside out when the clothing has one sleeve or leg, for example, turned wrong side out.

2. Have students use magazines and catalogs to cut and paste pictures to illustrate a nursery rhyme or "Statue on the Curb," found on page T1a. Have students share their pictures by telling where each object is placed in their pictures.

• Use leading statements such as *I see a leaf and there is . . .* to encourage students to use position words to complete them. Have students point to the leaf in Exercise 1. Ask which insect is on top of the leaf (ladybug) and which is on the bottom (caterpillar). Tell students to draw an X on the insect that is on top of the leaf and circle the insect that is on the bottom.

• Tell students to look at the picture in Exercise 2 and tell where the birds are, inside or outside, in or out of the house. Have students mark an X on the bird that is inside the birdhouse and circle the bird that is outside the birdhouse. Have students point to the flower in Exercise 3 and mark an X on the caterpillar at the top of the stem. Tell them to circle the one at the bottom. Now, tell students to find the log in Exercise 4 and mark an X on the animal that is jumping over the log and circle the insect under the log.

Using page 2 Encourage students to use position words in sentences to tell about the pictures. Have students tell where the bird and caterpillar in Exercise 1 are. Then, tell students to mark an X on the animal that is off the board and draw a circle around the one that is on it. Have students find the mushroom in Exercise 2 and mark an X on the insect that is above the mushroom and circle the one below it. Tell students to find the pine cone in Exercise 3 and mark an X on the caterpillar that is on top of it and circle the caterpillar that is beside it. Now, tell students to look at Exercise 4 and circle the animal that is outside the flowerpot and mark an X on the insect that is inside it.

4 Assess

Give each student a container with a lid and one of each of the following objects: paper clip, crayon, block, and stick. Tell students to put the crayon in the container, the paper clip beside the container, the stick on the lid, and so on. Have students tell about each action in a sentence.

T2

1 Getting Started

Objective
• To describe position using *first, last, middle, between, front, behind, right,* and *left*

Vocabulary
first, last, middle, between, front, behind, right, left

Materials
*yarn or string; *3 boxes with lids; *hand stamp or stickers; blocks

Warm Up • Mental Math
Tell students to put their finger
1. under their chin
2. on top of their head
3. beside their nose
4. above their eye
5. below their knee
6. in their shoe
7. on the bottom of their shoe
8. over their lips

Warm Up • Application
Play Simon Says by telling students that they are to do what is said only if Simon says to do it. Give a few directions by Simon and then one not by Simon to be sure students understand how to play. Use directions such as those in the above Mental Math activity.

2 Teach

Develop Skills and Concepts Give each student a block to play the game of "Directions." Tell students to put the block on top of their head, beside their left foot, and so on. After each direction, have students tell in a sentence where the block is.

• Form a yarn circle on the floor. Tell a student to demonstrate as you give directions to stand inside, outside, in, on, and beside the circle. Also, direct the student to put his or her left or right foot in the circle.

• Read "Statue on the Curb" on page T1a. Have students act out the poem and describe their actions.

• Place three closed boxes in a row on a table. Point to the box on the left and tell students that this box is the first box. Identify the last box similarly. Point to the box in the middle and tell students that this box is in the

middle and is between the other boxes. Have them tell about each box in a sentence such as "This is the last box" or "This box is between the other boxes."

• Put two boxes on a table for students to identify which one is on the left and which one is on the right. Repeat using the students' hands, groups of students, and other objects. Use the same activities for the positions in front of and behind.

3 Practice

Using page 3 Tell students that we can mark a picture to show it is the one we are talking about. Draw an object on the board and put a check mark on it as you describe what you are doing. Now, draw three objects on the board and have volunteers draw a check mark on the first, an X on the last, and a circle around the one in the middle. Discuss the positions of the animals in each exercise on this page with students. Then, give them directions to mark various ones with a check mark, an X, or a circle.

Use the vocabulary *first, last, middle,* and *between*.

4 Lesson I-2 • Position

Use the vocabulary *front*, *behind*, *right*, and *left*.

For Mixed Abilities

Common Errors • Intervention

Some students will need additional practice with the vocabulary. Place a stamp, sticker, or string on each student's right hand and have students identify that hand. Then, have students identify objects to their left or right.

Enrichment • Application

1. Have students use magazines and catalogs to cut out three pictures of cars. Tell students to draw a road and place the cars on the road so that one car is between the first and last cars.

2. Give students a paper prefolded into fourths to look like a book. Have students illustrate the front and back covers and the inside left and right pages. Then, have students open the paper and lay it flat to show the pictures. Have students tell about the positions of their pictures.

ESL/ELL STRATEGIES

Introduce the position words by giving several concrete illustrations using the position words in context. For example, point to different students and say, *Ali is sitting beside Pedro. Lisa is sitting between Frank and Sue.* Have students repeat each example.

Using page 4 In preparation, you may want to mark students' right or left hand with a stamp or sticker. Have students point to the picture on the page that is at the top, on the right, and so on, and tell about it. (Exercise 2) Ask them to point to the insect that is first, in the middle, and last. Next, have students identify the picture that is on the bottom left and describe it. (Exercise 3) Then, have students identify and describe other pictures on the page. After they have had enough practice describing these pictures, tell them to mark various animals with a check mark, an X, or a circle. Be sure your directions include position words and the words *right* and *left*.

4 Assess

Ask a student to stand at the front of the room. Tell another student to go up and stand behind the first child. Ask students to raise their left hand if they know who is standing in front, who is next, who is last, and so on. Have them use their classmates' names in sentences to identify their positions.

T4

1 Getting Started

Objectives
- To identify red, yellow, blue, orange, green, purple, brown, and black
- To recognize the color words *red, yellow, blue, orange, green, purple, brown,* and *black*

Vocabulary
red, yellow, blue, orange, green, purple, brown, black

Materials
*several color name cards for each color; *tape; *colored paper; Reproducible Master RM3; magazines or catalogs; small objects; crayons of different colors

Warm Up • Mental Math
Have students tell the name of
1. the first meal of the day (breakfast)
2. the person on their left
3. something above them
4. something inside a car
5. who jumped over the moon in a nursery rhyme (cow)

Warm Up • Application
Discuss two meanings of the word *right*, the right side compared to the left side and the right side of clothing compared to the wrong side of clothing. Use the word in several sentences and ask students to tell which meaning is used.

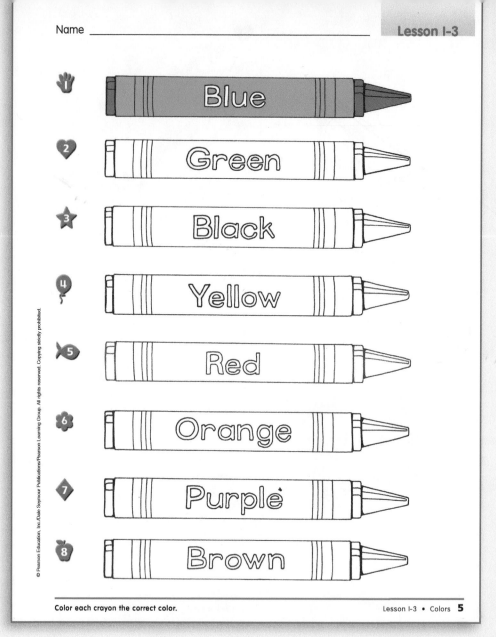

Color each crayon the correct color.

Lesson I-3 • Colors **5**

2 Teach

Develop Skills and Concepts Introduce each color and its color word separately by laying out several small objects of that color. Say the color word as you print the word on the board. Have students say the word as you point to it. Have them identify other objects in the room that are that color. Encourage students to find the color on their clothing, in books, and so on. Have them tell about things they have seen that are that color.

- Have students use magazines or catalogs to cut out pictures that show a particular color. Have students paste their pictures on a large sheet of paper to make a color collage.

- Direct three students who are wearing the same color to stand together in a group. Have a fourth student who is wearing a different color join them. Ask, *Who does not belong in this group? How do you know?* Encourage students to use the terms *same color* and *different color* to describe the group.

- Place a tape line on the floor. Give each student a sheet of colored paper. Have all students with yellow paper stand on the line, those with green stand behind a student with yellow, those with brown stand in front of a yellow, and so on. Have students then tell their color and use other color names to tell about where they are standing. (Possible answer: I am holding black and I am standing between red and orange.) Repeat the activity to combine the identification of color names with various position words.

- After all color words have been introduced, distribute one copy of Reproducible Master RM3 to each student. Encourage students to color each object the color they think best.

Color the picture.

For Mixed Abilities

Common Errors • Intervention

Some students may need additional practice recognizing colors and color names. Have them use color name cards to label objects in the classroom.

Enrichment • Application

1. Have students find pictures of houses and tell the color of each. Help students decide which colors are and are not commonly used for the outside color of houses. Discuss why.

2. Have students find pictures of different shades of several colors. Have students experiment with crayons to make different shades by applying different amounts of pressure.

• Have students work in pairs with a collection of colored objects. Tell students to sort their collection into groups by color. Ask students to tell why the objects in each group are the same and why the groups are different. Have students then identify each group by color.

3 Practice

Using page 5 Have students lay out their crayons. Write the eight color words on the board and have students hold up the correct crayon as you point to and say each color word. Now, have students point to the first picture of a crayon on the page. Ask what color the first crayon should be. (blue) Tell students to lay their blue crayon on top of the picture of a blue crayon. Have them look at the next crayon picture and lay the correct one on it. (green) Continue until all eight crayons are laid out and correctly matched to the pictures. Tell students that they are to look at each color word and color the crayon on the page that color. Have them color the first crayon as you watch. Then, have students complete the page.

Using page 6 Have students complete the page independently.

Now Try This! Have students talk about the picture, identifying each fruit. Have students lay out their eight crayons and tell which color might be used to color the mouse. (black or brown) Ask what fruit the mouse is eating and what color it would be. Tell students to color the fruit and the mouse and then use any colors left to color the design on the fruit bowl and the bowl itself. Have them complete the page independently and then tell about their color choices. Encourage them to use position words in complete sentences as they share their work. (Possible answers: My brown mouse is sitting in front of the blue and yellow bowl.)

4 Assess

Lay out the color name cards and have students sort crayons into piles beside the correct color words.

T6

1-4 Size

pages 7–8

1 Getting Started

Objective
- To compare using *small, smaller, smallest, big, bigger, biggest, large, larger,* and *largest*

Vocabulary
small, big, large

Materials
*containers; objects of various sizes; large and small blocks

Warm Up • Mental Math
Ask students to answer yes or no.
1. Dogs can be green. (no)
2. Many houses are white. (yes)
3. The sky is above us. (yes)
4. Your house is green.
5. Cars can be parked beside each other. (yes)
6. Each person has a right foot and a wrong foot. (no)
7. People who are bald have no hair on top of their head. (yes)

Warm Up • Application
Play Hokey Pokey with or without music. Have students stand in a circle and put the named part of their body inside the circle as you say or sing, *Put your right hand in, take your right hand out, put your right hand in, and shake it all about. Do the hokey pokey and turn yourself around. That's what it's all about.* Repeat for left hand, foot, head, and so on.

2 Teach

Develop Skills and Concepts Have a small group of students sit on the floor with you. Give each student a container of objects. Tell students to lay a small object on the floor in front of them. Tell them to lay a smaller object beside the first one and then to take out another object that is larger or bigger than the other objects. Then, have students identify the smallest and largest or biggest objects and place the objects in order from the largest to the smallest or the smallest to the biggest. Have students tell about their objects by size and color.

- Lay out a container of large and small blocks. Have students sort the blocks into piles of small blocks and large blocks. Ask a student to take a block from each pile and tell which block is smaller or larger. Then,

Use the vocabulary *smaller/smallest* and *bigger/biggest*.

Lesson 1-4 • Size **7**

have other students place one block on top of, behind, and beside the other. Have students tell about the sizes, positions, and colors of the blocks.

- Read the poem "Tall and Small" (anon.):

  Here is a giant who is tall, tall, tall;
  Here is an elf who is small, small, small;

  The elf who is small will try, try, try
  To reach to the giant who is high, high, high.

  Have students stand in a circle and act out the poem as you reread it. Encourage students to say the poem with you as they act it out again.

3 Practice

Using page 7 Encourage discussion about the pictures by asking students if in Exercise 3 the big flowerpot is on the right side of the mouse (yes) and if in Exercise 1 the small balloon is in the mouse's left hand (no), and so on. Encourage students to use position words in a sentence to answer each question.

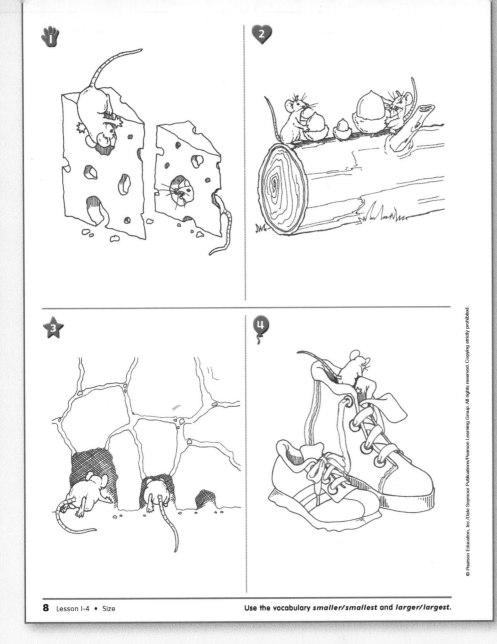

8 Lesson I-4 • Size   Use the vocabulary *smaller/smallest* and *larger/largest*.

For Mixed Abilities

Common Errors • Intervention

For students who find size difficult to understand, use physical objects for them to experience. For example, lay out three containers of different sizes and ask students to identify the largest and the smallest containers. Point to the largest container and ask students to point to one that is smaller and then to compare the two smaller containers for size.

Enrichment • Application

1. Have students tell about the biggest building they have visited and about the largest and smallest rooms in the school or in their homes. Then, have students find pictures of buildings larger and smaller than the school.

2. Post a picture of an animal and have students find pictures of animals that are larger and smaller than the animal whose picture is posted. Students may then wish to sort the animal pictures into groups of those that are small, those smaller, and those smallest.

• Have students point to the larger of the two balloons in Exercise 1. Tell them to circle the smaller balloon and put an X on the larger balloon.

• Ask students if the mouse in Exercise 2 is bigger or smaller than the cars in the picture. (bigger) Ask which car is closer to the mouse's tail. (the smaller car) Tell students to circle the small car and put an X on the bigger car.

• Have students look at the flowerpots and the mouse in Exercise 3 and tell what is in the middle. (the mouse) Ask if the mouse is larger than the flowerpots. (no) Tell students to mark an X on the biggest object in the picture and to circle the smallest object.

• Tell students that in Exercise 4, they are to circle the smallest mouse and mark an X on the biggest mouse.

Using page 8 Ask students questions that require answers in full sentences, such as *What is the mouse behind? Where are the mice going at the bottom left?* and so on.

• Tell students that they are to circle the smaller piece of cheese in Exercise 1 and to mark an X on the larger piece. Have students look at Exercise 2. Tell them to point to the largest acorn and mark an X on it. Then, have them point to the smallest acorn and circle it. Finally, have students mark an X on the largest hole in the wall in Exercise 3 and circle the smallest. For Exercise 4, have students point to the smaller shoe and tell them to circle it. Have students point to the larger shoe and mark an X on it.

4 Assess

Tell students that they are to listen carefully and follow the directions you give. Play cheerful music as you tell students to walk with small steps, make themselves large, make themselves smaller and smaller, and so on. Continue for more directions, using variations of *small*, *large*, and *big*.

T8

1-5 Size

pages 9–10

1 Getting Started

Objective
- To compare using *shorter, shortest, taller, tallest, longer,* and *longest*

Vocabulary
short, tall, long

Materials
*tape; picture books in assorted sizes; 12-in. ruler; yardstick; magazines or catalogs; crayons in assorted lengths; string

Warm Up • Mental Math
Ask students which is smaller.

1. ant or mouse (ant)
2. the school or their house (house)
3. a watch or the clock on the wall (watch)
4. their dad's shoe or their shoe (their shoe)
5. their house or their city (house)
6. their thumb or index finger (thumb)

Warm Up • Application
Have three students stand in front of the class. Have each hold a different-sized picture book. Have the first student say, "I have a big book," the second student say, "I have a bigger book," and the last student say, "I have the biggest book." Continue for using small, smaller, and smallest books.

2 Teach

Develop Skills and Concepts Have one student hold up a 12-in. ruler and another student hold up a yardstick. Ask students who has the taller stick and who has the shorter stick. Have another student join the group and hold up a pencil. Ask who now has the tallest stick and who has the shortest. Help students use position words along with size words to make more comparisons to describe the group.

- Have students cut out a picture of a tall person and a short person from a magazine or catalog. Have them show their pictures and tell which person is taller and which is shorter. Vary the activity by having students compare the size of animals in pictures.

- Give each student a sheet of paper and assorted lengths of crayons. Have students lay a red crayon on their papers and then lay a longer crayon next to it. Ask which crayon is shorter and which is longer. Tell students to lay another crayon between the first two and tell which is the longest and which is the shortest. Repeat for different crayons, having students name the color of each as they make length comparisons.

- Draw three lines of different lengths on the board, one under the other. Compare the lines by pointing to the shortest first and then each in turn as you say, *This line is long, but this line is longer and this line is the longest.* Then, reverse the procedure, beginning with the longest line, and say, *This line is short, but this line is shorter and this line is the shortest.* Help several students compare the lengths verbally. Then, have students lay out crayons or other objects of different lengths to make more comparisons.

Use the vocabulary *shorter/shortest* and *taller/tallest*.

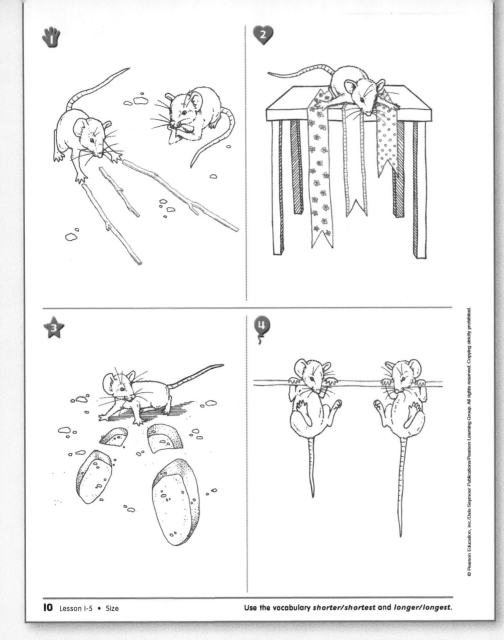

10 Lesson I-5 • Size

Use the vocabulary *shorter/shortest* and *longer/longest*.

For Mixed Abilities

Common Errors • Intervention

Some students may need additional practice with the vocabulary. Draw a line across the board for a starting point. Have students draw tall, short, and long sticks on the board to represent taller, longer, and shorter.

Enrichment • Application

1. Pair students. Have each pair draw a playground scene. Then, have pairs exchange pictures and tell three position facts about the new picture. (Possible answer: The girl is on the slide.)

2. Have students draw an object in the middle of a sheet of paper and then draw something shorter on the left and something taller on the right.

ESL/ELL STRATEGIES

Practice the comparative and superlative forms of words. Ask three students to line up. Describe the differences in their height using *tall*, *taller*, and *tallest* and then have students repeat. Repeat the activity with short pencils.

---

3 Practice

Using page 9 Have students look at the mouse in Exercise 1. Ask what the mouse is looking at. (cacti) Ask if anyone knows where cacti are found. (desert) Tell students to circle the shortest cactus and tell if it is on the left or right. (left) Tell students to put an X on the tallest cactus and tell if it is on the left, on the right, or in the middle. (middle)

• Use the same strategy with Exercises 2 to 4, using the concepts of shorter, shortest, taller, and tallest.

Using page 10 Tell students to look over the pictures and find the one where a mouse is on top of something. (Exercise 2) Ask a student to use the words *top* or *bottom* and *right* or *left* in a sentence to tell about that picture. Then, have a student similarly tell about Exercise 3. (Possible answer: The mouse is behind the footprints.)

• Have students point to the mice in Exercise 1. Ask them if the sticks are the same length. (no) Ask if the middle stick is the longest. (no) Tell them to then mark an X on the stick that is the longest and to circle the stick that is the shortest.

• Similarly, discuss the other exercises on this page, using the concepts of shorter, shortest, longer, and longest.

4 Assess

Have students work in pairs. Put a line of tape on the floor. Give each pair a piece of string longer than, shorter than, and the same length as the tape. Tell students to find the piece of string that is the same length as the tape. Repeat to have them identify the piece that is longer and the piece that is shorter.

T10

pages 11–12

1 Getting Started

Objectives
- To identify a circle, rectangle, square, and triangle
- To identify and compare the attributes of plane shapes

Vocabulary
circle, rectangle, square, triangle

Materials
*paper bag; cardboard circles, squares, triangles, and rectangles

Warm Up • Mental Math
Have students name something that is
1. longer than their arm
2. taller than they are
3. bigger than an elephant
4. smaller than their bed
5. shorter than their leg
6. larger than their hand

Warm Up • Application
Play Simon Says using any of the position words from Lessons 1-1 and 1-2.

Identify circles, squares, triangles, and rectangles.
Color the same shapes the same color.

Lesson 1-6 • Shapes   **11**

2 Teach

Develop Skills and Concepts Build visual identification of shapes by introducing each shape separately over a period of several days. Show students a cardboard circle and say, *This is a circle.* Give each student a cardboard circle. Help students see that circles have round edges and can be different sizes and colors. Ask students to identify objects in the classroom that have the shape of a circle. Introduce the square, rectangle, and triangle in the same manner.

- Place one of each cardboard shape in a paper bag. Tell students that you will describe a shape and then they will reach into the bag and select that shape without looking. Describe the shape by telling the number of sides and corners, whether it is round or pointy, whether it is large or small, and so on. Repeat several times with different shapes and descriptions.
- Place one object of each of the four shapes in various positions. Have students tell about each object using shape, color, and position words.

3 Practice

Using page 11 Have students point to any circle on the page and identify objects that are the shape of a circle. Ask if there are more circles on the page. (yes) Ask students to point to the circle in the middle of the page. Repeat for the circles at the top and bottom. Tell students to color all three circles the same color.

- Now, have students point to a rectangle on the page and name some objects that are the shape of a rectangle. Tell students to find other rectangles on the page and color each of the four rectangles one color. Repeat the procedure for having students find and color all the squares and all the triangles.
- Now, have students talk about the shapes with a partner, each describing the color he or she used for each shape. As an extension, have students name a classroom object shaped like a circle, rectangle, square, or triangle.

T11

**Identify circles, squares, triangles, and rectangles.
Color the same shapes the same color.**

For Mixed Abilities

Common Errors • Intervention

Give students additional practice identifying the shapes in the lesson by taking them outside to identify the shapes of things such as wheels on vehicles, sidewalk sections, rungs on the ladder of a playground slide, windows on the school building, and so on.

Enrichment • Geometry

1. Distribute outlines of a rectangle and an equilateral triangle along with cardboard shapes of squares and triangles. Have students explore putting shapes together to fill the outlined shapes. For example, students could put two squares together to make the rectangle. You could reverse the activity by having students cut large shapes into other shapes.

2. On the board, demonstrate a slide by sliding a cardboard shape in a straight line from one place to another. Then, demonstrate a flip by flipping a shape from one side to the other. Have pairs of students use the cardboard shapes to explore moving the shapes in flips and slides.

Using page 12 Ask questions to encourage students to talk about this picture. Be sure they use position and shape to discuss the robot and the mice. For example, ask students where the circles are on the robot and encourage them to use the vocabulary that they have learned, *above, below, to the left*, and so on, to describe their positions. Do the same with the other shapes. Have students choose a color for all of the circles. Use different colors for each of the other kinds of shapes. Some students may wish to color the mice as well.

4 Assess

Display many objects for students to sort into four groups according to shape. Have students use shape words to tell about the objects in each group.

1-7 Sorting and Classifying

pages 13–14

1 Getting Started

Objectives

- To sort and classify a set of shapes by one attribute
- To sort and classify a set of shapes by more than one attribute

Materials

objects that are the same color; attribute blocks; 2 yarn loops; set of animal pictures

Warm Up • Mental Math

Ask students to answer yes or no.

1. A circle has four sides. (no)
2. A door is shaped like a rectangle. (yes)
3. A triangle is round. (no)
4. A square has four corners. (yes)
5. A circle is round. (yes)

Warm Up • Geometry

Have students work in pairs to review shapes, sizes, and colors. Give each pair three classroom objects that are all different shapes and have students name the shapes. Repeat for color. Then, have students organize the objects according to size, from smallest to largest.

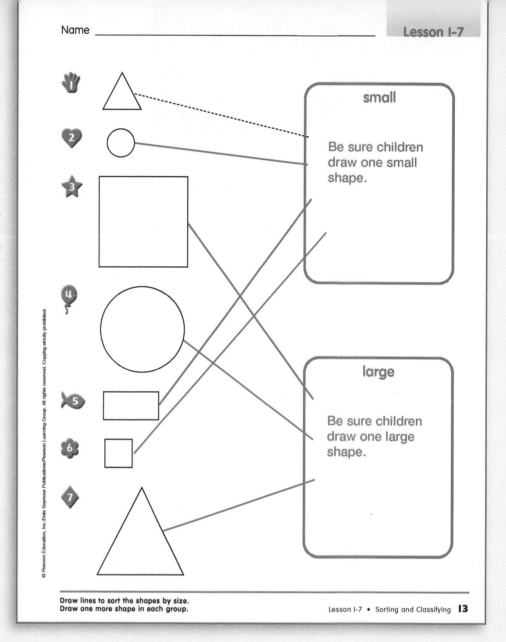

small

Be sure children draw one small shape.

large

Be sure children draw one large shape.

Draw lines to sort the shapes by size.
Draw one more shape in each group.

2 Teach

Develop Skills and Concepts Give students a set of objects that are all the same color, for example, red hat, red crayon, red fire truck, and red apple. Have students describe each object. Make sure they use the color word in their descriptions. Point out that even though the objects are different they are all the same color. Repeat with a set of objects that are the same shape but are different sizes and colors.

- Sort students into groups by different rules. First, sort by the color of clothing the students are wearing. Ask three students who are wearing black pants to stand and form a group. Then, ask three students wearing blue pants to stand and form a group. Have students look at each group and decide what the members have in common. Continue by sorting the students into two groups by attributes such as glasses/no glasses, blond hair/brown hair, and so on.

- Organize the class into small groups. Give each group five sheets of paper and a set of attribute blocks. Have students sort the blocks by one attribute. First, have them sort the blocks onto two sheets of paper by size. Next, have them sort the blocks onto three sheets of paper by color. Then, have them sort the blocks onto five sheets of paper by shape.

- Give students sets of attribute blocks. Have them pick out all of the large, red shapes. Ask, *How could you describe this group?* (large and red) Then, have students pick out all of the small triangles. Ask, *How can you describe this group?* (small and triangles) Have students find other ways to classify shapes by two attributes.

- Organize the class into pairs. Give each pair two yarn loops and a set of animal pictures that includes large and small animals, wild and domestic animals, mammals and reptiles, and so on. Ask students to take turns sorting the animal pictures into two groups. The partner not sorting the pictures has to name the group and explain how they are sorted. Have students switch

T13

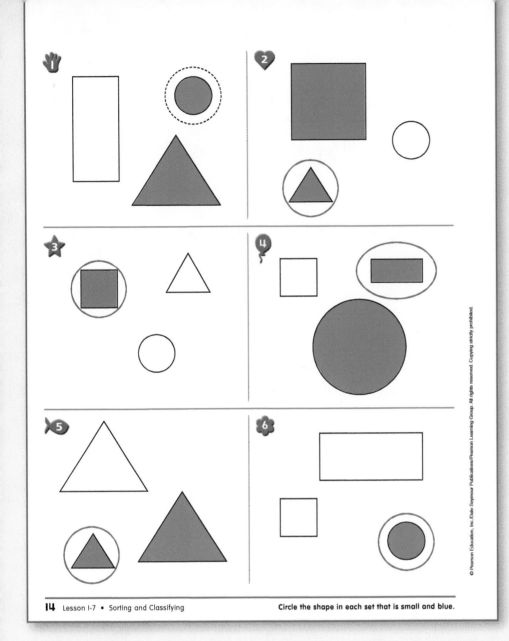

Circle the shape in each set that is small and blue.

For Mixed Abilities

Common Errors • Intervention

Some students may only be able to classify by one attribute at a time. Give those students more practice describing multiple attributes of shapes. Have students work in pairs. Give each pair cutouts of circles, squares, rectangles, and triangles in different colors and sizes. Have one partner hold up a cutout while the other partner describes the color, shape, and size. Have partners reverse roles and repeat.

Enrichment • Application

Have students use paper shapes to make pictures of objects, such as a house made from a square with a triangle roof, a rectangle door, square windows, and so on. Have students glue the shapes onto paper for a final product. Then, have them make a frame around the picture using shapes to create a pattern or a design.

roles and continue sorting the same set in as many different ways as they can.

3 Practice

Using page 13 Have students look at the shapes on the left side of the page. Ask them to name each shape, starting with the triangle at the top. Call students' attention to the two sorting boxes on the right side of the page. Explain that one is for small shapes and the other is for large shapes. Ask, *What size is the triangle in Exercise 1?* (small) Have students trace the line to match the triangle to the sorting box for small shapes. Have students complete the page independently by drawing a line to match by size each shape to the correct sorting box.

Using page 14 Ask students to look at the first exercise and point to any small shapes in the group. (circle) Then, have them point to any blue shapes. (circle and triangle) Ask, *Which shape is both small and blue?* (circle) Point out

that the triangle is blue, but it is not small. Have students trace the circle around the circle because it is both small and blue.

• If students need extra help, encourage them to put a check mark next to each shape that is small and then do the same for each shape that is blue. The shape in each group with two check marks is the one to be circled. Have students complete the page independently.

4 Assess

Ask students, *How do you sort groups of objects at home?* (Possible answers: We sort silverware into forks, knives, and spoons; We sort my toys from my siblings' toys.)

T14

1-8 Solid Figures

pages 15–16

1 Getting Started

Objectives

- To identify a sphere, cone, cube, and cylinder
- To identify and compare the attributes of solid figures
- To match objects to outlines of their shapes

Vocabulary

sphere, cone, cube, cylinder

Materials

classroom objects shaped like plane shapes; large models of solid figures; modeling clay; building blocks shaped like spheres, cubes, cylinders, and cones

Warm Up • Mental Math

Hold up a picture of a very large red circle. Ask students to answer yes or no.

1. The shape is yellow. (no)
2. The shape is a circle. (yes)
3. The shape is small. (no)
4. The shape is red. (yes)

Warm Up • Geometry

Have students work in pairs to review the plane shapes. Give each pair classroom objects shaped like the plane shapes. Have one partner hold up an object and the other partner name the shape. Have students reverse roles and continue.

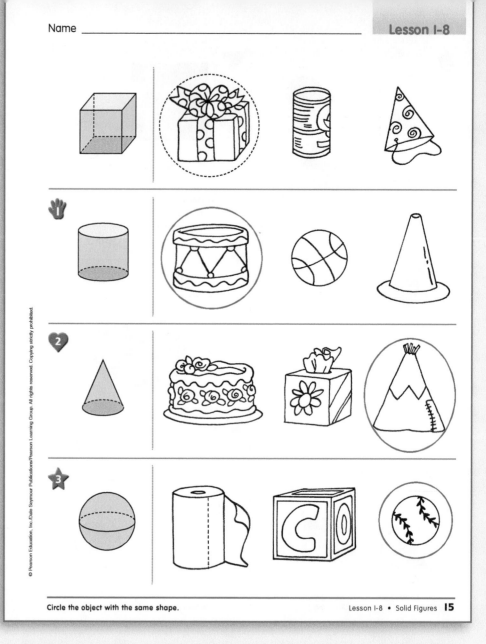

Circle the object with the same shape.

2 Teach

Develop Skills and Concepts Display large models of a cube, a cylinder, a cone, and a sphere. As you hold up each one, tell students its name and have them repeat the name after you. After naming the figures, have students go on a classroom hunt to find objects that are shaped like the solid figures. Place the models on a table and have students place the object they found with the model it matches.

- Distribute models of the solid figures and clay to students. Starting with the cube, have students press a flat surface of the figure into clay. Ask, *What shape do you see in the clay?* (square) Repeat with the cone and the cylinder, pointing out that there is only one flat surface on the cone and two on the cylinder. Make sure

that students see that the cone and cylinder make circles in the clay. Lastly, show the sphere and ask students, *Do you see any flat surfaces on this figure?* (no) Point out that a sphere is curved, so none of the surfaces can be flat.

- Distribute building blocks shaped like spheres, cubes, cylinders, and cones to students. Have students try to build towers with the blocks to determine which figures can stack. (cubes and cylinders) Then, have students make a ramp to see which figures will roll. (sphere, cylinder, and cone) Have students push each block on a table to see which will slide. (cube, cone, and cylinder)

- Organize the class into pairs. Have one partner hold a solid figure so the other partner cannot see it. Have partners describe their figures by telling whether it is curved and smooth, whether it has corners and is pointy, whether it stacks, slides, or rolls, and so on. The other partner then tries to guess the figure based on the description.

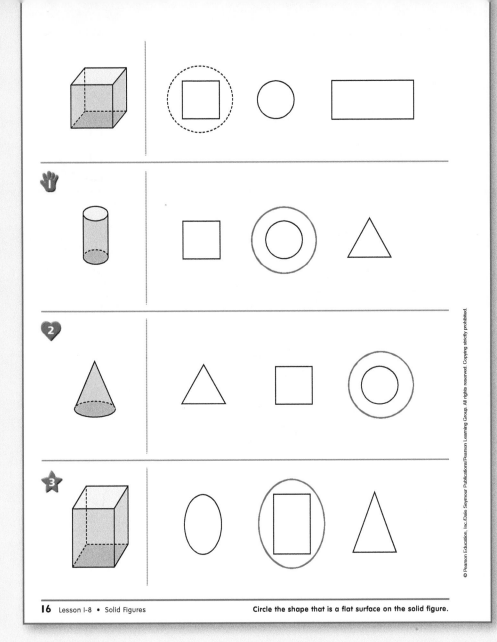

Circle the shape that is a flat surface on the solid figure.

For Mixed Abilities

Common Errors • Intervention

Some students may not be able to look at a drawing of a solid figure and understand if it can slide or roll. Have them work with models to see which ways each solid figure can move.

Enrichment • Geometry

1. Have students make shape patterns by dipping the flat surfaces of solid figures in paint and pressing them onto paper. Students can ask a classmate to guess which figures were used to make the pattern. Then, have the partner extend the pattern to show what comes next.

2. Have students look through supermarket flyers and cut out pictures showing packaged goods. Discuss how packages are shaped to hold certain foods. For example, most crackers are packaged in boxes with corners, but most liquids are packaged in rounded containers because they can flow into all of the space. In addition, most food containers are made to stack, which is why there are no containers shaped like spheres and cones.

3 Practice

Using page 15 Have students look at the pictures in the example. Point out that the solid figure on the left is a cube. Have students look at the gift box, the soup can, and the party hat and tell which object is shaped like the cube. (the gift box) Encourage students to look at attributes of the cube, such as corners, flat surfaces, and so on. Have them trace the circle around the gift box because it is about the same shape as the cube. Have students complete the page independently.

Using page 16 Have students look at the solid figure in the example. Ask a volunteer to name the figure. (cube) Encourage students to describe the shape of the flat sides on a cube. Ask, *Is there a square on a flat surface of the cube?* (yes) *Is there a circle?* (no) *Is there a rectangle?* (no) Have students trace the dashed circle around the square. If students need extra help, encourage them to use models of the solid figures. Have students complete the page independently.

• Make sure students understand that a sphere is curved and has no flat surfaces.

4 Assess

Ask students, *How are a cylinder and a sphere alike? How are they different?* (alike: both are round; different: a cylinder has two flat surfaces and a sphere has no flat surfaces.)

T16

pages 17–18

1 Getting Started

Objectives

- To recognize symmetry in the environment
- To identify shapes with a line of symmetry
- To identify the matching parts of a symmetrical shape

Vocabulary

symmetry, line of symmetry

Materials

*magazines; classroom objects shaped like solid figures; paper squares and circles; crayons; paint and paintbrushes; small mirrors

Warm Up • Mental Math

Ask students the following:

1. What figure rolls? (sphere, cylinder, cone)
2. What is the shape of a paper towel roll? (cylinder)
3. What shapes can roll and slide? (cone and cylinder)
4. Which figure has a square as a flat surface? (cube)

Warm Up • Geometry

Have students work in pairs to review solid figures. Give each pair classroom objects shaped like solid figures. Have one partner hold up an object while the other tells what solid figure it matches. Have students reverse roles and continue.

2 Teach

Develop Skills and Concepts Give each student a paper square. Have students fold the paper in half so that they have two equal halves. Check their work. Then, have students open the square and draw a crayon line on the fold. Explain that the fold is called the line of symmetry. Ask students to look at the square to see that the two parts are the same size and shape. Repeat with paper circles.

- Set out paper squares, paint, and paintbrushes. Have students fold the paper in half and then open it. Show students how to place a blob of paint on one side of the fold and then refold the paper, pressing gently. After

opening the paper, ask, *Is the paint blob on both sides of the fold the same?* Explain how the blobs are matching, so they are symmetrical.

- Give students crayons and a paper square and ask them to draw a large face on the paper. Show students how to fold the face vertically in half, between the eyes. Help each student place a mirror on the folded face so that the other half of the face is reflected in the mirror. Ask, *How does the mirror help you see a whole face?* Help students see that the mirror is showing the matching half of the face.

- Cut out symmetrical pictures from magazines and draw one line of symmetry in each picture. Have students take turns folding a picture in half along the line of symmetry and checking to see that the two parts match. Cut each symmetrical picture in half. Then, have students play a game to find the matching parts that go together to make whole pictures.

Name _____

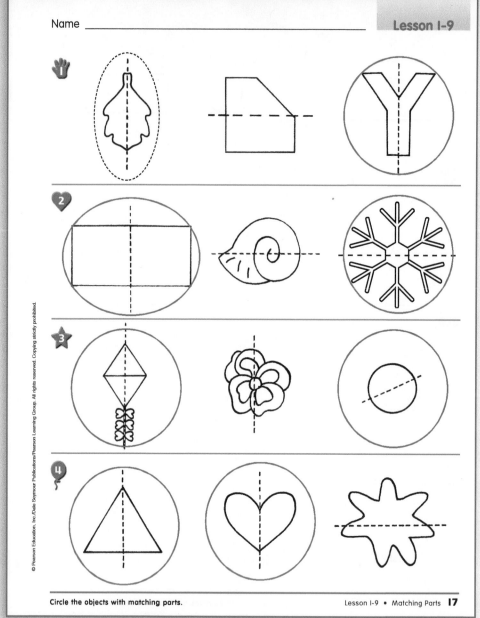

Circle the objects with matching parts.

Lesson 1-9 • Matching Parts **17**

18 Lesson 1-9 • Matching Parts

Circle the picture that shows the completed shape.

For Mixed Abilities

Common Errors • Intervention

Some students may need more practice identifying pictures with lines of symmetry. Have students work with partners. Give each pair large cutouts of all 26 capital letters of the alphabet. Have students fold the letters vertically to determine if they have symmetry. Encourage students to make separate piles of letters with and without symmetry. (letters without symmetry: F, G, J, K, L, N, P, Q, R, S, and Z)

Enrichment • Geometry

Point out to students that some shapes have more than one line of symmetry. For example, a circle has an unlimited number of lines of symmetry. Give students four paper squares and have them try to find all four lines of symmetry. Repeat the process with paper hexagons by having students try to find all six ways to make a line of symmetry.

3 Practice

Using page 17 Have students look at the pictures in Exercise 1. Explain that the line drawn through the leaf divides the picture into two matching parts. This line is called a line of symmetry. Have students look at the plane shape and ask a volunteer to tell if the line divides this shape into two matching parts. (no) Repeat with the letter Y. (yes)

• Tell students that in Exercises 2 to 4, there are two pictures in each row that show symmetry and one picture that does not. Allow students to work in small groups to complete the page.

Using page 18 Have students look at the folded paper on the far left in the example. Point out that the drawing on the paper shows half of a picture and that they need to predict what the whole picture would look like after

the paper is unfolded. If students need extra help, encourage them to draw the other half of the picture next to the half that is given on the page. Have students complete the page independently.

4 Assess

Ask students, *How can you tell if a picture shows symmetry?* (Possible answer: If you fold it in half and the same image is on both halves of the paper, then the picture shows symmetry.)

T18

1-10 Problem Solving: Look for a Pattern

pages 19–20

1 Getting Started

Objectives
- To identify patterns using color, size, or shape
- To complete a pattern
- To identify the shape that comes next in a pattern

Materials
*blue and green blocks; ribbon cut in different lengths; Reproducible Master RM4

Warm Up • Mental Math
Ask students what shape is the following:
1. the door going to the hallway (rectangle)
2. a wheel on a car (circle)
3. this page (rectangle)
4. the side of a cereal box (rectangle)

Warm Up • Application
Have three students stand in front of the class, each holding a different length of ribbon. Have them arrange themselves from shortest to longest ribbon and tell about their position. Encourage them to use complete sentences. (My ribbon is shorter than Bill's.)

Name _____

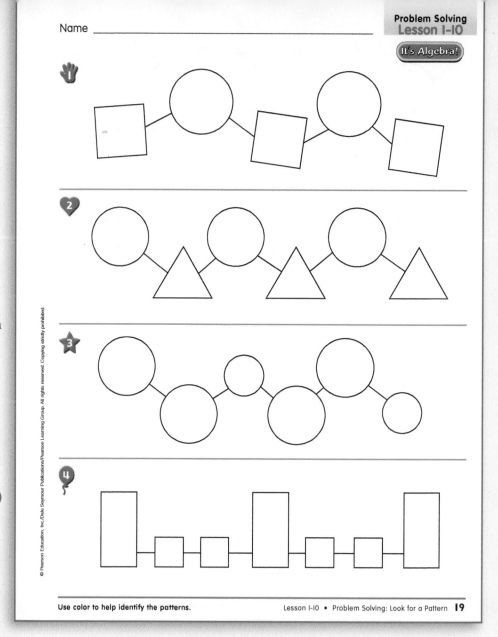

It's Algebra!

Use color to help identify the patterns.

Lesson 1-10 • Problem Solving: Look for a Pattern **19**

2 Teach

Develop Skills and Concepts Students can use a four-step plan to solve word problems or any exercise in a problem-solving lesson. This plan helps students break a problem into parts. The four steps are SEE, PLAN, DO, and CHECK.

- In SEE, students read the problem and look at the given art or graphs for information. Then, they list all the information.

- In PLAN, students think about how to use the information to solve the problem. Students then list the steps to be used with the information in the order they will be completed.

- In DO, students solve the problem using the plan.

- In CHECK, students make sure the answer is correct and check that they answered the question and that the answer is reasonable.

- Now, have four boys stand in a row facing the class. Tell students that the pattern is boy, boy, boy, boy. Substitute two girls for two boys and then continue to arrange various patterns for the students to describe.

- Lay out a pattern of small blue and green blocks to show green, green, blue, green. Have students say the pattern. Ask the color of the first block in the pattern. (green) Give students copies of Reproducible Master RM4. Tell students that they are going to color the first row of squares to look like this pattern of blocks. Ask, *What color should the first square be?* (green) Have students color the first square green. Help them decide which color to use for the next three squares. Create patterns using the same number of small blocks that is in each row on the reproducible master page. Then, have students color each pattern on each of the rows on the page.

It's Algebra! The concepts in this lesson prepare students for algebra.

T19

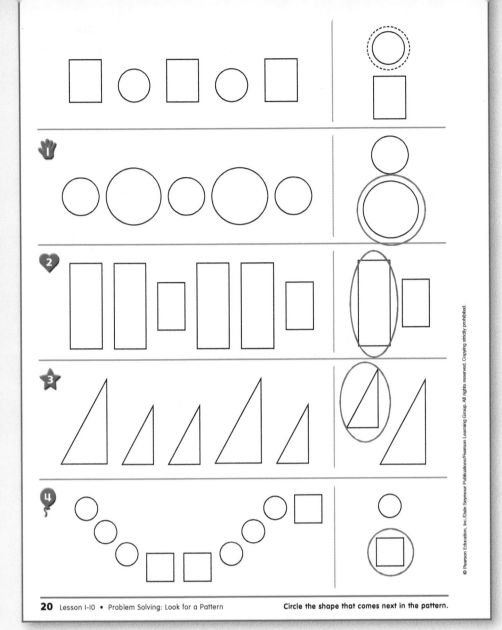

Circle the shape that comes next in the pattern.

For Mixed Abilities

Common Errors • Intervention

Some students may need additional practice with patterns. Have them clap patterns such as long-long-short-long-long-short with you. Then, have students draw sticks in lengths to show the pattern on the board. Have them discuss and demonstrate how the pattern would be continued in claps and sticks.

Enrichment • Application

1. Have students look at patterns in everyday life in order to predict the results of repeated activities. For example, when the light switch is moved up, then the light goes on, or when the school bell rings in the afternoon, then it is time to go home. Ask students to give their own examples of cause-and-effect relationships in their everyday lives.

2. Tell students that the first two patterns on page 20 are both AB patterns. This same pattern can be shown with sounds or movements. An example of the same AB pattern using sounds is loud clap, soft clap repeated. An example of the same pattern using movements would be stomp, clap repeated.

3 Practice

Using page 19 Have students tell what shapes they see on the page. (circles, squares, rectangles, triangles) Tell students that this page has some patterns for them to see and color.

• Now, have students look at Exercise 1 and apply the four-step plan. For SEE, ask, *What shapes do you see in this pattern?* (squares and circles) Tell students to color the first square blue, the first circle yellow, and the second square blue. For PLAN, ask students how they will decide which color the next shape should be. (color every square blue and every circle yellow) For DO, ask students what color the next shape should be colored. (yellow) Then, have students color only the first four shapes in the pattern using blue and yellow. Next, have them color the last square to continue the pattern. For CHECK, have students say the pattern with you. Continue this strategy in the rest of the patterns on this page.

Using page 20 Call attention to the vertical line that goes up and down the page from top to bottom. Tell students that they are to circle the shape at the right that will come next in each pattern. Have students look at the example and say the first pattern, being sure to stop at the vertical line. Tell students to color the rectangles green and then color the circles blue. Have students say the pattern aloud, green rectangle, blue circle, and so on. Have students look at the shapes to the right of the vertical line and tell which would come next in the pattern. (circle) Ask students what color the circle should be. (blue) Have students color the circle and then trace the circle around it to show it is the correct answer. Continue this method through the other patterns.

4 Assess

Lay out a pattern of small blue and green blocks, such as blue, blue, green, blue, blue, green. Ask students to name the next color. (blue)

T20

1-11 Making Patterns

pages 21–22

1 Getting Started

Objective
- To identify the object that comes next in a pattern

Materials
*large and small colored blocks; *two different colors of construction paper; string; beads or macaroni

Warm Up • Mental Math
Ask students the following:
1. What shape is round with no edges? (circle)
2. What is the shortest finger called? (thumb)
3. Is a bathtub larger or smaller than a sink? (larger)
4. Is a giant smaller or taller than an elf? (taller)

Warm Up • Language Arts
Say *glug, gurgle, glug* and have students repeat the words. Tell students that these words may sound like water or milk pouring from a container. Encourage students to experiment by repeating different patterns of the words such as *glug, glug, gurgle, glug* or *glug, gurgle.*

It's Algebra!

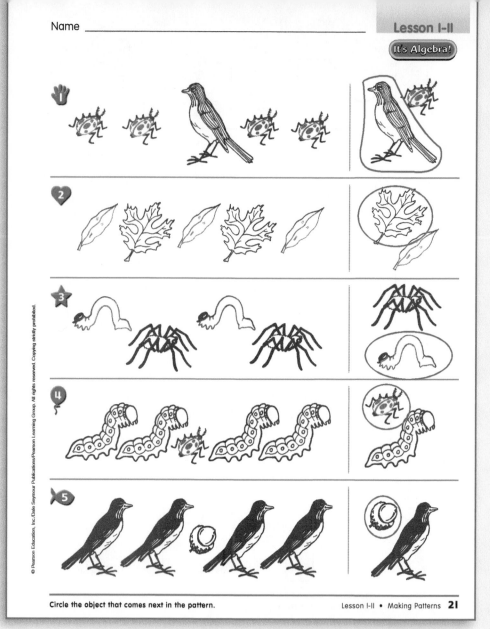

Circle the object that comes next in the pattern.

Lesson 1-11 • Making Patterns   **21**

2 Teach

Develop Skills and Concepts Use large and small colored blocks to lay out a pattern as follows: large yellow, small orange, large yellow, small orange, large yellow. Have students say the pattern aloud. Ask students what would come next in the pattern. (small orange) Have a student place the correct block in the pattern for all students to say the pattern.

- Lay out large and small colored blocks for students to work in pairs to build patterns.

- Have students string beads or macaroni to form patterns you show. Then, have students string patterns you dictate. Allow time for them to work independently to create their own patterns to share verbally with classmates. You may want to display the patterns for students to identify each pattern during free time.

- Have students listen and watch as you pat your knees, clap your hands, pat, and clap again. Tell students that this is also a pattern. Verbalize the pattern as you repeat it. Then, have students say and make the pattern with you. Continue for the following patterns, verbalizing each: clap, clap, pat, clap, clap, pat; clap, pat, pat, clap; pat, clap, touch shoulders, pat, clap, touch shoulders; pat, pat, clap, touch shoulders, touch head, pat, pat, clap, touch shoulders, touch head. Repeat action patterns, varying the motions. You may want to group students into pairs with partners facing each other to do patterns such as pat, clap, right hand to partner's left hand, left hand to partner's right hand, clap, pat.

It's Algebra! The concepts in this lesson prepare students for algebra.

T21

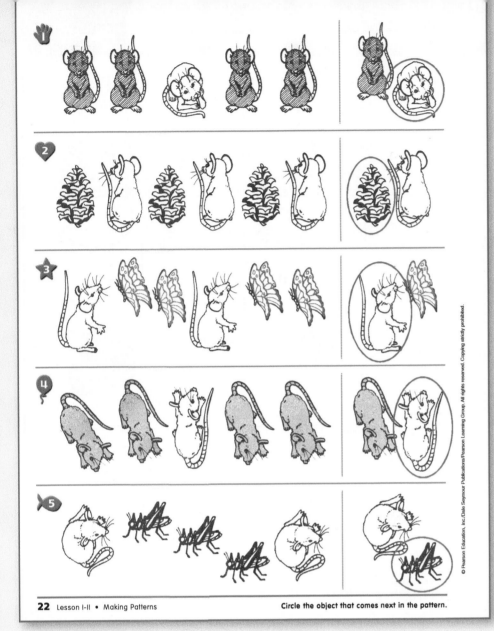

22 Lesson I-II • Making Patterns   Circle the object that comes next in the pattern.

For Mixed Abilities

Common Errors • Intervention

Some students may have difficulty continuing patterns. Lay out two erasers, a pencil, another two erasers, and another pencil. Have students work with partners to say the pattern and continue it by laying out the next several objects. Continue for more object patterns, helping students say each pattern rhythmically. Each example you start for the students should contain at least two sample pattern sets. Some students may want to create their own patterns. Ask them to explain how their patterns work.

Enrichment • Sequencing

1. Give students many small paper squares or rectangles in different colors. Show students how to lay out the shapes to form patterns such as they see on floors, walls, and so on.

2. Discuss the routines we follow each day and tell students that these routines are patterns. (eat, brush teeth, play, eat, brush teeth, play; get up, go to school, go to bed, get up, go to school, go to bed, and so on)

3 Practice

Using page 21 Have students identify the animals and objects they see on this page. Tell students that again they are to look at the choices on the right and choose the picture that comes next in each pattern. Have students say the pattern shown in Exercise 1 and tell which picture should come next. (bird) Have them say the pattern again to check their work. Then, have students say the pattern in Exercise 2, and choose and circle the picture that continues it. (big leaf) Have students repeat the pattern orally. Continue for Exercises 3 to 5, having students say each pattern, choose and circle the correct answer, and then repeat the pattern orally. You may then want to have students color the animals and objects so that a different color is used for each different object within a pattern.

• You may want to have students compare all the patterns on the page. Ask, *How are the patterns the same? How are they different?* Point out that all of the patterns

use objects and animals. The difference is that in Exercises 1, 4, and 5, they are all AAB patterns, and in Exercises 2 and 3, they are AB patterns.

Using page 22 Ask what pattern is used in Exercise 1. (standing mouse, standing mouse, sitting mouse) Have students point to the standing mouse and the sitting mouse in the answer column. Have them say the pattern aloud and circle the picture that should come next. (sitting mouse) Have students say the pattern again to check the answer. Repeat the procedure for Exercises 2 to 5. Have students say the pattern aloud to check answers. You may want them to color the objects in each pattern.

4 Assess

Create a patterned paper chain using two different colors of construction paper. Have students tell what color the next link in the chain should be and then attach a link of that color. Repeat until all students have added a link to the chain.

T22

Chapter 1 Test and Challenge

pages 23–24

1 Getting Started

Test Objectives

- **Item 1:** To describe position using *top, bottom, in, out, over, under, on top of, off, above, below, beside, outside,* and *inside* (see pages 1–2)
- **Item 2:** To describe position using *first, last, middle, between, front, behind, right,* and *left* (see pages 3–4)
- **Item 3:** To identify red, yellow, blue, orange, green, purple, brown, and black; to recognize each color word (see pages 5–6)
- **Item 4:** To compare using *shorter, shortest, taller, tallest, longer,* and *longest* (see pages 9–10)
- **Item 5:** To identify a circle, rectangle, square, and triangle; to identify and compare the attributes of plane shapes (see pages 11–12)
- **Item 6:** To sort and classify a set of shapes by more than one attribute (see pages 13–14)
- **Item 7:** To identify a sphere, cone, cube, and cylinder; to identify and compare the attributes of solid figures; to match objects to outlines of their shapes (see pages 15–16)
- **Item 8:** To identify shapes with a line of symmetry (see pages 17–18)
- **Item 9:** To identify patterns using color, size, or shape; to complete a pattern; to identify the shape that comes next in a pattern (see pages 19–20)

Challenge Objective

- To use vocabulary to describe the position, color, and size of objects

Materials

*2 different-sized tissue boxes

2 Give the Test

Prepare for the Test Bring in two different-sized tissue boxes. Put them under a book on top of a table in the front of the room. Have students describe the position and colors used in the tissue boxes. Be sure to listen for the correct vocabulary used by the students. (Possible

Circle the bug that is over the flower. Circle the bird that is first. Color each crayon the correct color. Circle the ribbon that is longest. Color the circles blue. Circle the square that is small and blue. Circle the object with the same shape. Circle the object with matching parts. Circle the shape that comes next in the pattern.

Chapter 1 • Test **23**

answers: The tissue boxes are under the book; the tissue boxes are on top of the table.)

- On the board, draw a rectangular prism resembling each tissue box. Label the rectangular prisms A and B. Ask, *Which tissue box has the same shape as solid figure A?* (Answers will vary.) Repeat this for solid figure B.
- After students have mastered the vocabulary needed to describe the position, color, length, and size of objects, sorting and comparing plane shapes and solid figures, as well as the ability to identify and complete patterns, have them complete the Chapter 1 Test. You may wish to review test-taking rules with students. Remind them that they are not to talk to each other and they are to finish the test on their own.

Using page 23 Students should be able to complete the test independently. For Exercise 1, students must know position vocabulary to choose the ladybug that is over the flower. For Exercise 2, they must know position vocabulary to choose the bird that is first. For Exercise 3, students will need to recognize the color words, identify

For Mixed Abilities

Common Errors • Intervention

Some students may have difficulty recognizing a pattern when asked to draw the shape that should come next. Suggest that students work in pairs and talk through the pattern together. Each partner should tell the other the shapes in the pattern, being sure to describe the colors. Then, each partner should decide on the next shape in the pattern. Both students must agree on the shape, helping each other if necessary.

Enrichment • Application

1. Give each student a sheet of paper listing the names of classmates who own a pet. Tell students to interview one classmate to find out as many details about this pet as possible, such as color and size. Then, have students draw these pets.

2. Have students bring in newspapers and magazines. Then, have them find similar-shaped objects in the newspapers and magazines that differ by size or color. Provide sheets of heavy paper for students to paste their selections on. Then, have them present their work to the class.

Color the picture. Then use the correct vocabulary to describe the position, color, and size of objects in the picture.

24 Chapter I • Challenge

the matching color, and color the appropriate crayon. For Exercise 4, students must know size vocabulary to choose the longest ribbon. For Exercise 5, students will identify and color two circles from a group of varied plane shapes. For Exercise 6, students will sort the set of shapes using two attributes to find the one plane shape that is both small and blue. For Exercise 7, students will identify and recognize the attributes of solid figures in order to find the object that matches the shape of the given solid figure. For Exercise 8, students will identify the shape with matching parts. For Exercise 9, students will identify the shape that comes next in the pattern.

3 Teach the Challenge

Develop Skills and Concepts for the Challenge Have students list the vocabulary they learned in this chapter. Create a vocabulary list on the board. Then, sort the words into groups of position, color, size, shapes, and solid figures. Have students volunteer sentences that use both the color and position vocabulary in one sentence. (Possible answer: My *yellow* pencil is *on top of* my desk.)

Using page 24 Allow students to color the picture first, using only blue, red, green, black, yellow, orange, purple, and brown crayons. Then, allow students to use the vocabulary listed on the board to describe the position, color, and size of objects in the picture.

4 Assess

Alternate Chapter Test You may wish to use the Alternate Chapter Test on page 205 of this book for further review.

Challenge To assess students' understanding of vocabulary needed to describe position, color, and size, lay three different-sized crayons on top of a book and place all of these objects on top of a desk. Have students describe the crayons' location, color, and size. (Possible answer: The red crayon is smaller than the blue crayon.)

pages 25–44

Chapter Objectives

- To use one-to-one correspondence to compare groups of objects without counting
- To count and compare one through five objects
- To associate the numbers 1 to 5 with the corresponding number of objects
- To define *one*, *two*, *three*, *four*, and *five*
- To write the numbers 1 through 5
- To connect numbered dots in order from 1 to 5
- To compare one to five objects using *as many as*, *more*, *fewer*, *more than*, *fewer than*, and *less than*

Materials

You will need the following:

*small bowls and spoons; *number cards 1 through 5; *three cardboard circles; *sets of 3 objects; *4 empty cans labeled 1, 2, 3, and 4; *ice-cream sticks; *5 cans or boxes; *rope or string about 10 feet long with a knot at every 2-foot interval; *tape

Students will need the following:

Reproducible Masters RM5–RM8; small objects for sorting and counting; yarn; counters in several colors; red and green sheets of paper; number cards 1 through 5; number cards 1 through 5 in relief; yarn; modeling clay; ice-cream sticks; green, red, and black crayons; pictures of groups of 1 through 5 objects

Bulletin Board Suggestion

Place large numerals on the bulletin board. Have students bring in pictures showing one, two, three, four, or five objects so that two pictures are displayed under each large number as it is learned. Have students connect yarn pieces from each picture to its number to show how many objects are shown.

Verses

Poems and songs may be used any time after the pages noted beside each.

Things in Twos (pages 29–30)

2 arms for reaching
2 legs that walk
2 nostrils for smelling
2 lips to talk

2 hands for holding
To hug or to touch
2 ears for listening
And washing too much

Twos by John Drinkwater (pages 29–30)

Why are lots of things in twos?
Hands on clocks and gloves and shoes,
Scissor-blades and water-taps,
Collar studs and luggage straps,
Walnut shells and pigeon's eggs,
Arms and eyes and ears and legs—
Will you kindly tell me who's
So fond of making things in twos?

If All the Seas Were One Sea (pages 29–30)

If all the seas were one sea,
What a BIG sea that would be!
And if all the trees were one tree,
What a BIG tree that would be!
And if all the axes were one axe,
What a BIG axe that would be!
And if all the humans were one human,
What a BIG human that would be!
And if that BIG human took that BIG axe
And chopped down that BIG tree,
And let it fall into that BIG sea
Hmmm! What a BIG splash THAT would be!

One for Me (pages 39–40)

We have five seat belts in our car.
One for Dad, one for Mom,
One for Grandpa, one for Grandma
And . . . one for me.

This Old Man (pages 41–42)

This old man, he played one.
He played knickknack on my thumb,
With a knickknack paddy whack,
Give the dog one bone.
This old man came rolling home.

You may wish to extend the song.
(2/shoe/2 bones; 3/knee; 4/door; 5/hive)

We Have a Whole Lot of Numbers (pages 41–42)

We have a whole lot of numbers in our hands.
We have a whole lot of numbers in our hands.
We have a whole lot of numbers in our hands.
Let's see what we have.

We've got the number 1 in our hands.
We have 1 more than zero in our hands.
We have 1 less than 2 in our hands.
We have a 1 in our hands.

You may wish to extend the song as follows:
2: 1 more than 1;   1 less than 3
3: 1 more than 2;   1 less than 4
4: 1 more than 3;   1 less than 5

Three Nice Mice

Mr. and Mrs. Mouse had a house. The house was at the *bottom* of a tall wall. Mr. and Mrs. Mouse had three little mice. The three little mice were nice. All five of the mice lived together at the *bottom* of the tall wall.

The three little nice mice liked to run and squeak and play. They squeaked and ran *in* and *out* of the house. They squeaked and ran *over* each other. They squeaked and ran *under* each other. They squeaked and ran *beside* each other. And they squeaked and ran—and crashed—into each other!

Sometimes they ran over the *top* of things. Sometimes they ran under the *bottom* of things. But mainly they squeaked and ran *in* and *out* of the house.

"Please, stay *inside* or *outside*!" Mother Mouse said.

"And stop running!" scolded Father Mouse. "Your legs can walk as well as run. Your mouths can close as well as open. I'm tired of all the squeaking and running! For once, please walk and be quiet!"

Then one day the three nice little mice all squeaked and ran to the door at once. They all got to the door at the same time. They all tried to run *out* of the door at the same time. And they all got stuck!

One was stuck on the *top*. Two were stuck on the *bottom*. The two on the *bottom* were stuck *beside* each other! And they were all stuck both inside and outside the door! And for once, all three of the nice little mice had stopped squeaking and running.

Mr. and Mrs. Mouse laughed and laughed and laughed!

But as soon as the three little nice mice were unstuck, they squeaked and ran *off* to squeak and run some more!

RM5 Chapter 2 • "Three Nice Mice"   Describe the picture using position words.

"Three Nice Mice," p. RM5

Activities

Use Reproducible Master RM5, "Three Nice Mice," to help students relate the story to the mathematical ideas presented.

- Reread the story. Have students use stuffed animals, small toy cars, or similar toys in place of the three little mice to act out the doorway predicament. Have each student tell about his or her character's position. Allow time for several groups to participate. Have students then reenact the story, creating their own dialogue.

- Discuss with students the subject of taking turns and forming lines to wait for one's turn. Ask them what might happen if you brought a kitten into the classroom and you said every student could pet it. Students should see that everyone would crowd around and want to be first. Help students verbalize and act out ways to make sure that all students would get a turn. (form a line, form small groups, take the kitten around to each student, and so on) Help students name situations in which people form lines and wait for their turn. (in a bank, in a grocery store checkout, in traffic, going into a sports arena, and so on) Encourage students to include position words as they tell about times they have stood in line to wait their turn.

2-1 As Many As

pages 25–26

1 Getting Started

Objectives

- To match objects one-to-one without counting
- To compare using *as many as*
- To use one-to-one correspondence to compare the objects in two sets without counting

Vocabulary

as many as, one-to-one correspondence

Materials

*small bowls and spoons; objects for sorting; Reproducible Master RM6

Warm Up • Mental Math

Have students respond to the following:

1. Is a mouse larger than an ant? (yes)
2. Who is sitting beside you?
3. Turn your head to your left.
4. Name something that is between you and the door.
5. Do you weigh more or less than a baby? (more)

Warm Up • Classification

Display objects and have students arrange them into groups according to whether or not they have the same size, shape, color, use, and so on. Have students tell why they have grouped the objects as they have. (Possible answers: Knives and forks are for eating; these objects are blue; these items have wheels.)

2 Teach

Develop Skills and Concepts Place 4 bowls and 4 spoons randomly on a table. Have students arrange the objects into groups having the same shape or use. Ask students to explain their groupings. (These are bowls and these are spoons.) Tell students that they are to find out if there are as many spoons as bowls. Have students place a spoon in each bowl. Ask if there are any spoons left over. (no) Ask if any bowls do not have spoons. (no) Ask if there are as many spoons as there are bowls. (yes) Ask students to repeat that answer in a sentence with you. (There are as many spoons as bowls.)

Lesson 2-1

It's Algebra!

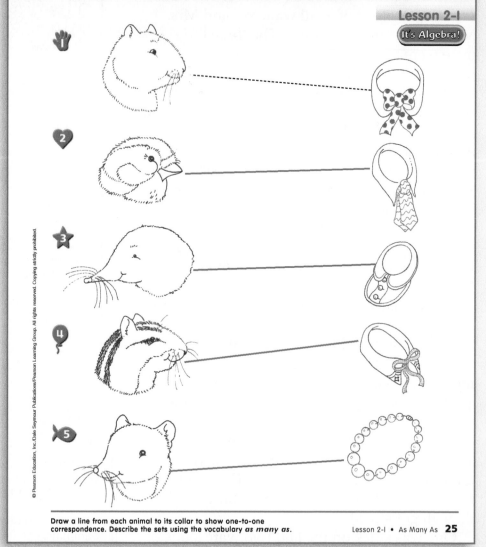

Draw a line from each animal to its collar to show one-to-one correspondence. Describe the sets using the vocabulary *as many as*.

Lesson 2-1 • As Many As **25**

- Place 5 small toys on a table and call 5 students' names to stand beside the table. Tell students that they are to find out if there are as many toys as students standing beside the table. Have each student pick up a toy. Ask if there any toys left over. (no) Ask if any student does not have a toy. (no) Ask if there are as many toys as students. (yes) Tell students to repeat the answer in a sentence with you. (There are as many toys as students.) Note: Students are not asked to count objects at this time, although some may do so.

- Draw 3 stick people in a group and 3 balls in another group on the board. Ask students how they might find out if each stick person can have a ball. Some students may relate to earlier activities and want to move the drawings by erasing and redrawing a ball beside a stick person. Help them realize that drawing a line from a stick person to a ball quickly pairs the stick person to the ball. Help students make statements to compare the two groups. Note that counting is not done at this time.

T25

26   Lesson 2-1 • As Many As

Draw a line from each animal to its treat to show one-to-one correspondence. Describe the sets using the vocabulary *as many as*.

For Mixed Abilities

Common Errors • Intervention

Some students may need help understanding the expression *as many as*. Have 4 students stand and 4 sit facing them. Give each sitting student a piece of yarn. Have each piece of yarn be extended from a sitter to a standee to see if there are as many standing students as sitting students. Have students extend the yarn. Ask, *Are there as many standing as sitting?* (yes) *Are there as many sitting as standing?* (yes) Repeat the activity.

Enrichment • Application

Have students draw a picture that shows a boy and his brother playing ball but not sharing the same bat.

ESL/ELL STRATEGIES

Throughout this chapter, demonstrate each new instruction: *draw a line, draw a circle, trace,* and so on. Then, have students use their fingers to practice the action on the appropriate page without making a mark. Check that all students understand before assigning the page.

• Have students complete Reproducible Master RM6.

It's Algebra! The concepts in this lesson prepare students for algebra.

3 Practice

Using page 25 Help students identify the animals on the left side of the page. (squirrel, bird, mole, chipmunk, weasel) Tell students that each animal wants a collar and we need to find out if there are as many collars as there are animals. Remind them that to find out if there are enough collars, we match an animal to a collar until all the animals have collars or until there are no collars left. Tell students to place their pencil point on the first animal and trace the line to the collar. Tell them to draw lines to match the other animals to their collars. Ask if there are any collars left over. (no) Ask if any animal does not have a collar. (no) Have students say, "There are as many collars as animals," and "There are as many animals as collars."

Using page 26 Have students identify the animals on the left side of the page. Tell them that each animal now wants a treat and that we need to see if there are as many treats as there are animals. Have students draw a line from the first animal to the first treat. Tell students to continue to match each animal to a treat until all the animals have treats or until all the treats are gone. Ask if there are any treats left over. (no) Ask if any animal does not have a treat. (no) Have students say, "There are as many treats as animals," and "There are as many animals as treats."

4 Assess

Draw 5 stick people in a vertical row on the board. In another vertical row, draw 5 cookies. Have students determine if there are as many cookies as there are people. (yes)

T26

2-2 More and Fewer

pages 27–28

1 Getting Started

Objectives

- To match objects from two groups to compare for more or less
- To compare using *more, fewer, more than, fewer than,* and *less than*

Vocabulary

more, fewer, more than, fewer than, less than

Materials

counters in several colors; red and green sheets of paper

Warm Up • Mental Math

Have students do the following:

1. put their hand on their right eye
2. put both thumbs on their left ear
3. make a circle in the air with their right hand
4. put their head between their hands
5. point to the body part between their shoulders

Warm Up • Application

Lay out 3 green and 2 red sheets of paper on a table. Have 5 students stand and individually approach the table to take a sheet of colored paper. Discuss whether there was enough paper for each standing student and why.

Draw a line from each animal to a chair to show one-to-one correspondence. Circle the set that has more.

Lesson 2-2 • More and Fewer **27**

2 Teach

Develop Skills and Concepts Lay out a precounted stack of red and green paper that has at least one more green than red. Have the rest of the students approach the table one by one to get a sheet. Have students who have red paper stand together in a group and those who have green stand together. Ask if anyone knows a way to find out if there are as many students having red as those having green. Students should volunteer pairing reds and greens until no more can be paired. Have matched pairs stand together and ask if there are as many that have red as green (no) and as many that have green as red (no), and if there are any greens left over (yes). Then, have students say, "There are more greens than reds." Tell them that there are fewer reds than greens and have them repeat the sentence with you.

Then, ask which group has more and which group has fewer, and have several students make statements to that effect.

- Draw 2 bare trees on the board. Have a student draw an apple on each tree. Ask if one tree has more apples than the other. (no) Have another student draw an apple on each tree and repeat the question. (no) Continue having students draw an apple on each tree until the left tree has 7 apples and the right tree has 8 apples. Now, have a student draw an apple on the tree on the right only. Ask students to tell about the trees now. (This tree has more apples or this tree has fewer apples.) Help them see that if this last student draws another apple on the left tree, the trees will have the same number of apples.

It's Algebra! The concepts in this lesson prepare students for algebra.

T27

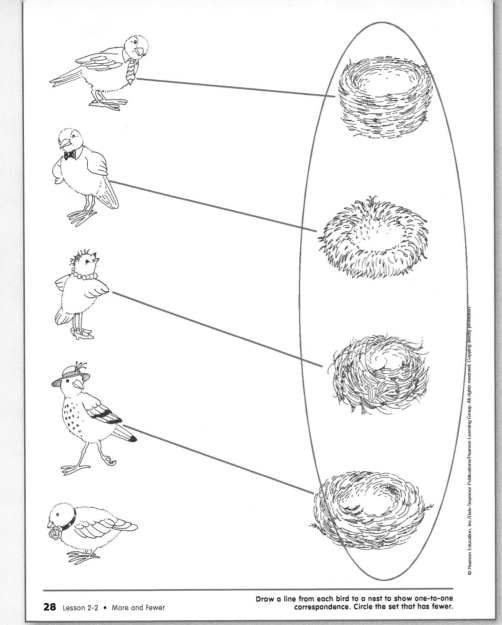

Draw a line from each bird to a nest to show one-to-one correspondence. Circle the set that has fewer.

28 Lesson 2-2 • More and Fewer

For Mixed Abilities

Common Errors • Intervention

Some students may need additional help in understanding the vocabulary. Have them work with partners and a group of 5 red and a group of 7 blue counters. Have the students match 1 red counter with 1 blue counter and continue to match the counters one to one. Ask, *Do you have any counters left over?* (yes) *What color are they?* (blue) *So, are there more blue counters or more red counters?* (more blue) Change the number of counters and repeat.

Enrichment • Application

Have students trace their hands on a sheet of paper and decorate their fingers with rings so that the left hand has more rings.

ESL/ELL STRATEGIES

Students may be very curious about the English names for the plants, animals, and objects that appear throughout this chapter. Try to avoid spending a lot of time on vocabulary. Instead, substitute words such as *birds*, *animals*, and *things* when discussing the objects.

3 Practice

Using page 27 Tell students that now that the animals have put on their collars and each has had a treat, they are tired and want to sit down. Tell students that they are to find out if there are as many chairs as animals. Ask how they can find out. (draw a line from each animal to a chair) Have them trace the line from the weasel to the chair. Supervise students in drawing a line from an animal to a chair. Ask if there are enough chairs for each animal to have one. (no) Ask if there are more animals than chairs. (yes) Have students say, "There are more animals than chairs." Have students draw a circle around the animals to show that there are more animals than chairs.

Using page 28 Tell students that they are to find out if there are fewer birds or fewer nests on this page. Supervise students as they draw a line from a bird to a nest. Ask if there is a nest for each bird. (no) Ask if there

are fewer nests than birds. (yes) Have students say, "There are fewer nests than birds." Tell students to draw a circle around all the nests to show that there are fewer nests than birds.

4 Assess

Lay out a group of 5 counters. Ask 4 students to each take a counter. Ask the class, *Are there more students or more counters?* (more counters) Then, lay out 3 counters. Ask a different group of 4 students to each take a counter. Ask, *Are there fewer students or fewer counters?* (fewer counters)

T28

pages 29–30

1 Getting Started

Objectives

- To associate the number 1 with a set of one object
- To associate the number 2 with a set of two objects
- To define *one* and *two*

Vocabulary

one, two

Materials

*tape; small objects for sorting; yarn; number cards 1 and 2

Warm Up • Mental Math

Ask students which is smaller.

1. them or their house (them)
2. their finger or their leg (finger)
3. their wrist or their head (wrist)
4. the whole school or this classroom (this classroom)
5. their hand or a baby's hand (baby's hand)

Warm Up • Application

Lay out 4 small objects and 4 yarn pieces. Ask students if there are as many yarn pieces as there are small objects. (yes) Repeat this activity by adding 2 more small objects and asking if there are more small objects than yarn pieces or fewer small objects than yarn pieces. (more small objects than yarn pieces)

2 Teach

Develop Skills and Concepts Hold up one finger and tell students that we say the word *one* to mean one thing. Tell students to watch you use one finger as you point to your arm, head, eye, and so on as you recite this poem:

One finger, one arm,
One head, one eye.
One me, one you,
One nose, one sky.

Repeat the poem several times as students do the pointing and join in the recitation.

- Hold up both index fingers and tell students that we say the word *two* to mean two things. Tell students to listen for the word *two* in this poem as you use two

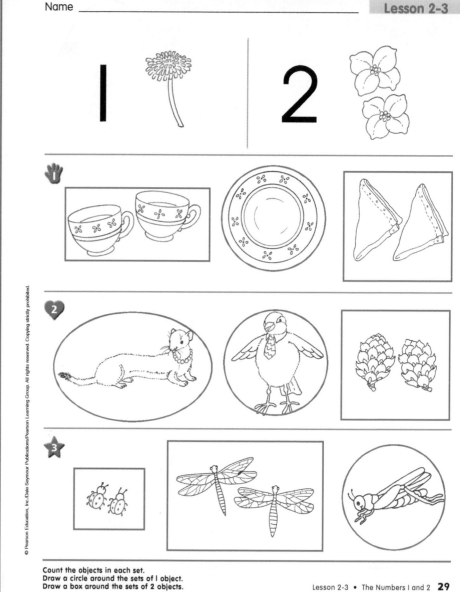

Count the objects in each set.
Draw a circle around the sets of 1 object.
Draw a box around the sets of 2 objects.

Lesson 2-3 • The Numbers 1 and 2 **29**

fingers to point:

Two arms, two legs,
Two shoulders, two hips.
Two nostrils, two feet,
Two shins, two lips.

- Distribute the number cards and have students show the number whose name is heard as they say each poem again.

- Place one or two objects on the table so that sets of one or two can be sorted by size, shape, color, or use. Use tape to make a large 1 and a large 2 each inside a yarn ring. Help students identify each number with its oral word name. Ask students to place one object in the 1 ring. Remove each group of objects before placing the next in the ring. Repeat for the 2 ring.

- Distribute the number cards and have a student hold up a card and say its number name. Then, have the same student name a group in the room that contains that many objects.

1   2   1   2

1   2   1   2

1   2   1   2

1   2   1   2

30 Lesson 2-3 • The Numbers I and 2   Circle the number that tells how many objects are in the set.

For Mixed Abilities

Common Errors • Intervention

Some students may have difficulty associating a number with a set. Have them practice by playing the Pop Up and Count game. Have a student sit or stoop behind a barrier. Tell the student to pop up when called by name. Call the student's name and say *one* when you see the student. Remind students that you did not count 1 until the student popped into sight. Repeat for two students and the number 2. Then, have students count as you pop up one or two items from behind the barrier.

Enrichment • Number Sense

1. Have students work in pairs to build a person with felt or magnetic pieces. Have one student position one trunk, two arms, and so on as the other student calls for the parts by name and number.

2. Have students cut pictures of groups of one and two from magazines and describe their choices to the class. Have students paste the objects on paper under the number 1 or 2.

3 Practice

Using page 29 Ask students the name of the number in the first example at the top of the page. (one) Tell them to look at the flower. Ask if this is one flower. (yes) Have students say, "There is one flower." Tell them to circle the flower to show there is one.

- Then, ask students the name of the number in the second example at the top of the page. (two) Tell them to look at the flowers. Ask if this picture shows two flowers. (yes) Have students say, "There are two flowers." Tell them to draw a box around the two flowers to show there are two.

- Repeat for the items in Exercise 1. Have students circle the set of one item and draw a box around the sets of two items. Have students complete Exercises 2 and 3 independently.

Using page 30 Have students name the numbers in Exercise 1. (1, 2) Ask them which number tells how many teapots there are. (1) Tell students to trace the circle around the 1 to show that there is one teapot. Continue similarly to talk through the next two exercises to be sure students understand to circle the number that tells how many objects there are. Then, have students complete the page independently.

4 Assess

Tell students that you will point to a part of your body and they are to tell if a human has one or two of that part by raising one or two fingers.

T30

2-4 The Number 3

pages 31–32

1 Getting Started

Objectives

- To associate the number 3 with a set of three objects
- To recognize that 3 is more than 2 or 1
- To define *three*

Vocabulary

three

Materials

*3 cardboard circles; *number cards 1, 2, and 3; *sets of 3 objects; counters

Warm Up • Mental Math

Ask students to tell how many there are.

1. the student and a friend (2)
2. ears on their head (2)
3. wheels on a bicycle (2)
4. their teachers
5. all the students in the class

Warm Up • Number Sense

Show 1 counter and ask students how many. (1) Ask them how many counters you would have if you have 1 more as you hold up a second counter. (2) Have students count the 2 counters. Repeat for more groups of two objects.

2 Teach

Develop Skills and Concepts Tell students to count with you as you count out 2 cardboard circles. Place 1 more circle with the 2 and tell students that there are now 3 circles. Write **3** on the board and have students say the number name. Point to each of the circles as you say the numbers 1, 2, and 3. Write **1** and **2** on the board in front of the 3. Have a student hand you 1 circle as you count it and point to the 1 on the board. Have the student hand you another circle as you count it and point to the 2. Repeat for the last circle and the number 3. Have students answer you in a sentence when you ask how many circles there are. (There are 3 circles.) Hold up 1, 2, or 3 circles and have students count them and point to the number on the board that tells how many there are.

- Give the number cards to three students. Have each student say the number and identify a set of objects in the room that has that number. Repeat until all students have correctly associated a number card with a set containing that number of objects.

- Lay out many counters on a table. Have a student count out 3 counters from the group on the table. Have another student count to check the first student's work and then count out another group of 3 counters. Continue until all counters are in groups of 3.

- Lay out many objects so that there is one of some items, two of some, and three of others. Ask a student to find all the large blue circles, for instance, and tell how many there are. Continue until all objects have been grouped by kind, size, and so on. Then, have students name a group that has more than, fewer than, or the same as a group you name. Have them tell why. Continue until many comparisons have been made.

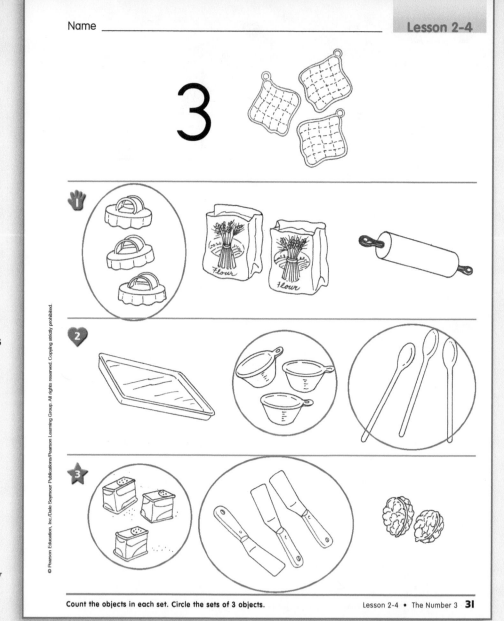

Count the objects in each set. Circle the sets of 3 objects.

Lesson 2-4 • The Number 3 **31**

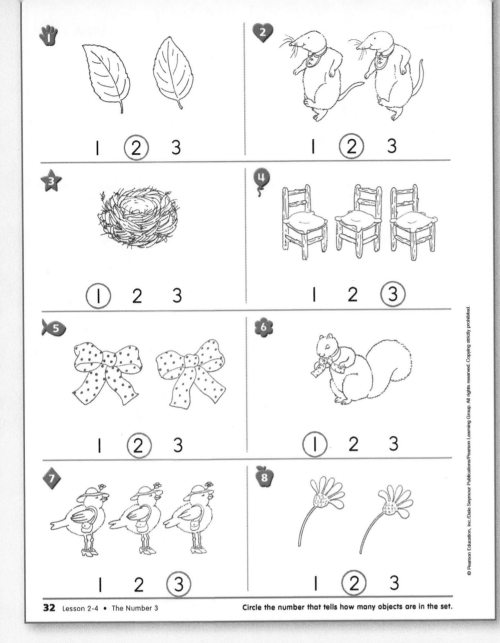

1. | 2 | 3

2. 1 | (2) | 3

3. (1) | 2 | 3

4. 1 | 2 | (3)

5. 1 | (2) | 3

6. (1) | 2 | 3

7. 1 | 2 | (3)

8. 1 | (2) | 3

32 Lesson 2-4 • The Number 3

Circle the number that tells how many objects are in the set.

For Mixed Abilities

Common Errors • Intervention

Students may need more practice with the number 3. Draw a large 2 and 3 on the board. Have one student take 3 objects to the board to hold beside the 3 and another student take 2 objects to hold beside the 2. Then, have them draw 2 large dots under the 2 and 3 large dots under the 3.

Enrichment • Number Sense

1. Have students cut pictures of 1, 2, or 3 objects from magazines. Provide three boxes, each with a large 1, 2, or 3 on it. Mix up the pictures and have students choose one and place it in the correct numbered box.

2. Give students three 3 × 5-in. cards that have a 1, 2, or 3 on them. Have students color one, two, or three objects on another set of three cards. Have students use the cards to play a matching game.

3 Practice

Using page 31 Have students name the number in the example. (3) Ask students to count the hot pads and tell how many there are. (3) Tell students to circle the group of hot pads because there are 3 of them to go with the number 3 on this page. Tell students that they are to count each group of objects and circle the group if there are 3 in it. Ask them if they will circle a group of 1 or 2 utensils. (no) Have students complete the page independently. Then, have students tell about each group of objects in a sentence.

Using page 32 Tell students that they are to count the objects in the group and circle the number that tells how many objects there are. Have students complete the page independently. Then, ask if there are as many nests as chairs (no), more than one bow (yes), fewer squirrels than moles (yes), and so on.

4 Assess

Have students look through magazines to find pictures of three objects. Have them cut out those pictures and create a collage of the number 3.

T32

Name _____

1 Getting Started

Objectives

- To count one, two, three, or four objects
- To define *four*
- To recognize that 4 is more than 3, 2, or 1

Vocabulary

four

Materials

*counters; *4 empty cans labeled 1, 2, 3, and 4; *ice-cream sticks

Warm Up • Mental Math

Ask students what number is
1. less than 2 (1)
2. the same as 2 (2)
3. less than 3 (1 or 2)
4. the same as the number of their hands (2)
5. the same as the number of clocks in the room

Warm Up • Number Sense

Write **3**, **1**, and **2** on the board and have a student point to each number you say. Have students count to that number. Now, have them draw dots to show each number.

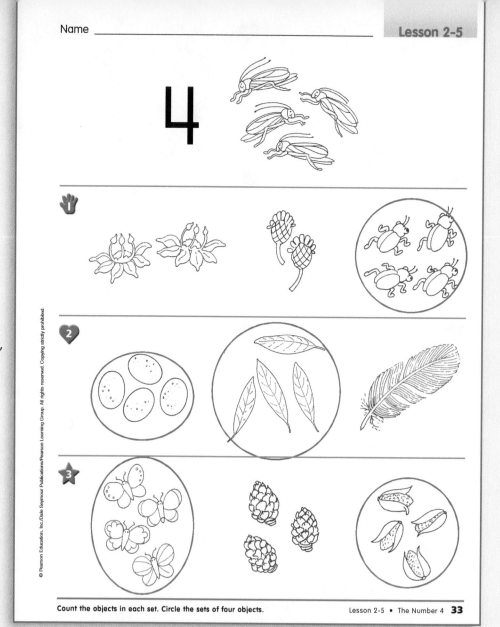

Count the objects in each set. Circle the sets of four objects.

2 Teach

Develop Skills and Concepts Have students count as you lay out 3 counters. Write **1**, **2**, and **3** on the board. Put 1 more counter with the 3 and say *four* as you do so. Write **4** on the board and tell students that 4 comes after 3 because 3 and 1 more is 4. Have them count as you draw 1, 2, 3, or 4 large dots under each number on the board.

- Draw 4 large circles on the board. Draw 3 houses, trees, or similar objects in the first circle, 1 in the second, 4 in the third, and 2 in the remaining circle. Have students count the number of objects in each circle and then say a sentence to tell how many objects there are. (There are three houses in this circle.)

- Now, write the numbers **1**, **2**, **3**, and **4** in random order on the board. Point to the first circle of objects and ask students which of the numbers tells how many are in the circle. (3) Have students draw a line from the circle of objects to the 3. Repeat to match the other groups with their correct numbers. Erase the circles and drawn lines and have students circle the set of 4 and draw a line to the 4. Repeat for the other groups of objects.

- Show students the number 4 can. Ask what number is on the can. (4) Have students count with you until 3 ice-cream sticks are placed in the can. Ask them if the number of sticks in the can is the same as the number on the can. (no) Ask if it is more. (no) Ask if it is less. (yes) Ask what is needed to be the same number. (put 1 more in) Have a student put 1 more stick in the can and count to check the number. Repeat the activity for the other cans.

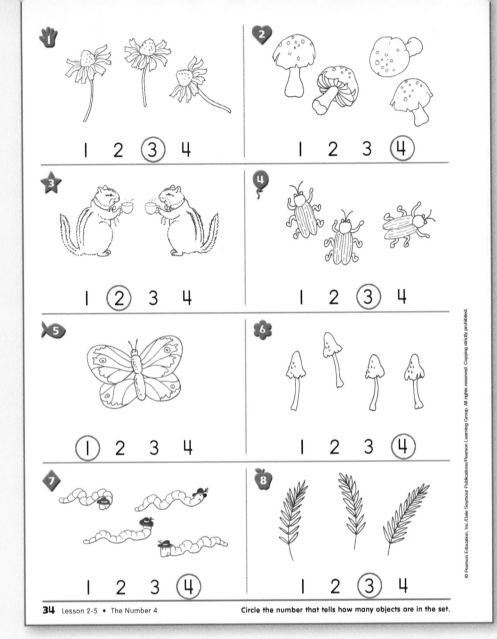

1   2   ③   4

1   2   3   ④

1   ②   3   4

1   2   ③   4

①   2   3   4

1   2   3   ④

1   2   3   ④

1   2   ③   4

34 Lesson 2-5 • The Number 4   Circle the number that tells how many objects are in the set.

For Mixed Abilities

Common Errors • Intervention

Some students may need more practice with the numbers 1 through 4. Begin with a fist and raise one finger at a time as you say:

First comes 1, and then comes 2. This is how we count to 2. Next is 3, and then comes 4. Count four things—1, 2, 3, 4.

Then, have students say the poem with you as they show the number of fingers or objects.

Enrichment • Number Sense

1. Have students cut out the numbers 1, 2, 3, and 4 from a sheet of numbers and paste one number onto each of four paper bags. Tell students to fill their bags with pictures of groups of one, two, three, or four objects cut from magazines or newspapers.

2. Tell students a story of a boy, his sister, their dad, and their grandpa going for ice-cream cones. Tell students to draw the ice-cream cones to show that the children had two scoops each, whereas the adults had one scoop each.

3 Practice

Using page 33 Have students point to the number 4 in the example. Ask students to count how many grasshoppers there are. (4) Tell them to circle the group of 4 grasshoppers. Tell students that they will circle all the groups that have 4 in them in each exercise. Have students complete the page independently. Now, ask students if there are more butterflies than pine cones (yes), more eggs than leaves (no), fewer pine cones than flowers (no).

Using page 34 Tell students that they are to count the objects in the set and circle the number that tells how many objects there are. Have them complete the page independently. Then, ask a student to choose a group of objects and use a sentence to tell about the number of objects in that group. Continue for more students to tell about groups of objects.

4 Assess

Draw four groups of flowers on the board. Draw 3 flowers in the first group, 2 in the second, 4 in the third, and 1 in the last. Point to each group and have students hold up 1, 2, 3, or 4 fingers to show how many are in each group.

1 Getting Started

Objective

• To write 1 and 2

Materials

number cards 1 and 2 in relief; Reproducible Master RM7; 4 yarn rings; counters; clay; ice-cream sticks; green, red, and black crayons

Warm Up • Mental Math

Have students answer yes or no.

1. 2 is less than 3. (yes)
2. You are larger than an elephant. (no)
3. 3 is more than 2. (yes)
4. 2 is less than 1. (no)
5. More than 2 is 3. (yes)
6. I live outside my house. (no)
7. The stomach is in the middle of the body. (yes)

Warm Up • Number Sense

Have students form groups of 1, 2, 3, and 4 counters in yarn rings according to numbers you write on the board. Have students check each other's work.

Trace the paths for writing 1 and 2.

Lesson 2-6 • Writing 1 and 2 **35**

2 Teach

Develop Skills and Concepts Write **1**, **2**, **3**, and **4** across the board and draw a green dot under the 1 and a red X under the 4. Tell students that the green dot tells us where to start and the red X tells us where to stop. Draw a line between the dot and the X. Have students count with you from the green dot as you point to each number, ending at the red X. Now, have students take turns pointing as classmates count from 1 to 4.

• Display large numbers 1 and 2 that have directional arrows along with a green dot at the starting point and a red X at the end. Have a student start at the green dot and trace each number. Provide time for all students to trace the numbers.

• Trace the large numbers 1 and 2 on the number cards in relief as you sing the following song to the tune of "Mulberry Bush":

We start at the top and go straight down,
We start at the top and go straight down,
We start at the top and go straight down
To write the number 1.

We curve around, then straight across,
We curve around, then straight across,
We curve around, then straight across
To write the number 2.

Then, give each student a card with 1 and 2 in relief to trace and sing again.

• Give students copies of Reproducible Master RM7 to practice writing 1s and 2s.

• Give each student a piece of clay. Tell students to roll or pat the clay to make it flat. Have them use a finger or an ice-cream stick to write the numbers 1 and 2 on the clay.

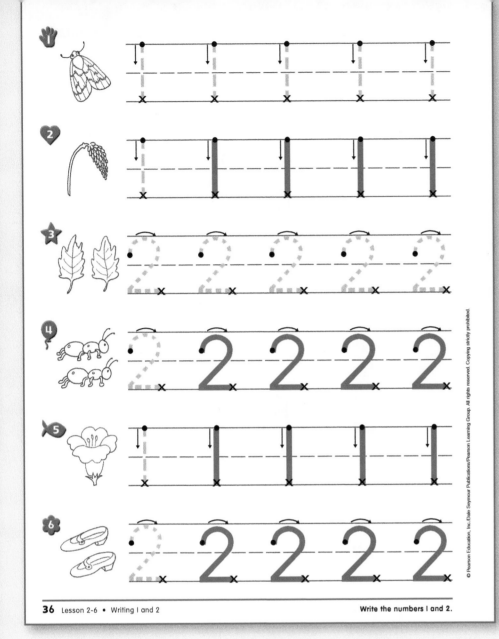

36 Lesson 2-6 • Writing 1 and 2

Write the numbers 1 and 2.

For Mixed Abilities

Common Errors • Intervention

Some students may need additional practice writing 1 and 2. Take students outside with a pail of water, a paintbrush, and chalk. Use the chalk to make a dot on the sidewalk to show where to begin, arrows to show in which direction to go, and an X to show where to end. Note: See page 35 for a sample. Use the paintbrush and water to make a large 1 from the dot to the X. Have students take turns using the paintbrush to go over your drawing. Repeat for the number 2.

Enrichment • Number Sense

1. Have students bring in two empty food containers or labels with the number 1 or 2 on them. Have students circle the numbers and then display.

2. Make a number tree. Provide a bare tree branch. Have students write 1 or 2 on a card to decorate the tree with number leaves.

3. Have students make their own sandpaper numbers 1 and 2 by pasting the outlines on 3 × 5-in. cards and covering the outlines with sand.

3 Practice

Using page 35 Discuss the groups shown around the number 1. Tell students to lay out their green and red crayons. Tell them to use the green crayon to color the dot at the top of the 1 and then use the red crayon to color the X at the bottom. Remind students that green on a stoplight means to go and red means to stop. Tell students to place a finger on the green dot and trace the dotted line down to the red X. Repeat as you sing the song for writing 1. Repeat this as many times as necessary before having students use a black crayon to write 1.

- Have students point to the large 2 and use the number to tell about the groups of objects. Tell them to color the starting dot green and the ending X red. Have students trace the number 2 several times with their finger before they use the black crayon to write 2.

Using page 36 Tell students that the number beside each group tells how many objects there are. Tell students to find the butterfly in Exercise 1 and trace the 1s in the first row across the page. Remind them to begin at the dot and end at the X to trace the 1s. Tell students to trace the first 1 in the second row and then begin at the dot to write more 1s. Show students that there are two leaves in Exercise 3. Have students begin at the dot and end at the X to trace the first 2. Tell them to trace this row of 2s, to trace the first 2 in Exercise 4, and then to write more 2s. Have students complete the last two rows in the same way.

4 Assess

Tell students to draw a set with one object and write the number 1 under the set. Repeat for the number 2.

T36

1 Getting Started

Objective
• To write 3 and 4

Materials
number cards 3 and 4 in relief; yarn pieces; modeling clay; Reproducible Master RM7; red, green, and black crayons

Warm Up • Mental Math
Ask students which is more.
1. 4 or 2 (4)
2. 3 or 4 (4)
3. 2 or 1 (2)
4. 2 or 3 (3)
Ask students which is less.
5. 1 or 4 (1)
6. 4 or 3 (3)
7. 2 or 4 (2)
8. 3 or 2 (2)

Warm Up • Number Sense
Have students form a 1 and a 2 with their yarn pieces. Then, have each student check a classmate's number.

2 Teach

Develop Skills and Concepts Repeat the first activity from Lesson 2–6 on page 36, if necessary, to help students follow directional aids for making a 3 and a 4.

• Display number cards 3 and 4 that have directional arrows along with a green dot at the starting point and a red X at the end. Have a student start at the green dot and trace each number. Provide time for all students to trace the numbers. Note that the 4 should have two starting points and two ending points so that students trace the left side of the 4 first and stop and then trace the vertical line that crosses the first line.

• Trace the large numbers 3 and 4 as you sing the following song to the tune of "Mulberry Bush":

> We go around and around again,
> We go around and around again,
> We go around and around again
> To write the number 3.

> Straight down, across, and then straight down,
> Straight down, across, and then straight down,
> Straight down, across, and then straight down
> To write the number 4.

Trace the paths for writing 3 and 4.

• Give each student the number cards 3 and 4 in relief to trace as the class sings the song again.

• Have students write a 3 and a 4 on flattened clay.

• Have students practice writing 3s and 4s on copies of Reproducible Master RM7.

3 Practice

Using page 37 Discuss the groups of three shown around the large number 3. Tell students to use their green crayons to color the dot at the top of the 3 and their red crayons to color the X at the bottom. Then, have each student use a finger to trace the 3 from the green dot to the red X as the class sings the song for writing 3. Have students use black crayons to trace the 3.

• Have students point to the two dots at the top of the 4 and color them green. Have them color the two Xs red. Tell students to trace the dotted line with their finger

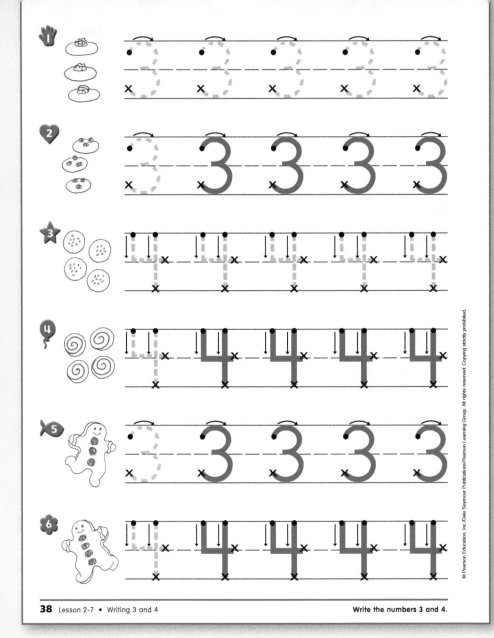

38 Lesson 2-7 • Writing 3 and 4

Write the numbers 3 and 4.

For Mixed Abilities

Common Errors • Intervention

Some students may need additional practice writing 3 and 4. Have them use the board and a pail of water. Draw a large 3 on the board. Dip your finger into the water and, beside the chalk-drawn 3, draw another 3 with the water. Have students take turns, wetting their finger and tracing over your slippery number 3. Repeat the activity for the number 4.

Enrichment • Application

1. Have students make cards for 3 and 4 to attach to the number tree.

2. Tell students to draw 3 balls, color them 3 different colors, and write the number 3 under them. Repeat for 4 scoops of ice cream in 4 different flavors.

3. Have students make number books with one page for each number, 1 to 4. Tell them to cut out and paste the correct number of objects on each page.

---

from the green dot on the left to the red X. Tell students to begin again at the green dot on the right and go straight down to the red X. Have students now sing the song for writing 4 with you. Repeat several times before having students trace the 4 with their black crayons.

Using page 38 Tell students to find the group of 3 treats in Exercise 1 and to trace the 3s across the page. Remind students to begin at the dot to trace the 3s. Tell them to begin at the dot to trace the first 3 in Exercise 2 and then to write more 3s. Have students begin at each dot and end at each X to trace the first 4 in Exercise 3. Tell students to trace the row of 4s. Have students trace the first 4 in Exercise 4 and then to write more 4s. Tell students that in Exercise 5, they are to trace the 3 to tell about the 3 buttons and then to write more 3s. In Exercise 6, tell them to trace the 4 to tell about the 4 buttons and then to write more 4s. Have students complete the last two rows independently.

4 Assess

Tell students to draw a set with three objects and write the number 3 under the set. Repeat for the number 4.

2-8 The Number 5

pages 39–40

1 Getting Started

Objectives
- To associate the number 5 with a set of five objects
- To define *five*

Vocabulary
five

Materials
*5 cans or boxes; pictures of groups of 1 to 5 objects; objects to count; yarn ring; counters

Warm Up • Mental Math
Ask students if they have the following:
1. more arms than legs (no)
2. 3 eyebrows (no)
3. the same number of eyes as ears (yes)
4. more cheeks than tongues (yes)
5. fewer hands than feet (no)
6. the same number of wrists as ankles (yes)
7. more knees than legs (no)

Warm Up • Application
Have students draw a picture showing a dinner table set for their family. Tell students to include placemats, plates, and so on in their drawings and to be sure that there is a one-to-one correspondence between the number of place settings and the number of people in their family.

2 Teach

Develop Skills and Concepts Place a yarn ring on a table and put 1 counter in it. Ask how many counters are in the ring. Continue putting 1 more counter in the ring and asking how many until there are 4 counters in the ring. Put in 1 more counter and write a large **5** on the board. Point to the 5 and tell students that the number 5 tells how many counters are in the ring. Ask if 5 is more or less than 4. (more than) Tell students that 4 and 1 more is 5.

- Set out five boxes or cans, each labeled with a 1, 2, 3, 4, or 5. Mix up the pictures of 1 to 5 objects and have students sort the pictures by placing them in the correct labeled container.

- Have each student hold up one hand and count the fingers and thumb together orally. Ask students how

many toes are on one foot. Ask if anyone has five buttons on a shirt or blouse. Have students find the 5 on the classroom clock.

- Use fingers or objects to show how to count one at a time as you say the "One for Me" poem found on page T25a. Then, have students show the correct number of objects as they say the poem with you.

- Lay out 4 objects and keep a fifth object concealed. Have students count the 4 orally with you. Prepare to lay the other object with the 4 as you say the following poem:

    If we have 4 things and then 1 more,
    We now have 5. 5 comes after 4.
    To count to 5, we start at 1,
    2, 3, 4, 5. This counting's done!

Point to each object as the poem says to count from 1 to 5.

Name _____

5

Count the objects in each set. Circle the sets of 5 objects.

Lesson 2-8 • The Number 5 **39**

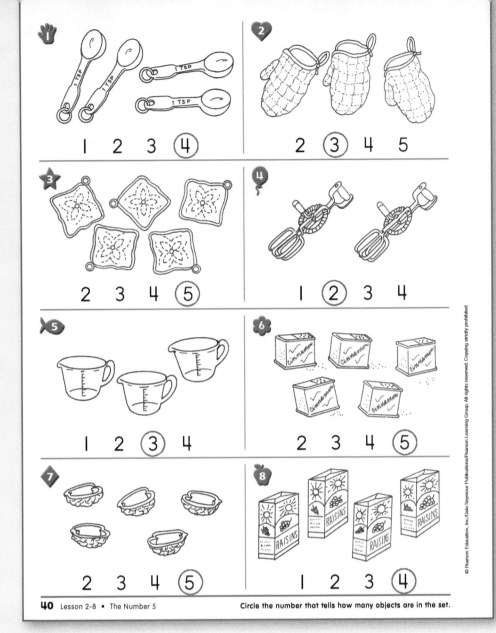

1.   1   2   3   ④

2.   2   ③   4   5

3.   2   3   4   ⑤

4.   1   ②   3   4

5.   1   2   ③   4

6.   2   3   4   ⑤

7.   2   3   4   ⑤

8.   1   2   3   ④

40   Lesson 2-8 • The Number 5   Circle the number that tells how many objects are in the set.

For Mixed Abilities

Common Errors • Intervention

Some students may need more practice with the number 5. Have them show 5 fingers on one hand. Then, have them match a finger on one hand to a finger on the other as they count from 1 to 5. Then, have students go around the classroom to identify and count, by one-to-one matching with the fingers on their hand, groups of up to 5 objects.

Enrichment • Application

1. Discuss with students the meaning of the expression "Give me 5." (shake hands) Have students find pictures to share where the saying could be an appropriate caption. Discuss the facial expressions of people who are shaking hands. Have students make up stories about the people and why they are shaking hands.

2. Have students look at each exercise on page 40 and tell how many objects would be in each set if there were 2 fewer objects shown. For example, in Exercise 1, there are 4 teaspoons. A set with 2 fewer would have 2 teaspoons.

3 Practice

Using page 39 Have students point to the number 5 in the example and say its name. (five) Have students count aloud the number of cookies on the cookie tray. (5) Tell them to circle the cookie tray.

• Tell students that they are to count the number of cookies on each cookie tray and circle the trays with 5 cookies. Ask them if they will circle a cookie tray that has 4 cookies on it. (no) Repeat the question for 1, 2, and 3 cookies so that students understand they are only to circle a tray of 5 cookies. Have them complete the page independently.

Using page 40 Have students name the pictured objects on this page. Tell them that they are to count the objects in the set and circle the number that tells how many objects there are. Have students complete the page as independently as possible. Then, have students use numbers in sentences to tell about each set of objects.

4 Assess

Ask students to name any groups of 5 objects they see around them.

T40

2-9 Numbers 1 Through 5

pages 41–42

Objectives

- To recognize sets of one to five objects
- To write 1, 2, 3, 4, and 5

Materials

Reproducible Masters RM7 and RM8; number card 5 in relief; counters; number cards 1 through 5; red, green, and black crayons

Warm Up • Mental Math

Have students name the number that

1. sounds like or rhymes with shoe (2)
2. is 1 more than 4 (5)
3. tells about their ears (2)
4. tells about their fingers on one hand (5)
5. is 1 more than 3 (4)
6. tells the number of colors on a stoplight (3)
7. tells how many tails a cat has (1)

Warm Up • Number Sense

Have students show counters and then number cards to represent bones as they sing "This Old Man."

Trace the path for writing 5.
Write the number 5.

Lesson 2-9 • Numbers I Through 5   **41**

Develop Skills and Concepts Have students follow dotted lines and arrows you provide on the board to write the numbers 1 through 4. Show students a large 5 that has directional aids. Tell them that to write a 5, they start at the dot and go down and around and stop. Then, show students that they start again at the dot and go across. Repeat the procedure as you trace the 5 again.

- Have students follow the arrows to trace the number card 5 in relief as you sing the following song to the tune of "Mulberry Bush":

    Straight down, around, and across the top,
    Straight down, around, and across the top,
    Straight down, around, and across the top
    To write the number 5.

    Have students sing the song with you as they trace the number. Stop abruptly after the word *around* to allow time to move to the starting dot.

- Have a student place the number card 2 on a table. Ask another student to write the number that comes next on the board. (3) Have a third student write the number that is less than 2 on the board. (1) Continue for other number cards and have students write the numbers that are the same as, 1 more than, or 1 less than each number.

- Give students copies of Reproducible Master RM7 to practice writing the numbers 1 through 5.

- Give each student a set of number cards. Have students hold up the number heard as you sing "We Have a Whole Lot of Numbers" from page T25a to the tune of "He's Got the Whole World."

- Give students copies of Reproducible Master RM8. Tell students that they are to find and color each number, 1 to 5. Note that there are actually two sets of the numbers 1 to 5.

42 Lesson 2-9 • Numbers 1 Through 5   Write the numbers 1 through 5.

For Mixed Abilities

Common Errors • Intervention

Some students may need additional practice writing the numbers 1 to 5. Either write large numbers on the board for students to trace in the air or provide water for them to trace the number on the board.

Enrichment • Application

1. Have students cut out and paste pictures of 1 to 5 objects, and write the number to tell how many are in each group.

2. Have students cut from magazines or newspapers several advertisements for food products they see in their homes. Tell students to count the number of cans, packages, and so on of each item they have in their home pantry and to write the number beside the picture.

3 Practice

Using page 41 Discuss the groups of 5 objects around the large 5 at the top of the page. Tell students that to write a 5 they need to use two motions, both starting at the same point. Have students use a green crayon to color the dot and a red crayon to color the two Xs at the end of the two motions. Have students then use a black crayon to trace the dotted 5 at the top of the page.

• Tell students to trace each 5 in the first row. For each of the last two rows, tell students to first trace the 5 and then write 5 fives. Have students complete the page as independent as possible.

Using page 42 Have students tell how many pens, name cards, strawberries, and balloons are shown. Tell them to trace the 2 and then write more 2s in the row. Then, tell students that they are to trace each number and write more 3s, 4s, and 5s. Tell students that they are then to count the strawberries in each group at the bottom of the page and write the number that tells how many there are. Have them complete the page independently.

4 Assess

Have students write the numbers 1 to 5 in order.

T42

Chapter 2 Test and Challenge

pages 43–44

1 Getting Started

Test Objectives

- **Items 1–6:** To associate the numbers 1 to 5 with sets of the appropriate number of objects (see pages 29–34, 39–40)
- **Item 7:** To write the numbers 1 to 5 (see pages 35–38, 41–42)

Challenge Objective

- To connect numbered dots in order from 1 to 5

Materials

*rope or string about 10 feet long with a knot at every 2-foot interval; number cards for 1 to 5

2 Give the Test

Prepare for the Test Set up 5 chairs in a row with a chair's space between each. Attach number card 1 to the first chair, number card 2 to the second, and continue for the 5 card on the last chair. Have 5 students sit on the 5 chairs. Ask how the students could be joined to make one long row. (hold hands) Have students do so in order from 1 to 5 as each student says the number on that chair.

- Now, move the chairs slightly so that the numbers are still in order but not in a definite row. Repeat the exercise to join two at a time and have students say the numbers in order from 1 to 5.

- Position 5 students in a row with about a foot of space between each. Tell each student a number from 1 to 5 in order from the first student to the fifth. Tell students that they are to say their number aloud when they are handed the rope or string. Hand an end knot of the rope to the first student. Wait for the student to say the number 1 and then place the second knot in the second student's hands. Wait for the student to say the number 2. Proceed until every student is holding the rope at a knot and the numbers 1 to 5 have been said in order. Ask students how many knots are in the rope. (5) Have all students count from 1 to 5 as each student who is holding a knot holds it up in order. Take out one knot and repeat for counting to 4.

- After students have mastered counting and writing numbers through 5, have them complete the Chapter 2 Test. You may wish to review test-taking rules with students. Remind them that they are not to talk to each other and they are to finish the test on their own. Decide ahead of time whether or not to allow use of manipulatives.

Using page 43 Tell students to count the acorns in Exercise 1 and circle the number that tells how many there are. (3) Have them complete the next five exercises independently. Then, tell students in Exercise 7 they are to count the butterflies in the group and write the number under the group. Have them complete the row independently.

Name _____

Circle the number that tells how many objects are in the set.
Write the numbers 1 through 5.

Chapter 2 • Test **43**

For Mixed Abilities

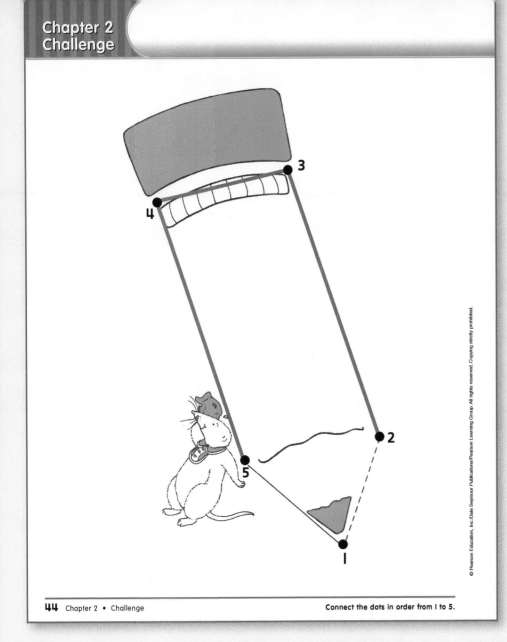

Connect the dots in order from 1 to 5.

Common Errors • Intervention

Students may have difficulty finding the correct number of objects in a set. They may count several objects from the set correctly and then rote count, skip an object when counting, or count an object twice. Allow students to use highlighters or crayons when counting. As they count, students should highlight or color each object and write the number above it. The number they end on is the number of objects in the set.

Enrichment • Number Sense

1. Have students write five 5s, four 4s, three 3s, two 2s, and one 1.

2. Have students expand their number books to include a page for 5.

3. Tell students to cut pictures from magazines to show things around them that have the numbers 1 to 5 in order. (book pages, telephones, computers)

4. Give each student a book with at least five numbered pages. Have students find page 4 and then tell if they will go backward or forward to find page 5. Repeat for other pages, changing the starting page.

3 Teach the Challenge

Develop Skills and Concepts for the Challenge Write the numbers 1 to 5 in order on the board but not in a row. Draw a large dot at each number. Have students start at 1 to connect the numbers from 1 to 5. Then, arrange dots and numbers again to form a design or picture and have students join the numbers. Repeat with other pictures. Explain that in connect-the-dot pictures, they are completing a picture by drawing lines between consecutive numbers.

Using page 44 Have students look at the picture on this page. Tell them that the picture needs to be completed and they can do so by joining the numbers from 1 to 5. Help students find the 1 and then tell them to trace the dashed line to the 2. Tell students to continue to join the numbers in order through 5.

4 Assess

Alternate Chapter Test You may wish to use the Alternate Chapter Test on page 206 of this manual for further review and assessment.

Challenge On the board, draw a connect-the-dot picture of a flower using the numbers 1 to 5. Have students join the numbers from 1 to 5 to complete the picture.

T44

Numbers 6 Through 9 and 0

pages 45–66

Chapter Objectives

- To count one to nine objects and choose the corresponding number
- To write 0 and 6 to 9
- To define *six*, *seven*, *eight*, *nine*, *zero*, *nothing*, *not any*, and *none*
- To use the try, check, and revise strategy to solve problems
- To estimate quantities of nine or fewer
- To associate no objects with the number 0

Materials

You will need the following:
*green crayons; *numbers 0 and 6 through 9 in relief; *tape; *number cards 0 through 9; *counters; *objects to be sorted by use in kitchen, bathroom, and bedroom; *pictures of kitchens, bathrooms, and bedrooms

Students will need the following:
Reproducible Masters RM7, RM9, and RM10; cards showing 0 through 7 objects; number cards 0 through 9; number cards 0 and 6 through 9 in relief; blank index cards; tubular pasta; string; 1-pound plastic container with a lid per pair of students; 9 real or play pennies per pair of students; beans; interlocking cubes; magazines; counters; dominoes; crayons

Bulletin Board Suggestion

Have students bring in pictures or small objects that can be posted. Post a different number of items each day and display matching number cards through the latest number learned. Have students count the objects and choose the correct number card to post for the day. Vary the activity by having several students write the number on 3×5-in. cards to post beside the objects.

Verses

Poems and songs may be used any time after the pages noted beside each.

St. Ives (anon.) (pages 47–48)

As I was going to St. Ives,
I met a man with seven wives.
Each wife had seven sacks,
Each sack had seven cats,
Each cat had seven kits.
Kits, cats, sacks, and wives,
How many were going to St. Ives?

Seven Names (pages 49–50)

One, two, three, four, five, six, seven;
Let's count names like Jan and Evan.
That's two so far, let's say some more;
Steve makes three and Sue makes four.
One more name now, Ted is five;
Ted makes one, two, three, four, five.
Five and one more will be six
And Trixie's nickname must be Trix.
But who will number seven be?
One more than six is Stephanie.
Jan, Steve, Sue, Ted, Trix, and Evan;
Stephanie makes number seven.

Zero (pages 61–62)

How many is nothing?
How many is none?
How many's not any?
Is it one less than one?

Yes, zero's not any.
Zero is none.
It's nothing at all!
Zero's one less than one.

None (pages 61–62)

Give me some things,
I'll hold every one.
Take them away
And I will have none.

Give me one thing
And I will have one.
Take it away
And I will have none.

We Have a Whole Lot of Numbers (pages 63–64)

We have a whole lot of numbers in our hands.
We have a whole lot of numbers in our hands.
We have a whole lot of numbers in our hands.
Let's see what we have.

We've got the number 1 in our hands.
We have 1 more than zero in our hands.
We have 1 less than 2 in our hands.
We have a 1 in our hands.

You may wish to extend the song as follows:

2:	1 more than 1;	1 less than 3
3:	1 more than 2;	1 less than 4
⋮	⋮	⋮
8:	1 more than 7;	1 less than 9

Scampy's Party

One day Scampy Squirrel decided to have four of her friends over for a party. She thought four friends would be just right because she had one table and four chairs. She also had four plates. So all Scampy had to do was to bake some cookies. And that is what she did.

As the cookies were cooling, Scampy rushed out to invite her favorite friends—Jenny, Benny, Winnie, and Flo—to her party. Everyone said, "Thank you," so Scampy ran home to set the table. She placed one plate in front of each chair. Then, she put four cookies on each plate.

Just then Scampy's friends arrived. She greeted them and showed them to the table. Jenny sat in one chair. Benny sat in one chair. Winnie sat in one chair. And Flo sat in one chair.

"Oh, dear," said Scampy. All the chairs are taken. "Where am I going to sit?"

Everyone laughed, and Benny asked, "Do you mean that you forgot to count yourself, Scampy?"

Scampy was embarrassed. "Yes!" she laughed. "I forgot to count myself. I invited four friends and I only have four chairs! So now I have no place to sit! I really need one more chair if we are all going to sit down. I need five chairs!"

"I'll tell you what," said Flo. "It's such a lovely day. Why don't we take our cookies outside and sit on the lawn?" So everyone agreed.

Once they were outside, Jenny, Benny, Winnie, and Flo each gave Scampy one cookie. "Wait!" said Scampy. "Now you each have only three cookies on your plate, but I have four cookies! I have one more than you do!"

"That's okay," said Winnie. "After all, you did all the work! So you deserve an extra cookie!"

Everyone agreed and had a wonderful time at Scampy's party.

"Scampy's Party," p. RM9

Activities

Use Reproducible Master RM9, "Scampy's Party," to help students relate the story to the mathematical ideas presented.

• Have a group of four students sit around a table or on the floor. Give another student 16 counters or small slips of paper. Tell the student to give one counter or paper at a time to each student until all are given away. Ask how many objects each student has. Now, have the student collect the objects. Repeat the activity so that the distributor receives some also. Discuss the outcome.

• After students have completed Reproducible Master RM9, ask if anyone seated the chipmunk between the squirrel and the bird. Continue to ask students to name the animal they seated between or beside other animals. Have students use the counters or other objects to set the table arrangement at a round table or on the floor. Include raisins or other snacks to represent the cookies. Have one student direct the activity. Let that student cast the characters and direct each animal to sit around the table according to the arrangement of animal pictures on the activity sheet. Allow time for more students to direct the activity to show the way they arranged the animals.

1 Getting Started

Objectives
• To count one to six objects and choose the correct number
• To define *six*

Vocabulary
six

Materials
cards showing 1 through 6 objects; number cards 1 through 6; tape

Warm Up • Mental Math
Have students name the number that comes

1. between 3 and 5 (4)
2. between 1 and 3 (2)
3. before 2 (1)
4. after 4 (5)
5. after 3 (4)
6. between 2 and 4 (3)

Warm Up • Number Sense
Write **2 5 1 4 3** on the board. Give students the number cards 1 through 5. Play the Simon Says game by telling students that Simon says to show the number card that is the same as the number you point to on the board. Continue for the number that is 1 more than or 1 less than the number on the board.

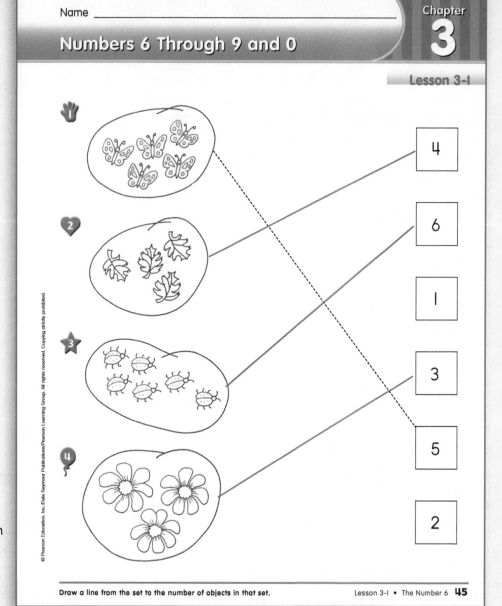

Draw a line from the set to the number of objects in that set.

Lesson 3-1 • The Number 6 **45**

---

2 Teach

Develop Skills and Concepts Draw a group of 5 objects on the board. Have students count the objects and tell how many there are. (5) Draw 1 more object and write a large **6** on the board. Tell students that 6 is the number we write for 5 and 1 more. Tell them to count aloud with you as you point to each object in the group of 6.

• Place the cards showing 1 through 6 objects on a table in random order. Have students identify the card that has 6 objects, 3 objects, and so on.

• Draw groups of 1 through 6 sticks across the board and circle each group. Tape the number cards 1 through 6 in random order along the board. Have students count the objects in each group and place each number card under its group. Repeat for more students to participate. Now, erase the sticks and mix the number cards. Have a student draw a card from the pile, place

it under a circle and draw the correct number of sticks in the circle. Continue for the other number cards.

3 Practice

Using page 45 Have students name the objects in each group and then tell how many groups of objects there are. (4) Have them name the numbers down the right side of the page and then tell how many numbers there are. (6) Ask students if there are more or less numbers than groups of objects. (more) Tell students to count the butterflies and tell how many there are. (5) Ask students to find the number 5 at the right and trace the line from the five butterflies to the number 5. Have them count the leaves and draw a line from the leaves to the number that tells how many there are. (4) Tell students that they are to count the ladybugs and flowers and draw a line from each group to the number that tells how many there are. Remind students that some of the numbers will not be joined to a group. Have them complete the page independently.

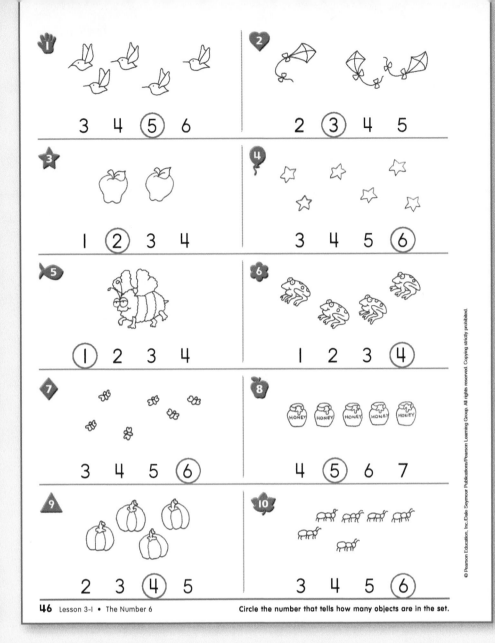

1.   3   4   (5)   6

2.   2   (3)   4   5

3.   1   (2)   3   4

4.   3   4   5   (6)

5.   (1)   2   3   4

6.   1   2   3   (4)

7.   3   4   5   (6)

8.   4   (5)   6   7

9.   2   3   (4)   5

10.   3   4   5   (6)

46   Lesson 3-1 • The Number 6   Circle the number that tells how many objects are in the set.

For Mixed Abilities

Common Errors • Intervention

Some students may need additional practice counting one to six objects. Have a student draw one stick on the board, say the number one, and then write the number **1** under it. Tell another student to draw one more stick, say the number of sticks now shown (two), and write the number **2**. Continue through five sticks and the number 5. Draw a sixth stick, say the number six, and write **6** under the stick. Circle the sticks and ask students to say the number that tells how many sticks are inside the circle.

Enrichment • Number Sense

1. Have students place number cards and objects cards in reverse order from 6 to 1.

2. Give students the number cards 2 through 6 from two suits of a deck of playing cards. Shuffle the cards and lay them facedown individually. Have a student turn over two cards and keep matched cards. Unmatched cards are turned facedown again for continued play until all cards are won.

ESL/ELL STRATEGIES

Explain that a set is several objects that form a group. You might hold up a key ring and say, *This is a set of keys*. Ask students to group classroom objects, such as books or chairs, and tell how many are in a set.

Using page 46 Have students count the birds in Exercise 1 and tell how many there are. (5) Ask them to name the numbers under the birds. (3, 4, 5, 6) Tell students to draw a circle around the number that tells how many birds are shown. (5) Help students work through the next several exercises before having them complete the page independently.

• Then, ask students to find all of the sets on the page that show the same number of objects as the group of stars in Exercise 4. (the butterflies and the ants) Point out that all of these groups show 6 objects.

4 Assess

Have students draw a picture of a house with six windows.

T46

3-2 The Number 7

pages 47–48

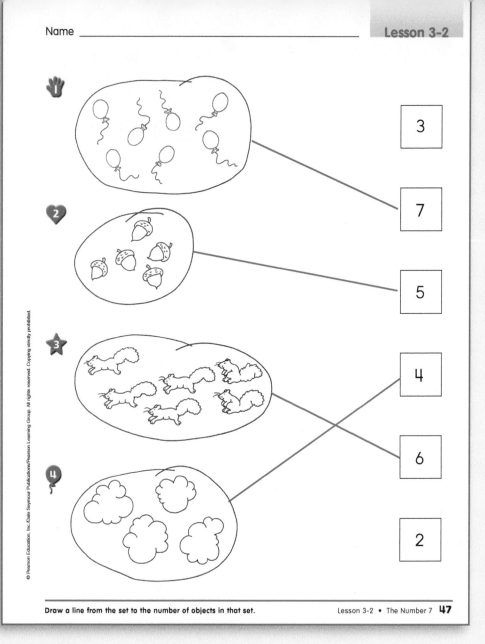

Name _____

Draw a line from the set to the number of objects in that set.

Lesson 3-2 • The Number 7 **47**

1 Getting Started

Objectives
- To count one to seven objects and choose the correct number
- To define *seven*

Vocabulary
seven

Materials
set of number cards 1 through 7 per pair of students; 28 counters per pair of students

Warm Up • Mental Math
Ask students who has more.

1. I have 3; Jan has 5. (Jan)
2. Nick has 6; Elly has 5. (Nick)
3. Fay has 6; I have 3. (Fay)
4. Kris has 4; Ed has 6. (Ed)
5. Li has 2; Al has 4. (Al)
6. Beth has 4; Joe has 5. (Joe)
7. Mia has 6; I have 3. (Mia)

Warm Up • Number Sense
Give students the number cards for 1 through 6. Ask them to show the number that comes between 3 and 5, before 6, after 2, after 4, is 1 less than 6, and so on.

2 Teach

Develop Skills and Concepts Place a set of 6 objects on a table. Tell students that you are putting 1 more object with the 6. Tell students that you now have 7 objects as you write **7** on the board. Have them say the number word in unison with you.

- Group students into pairs and give each pair a set of number cards 1 through 7 and at least 28 counters. Tell students that they are to build a group of 1, a group of 2, and so on, through a group of 7 counters. Have them place each number card beside the correct group of counters.
- Play the "Listen and Count" game. Tell students to clap seven times. Continue by having students wink seven times, take seven steps forward, hop seven times, show seven fingers, and so on. Now, vary the numbers from 1 to 7 for continued play.
- Read and then illustrate on the board the poem "St. Ives," found on page T45a. Help students see that careful listening is needed to answer the riddle. Repeat the poem for students to point to each of the seven objects mentioned.

3 Practice

Using page 47 Have students name the objects in each group and then tell how many groups of objects there are. (4) Have them name the numbers down the right side of the page and then tell how many numbers there are. (6) Ask students if there are more or less numbers than groups of objects. (more) Tell them to count the balloons and tell the number. (7) Ask students to find the number 7 at the right and draw a line from the seven balloons to the number 7. Have students count the acorns and draw a line from the acorns to the number that tells how many are shown. (5) Tell them that they are to count the squirrels and clouds and draw a line from each group to the number that tells how many there are. Remind students that some of the numbers will not be joined to a group. Have them complete the page independently.

T47

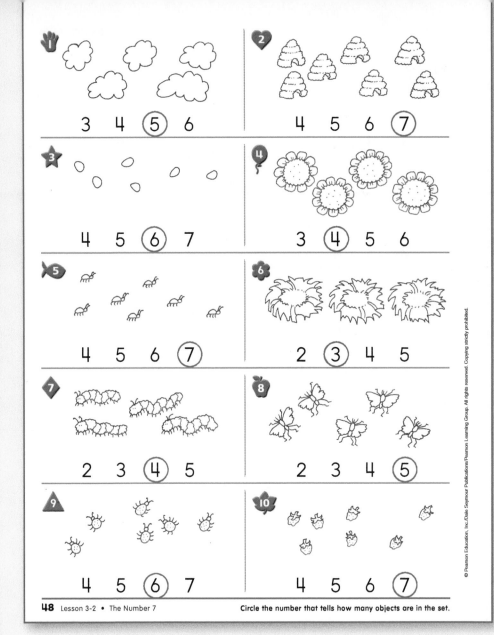

1. 3 4 ⑤ 6
2. 4 5 6 ⑦
3. 4 5 ⑥ 7
4. 3 ④ 5 6
5. 4 5 6 ⑦
6. 2 ③ 4 5
7. 2 3 ④ 5
8. 2 3 4 ⑤
9. 4 5 ⑥ 7
10. 4 5 6 ⑦

Circle the number that tells how many objects are in the set.

For Mixed Abilities

Common Errors • Intervention

Some students may need additional practice with the number 7. Give each student a set of number cards 1 to 7. Draw 7 groups of circles on the board. Draw 1 circle in the first group, 2 in the second, and so on until you have drawn a group of 7 circles. Moving from left to right, point to the groups one at a time. As you point, every student is to hold up the correct number card for that group. After you go through all seven groups in order, continue the activity by pointing to different circles at random as students hold up the appropriate number cards.

Enrichment • Application

1. Have students place from 1 to 7 counters in each cup of an egg carton so that no more than two cups have the same number of counters.

2. Read *Snow White and the Seven Dwarfs* to the students. Have students draw a face for each dwarf.

Using page 48 Have students count the clouds in Exercise 1 and tell how many there are. (5) Ask students to name the numbers under the clouds. (3, 4, 5, 6) Tell them to circle the number that tells how many clouds are shown. (5) Help students work through the next several exercises before having them complete the page independently.

4 Assess

Write the following on the board and have students match the columns by drawing a line from each number to that number of Xs:

4	x x x
6	x x x x x
5	x
7	x x x x
2	x x x x x x x
3	x x
1	x x x x x x

T48

3-3 Numbers 1 Through 7

pages 49–50

1 Getting Started

Objective
• To count one to seven objects and choose the correct number

Materials
number cards 1 though 7; cards showing 1 through 7 objects; tape

Warm Up • Mental Math
Have students count to 7 from the following numbers:
1. 2 (2, 3, 4, 5, 6, 7)
2. 6 (6, 7)
3. 1 (1, 2, 3, 4, 5, 6, 7)
4. 5 (5, 6, 7)
5. 4 (4, 5, 6, 7)
6. 3 (3, 4, 5, 6, 7)

Warm Up • Number Sense
Lay out the number cards 1 through 7 in random order. Ask a student to find the 6 card and place it on a table. Tell a student to find the 5 card and place it in front of the 6 card. Now, have a student give an instruction to properly place the 7 card. (Place the 7 card after the 6 card.) Continue to build the numbers 1 through 7 in proper order.

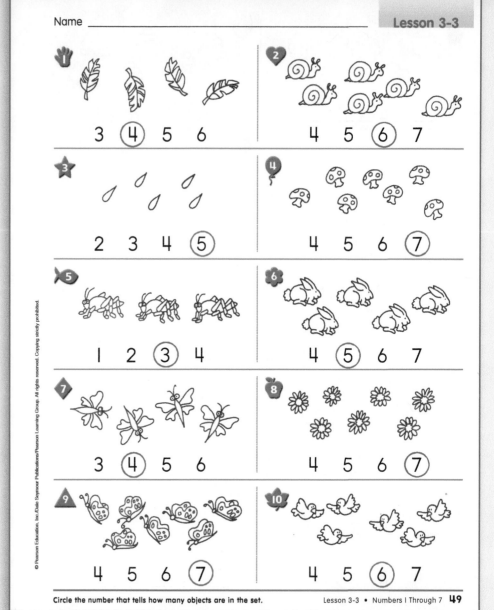

Circle the number that tells how many objects are in the set.

Lesson 3-3 • Numbers I Through 7 **49**

2 Teach

Develop Skills and Concepts Play the "Calling All Cards" game. Give each pair of students a set of number cards 1 through 7 and a set of cards with 1 through 7 objects on them. Tell students that you are calling all cards for the number that comes after 6. Have one student in each pair show the 7 card and the other show the 7-objects card. Continue the game, asking for the number that comes between 5 and 7 (6), the number before 3 (2), and so on.

• Give each student one number card. Tell students to build a group having that number of objects.

• Sing the poem "Seven Names," found on page T45a, to the tune of "Twinkle Twinkle Little Star" as you point to number cards whose names are heard in the poem. Then, have students pretend they are the names mentioned and stand as their pretend name is heard in the poem. Have the class count the standing students at the end. You may want to change the names in the poem to use names of students in the class.

3 Practice

Using page 49 Have students count the leaves in Exercise 1 and tell the number. (4) Tell them to find the 4 under the leaves and circle it to tell that there are four leaves. Work through two more exercises with students, if necessary, before having students complete the page independently.

Using page 50 Tell students that they are to complete the picture by joining the numbers 1 to 7 in order. Help students locate the 1. Tell students to join the dot beside the 1 to the dot beside the 2 by tracing the broken line. Supervise as they continue to join the numbers in order through 7. Discuss the story told by the picture by asking students how they think the bear is feeling. (scared, wondering, and so on) Ask why. (The bees might sting the bear to protect their honey.) Tell them to color the picture in colors of their choice.

T49

HONEY

Connect the dots in order from I to 7.

For Mixed Abilities

Common Errors • Intervention

Some students may need additional practice counting from 1 to 7. Have seven students hold the number cards 1 through 7 and arrange themselves in order from 1 to 7. Now, place students holding number cards 3 and 5 next to each other and ask if a number belongs between them. Have students count orally from 1 to 7, if necessary, to decide. Then, have the student with the 4 card stand between the students holding the 3 card and the 5 card. Continue building the numbers in order from 1 to 7 in this way.

Enrichment • Number Sense

1. Have students draw a picture and then count to solve this problem.

   Aunt Lucy has an egg carton in her refrigerator. The egg carton has seven eggs in it. How many eggs has Aunt Lucy used? (5 eggs)

2. Have students draw lines on a sheet of paper to divide it into seven parts. Have them write one number from 1 to 7 in each part. Have students then color each part a different color.

4 Assess

Give students number cards 1 through 7. Draw groups of 1 through 7 objects on the board and have students tape the correct number card to the board beneath each group.

T50

Trace the paths for writing 6 and 7.

1 Getting Started

Objective
• To write 6 and 7

Materials
*numbers 6 and 7 in relief; Reproducible Master RM7; counters; number cards 6 and 7 in relief; red, green, and black crayons

Warm Up • Mental Math
Have students tell the names of the following:
1. four friends
2. five toys they like
3. six classmates
4. three things behind them
5. two things above them
6. seven people they know
7. one thing that is to their left

Warm Up • Number Sense
Give each student a pile of counters. Review building groups of one to seven objects by saying a number and having students lay out that number of counters. Then, give each student a number from 4 to 7 and tell students to draw a picture to show the number.

2 Teach

Develop Skills and Concepts Have a student go to the classroom calendar and point to every number 6. Repeat for the number 7. Now, have students locate 6s and 7s in books, in room displays, on the clock, and so on. Ask if the 6s are all made alike. (yes) Repeat to observe the 7s.

• Display the large 6 in relief and sing the following song to the tune of "Mulberry Bush" as you trace the number:

    Curve down and around to make a ball.
    Curve down and around to make a ball.
    Curve down and around to make a ball
    To make the number 6.

Give each student a number 6 in relief to trace as they sing along with you. Repeat the procedure for the number 7, using the following lyrics:

    Go straight across and then slant down.
    Go straight across and then slant down.

Go straight across and then slant down
To make the number 7.

• Have students write the numbers 6 and 7 in the air and then on copies of Reproducible Master RM7.

3 Practice

Using page 51 Have students count the frogs at the top of the page and tell how many are shown. (6) Have students count the turkeys and tell how many there are. (6) Tell them to circle each group of 6 objects. Tell students to find the small circle at the top of the large number 6 and color the circle green. Remind students that green means to go. Tell them to find the X and color it red to remind them to stop at the X. Have students use their index finger to begin at the green circle and trace the path to the red X as they sing the song for writing a 6. Repeat the finger tracing several times before having students trace the 6 with a black crayon. Repeat the procedure to count and circle the groups of 7 before tracing the 7.

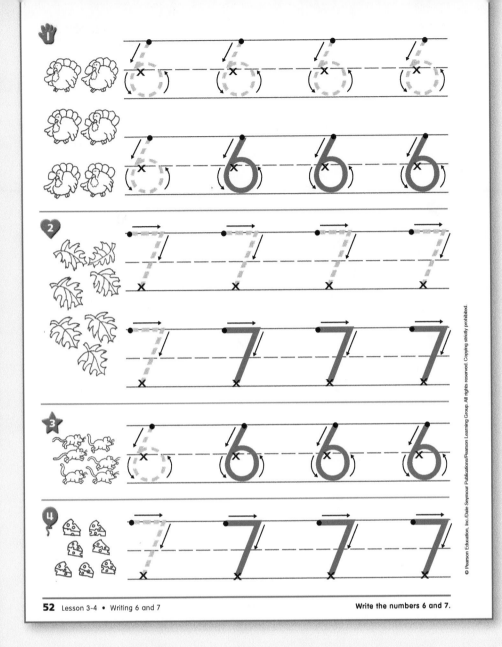

Write the numbers 6 and 7.

For Mixed Abilities

Common Errors • Intervention

Some students may need additional practice writing 6 and 7. Write a large 6 and 7 on the board, marking the beginning dot, directional arrows, and ending X. Have students trace the numbers in the air. Then, have them use clay to make the numbers 6 and 7.

Enrichment • Application

1. Have students make a list of seven things to buy at a grocery store. Tell them to write the numbers 1 to 7 down one side of a large sheet of lined paper. Have students cut out a picture of each item and paste it next to each number.

2. Tell students to make a list of seven things they do before arriving at school. Encourage students to use pictures to illustrate each activity and number them.

ESL/ELL STRATEGIES

Some students may know slightly different forms for the numbers 1, 4, and 7 from first language experiences. After they have traced the dotted numbers, check that they write the numbers 1, 4, and 7 using American English forms.

Using page 52 Have students count the turkeys and tell the number. (6) Tell them that they will now write the number 6. Have students trace the 6s and then start at the dots to write more 6s. Repeat the procedure to have students count the 7 leaves and trace and write the 7s to tell how many leaves are shown. Have students complete the last two rows independently.

4 Assess

Have students draw a group of six objects on one index card and a group of seven objects on a second index card. Have them switch index cards with a partner and count the number of objects in each group and write that number on the back of each index card.

T52

3-5 The Number 8

pages 53–54

1 Getting Started

Objectives
- To count one to eight objects and choose the correct number
- To define *eight*

Vocabulary
eight

Materials
*number cards 1 through 8; *counters; *8 green crayons

Warm Up • Mental Math
Have students name the number.
1. their age
2. 5 and 1 more (6)
3. how many children are in their family
4. how old they will be on their next birthday
5. 1 less than 7 (6)

Warm Up • Number Sense
Have a volunteer think of a number between 1 and 7. Tell the volunteer not to share the number with anyone. Have the student point to that number of different objects around the room. Have the other students count how many objects are pointed out and tell the secret number. Repeat for more secret numbers.

Name _____

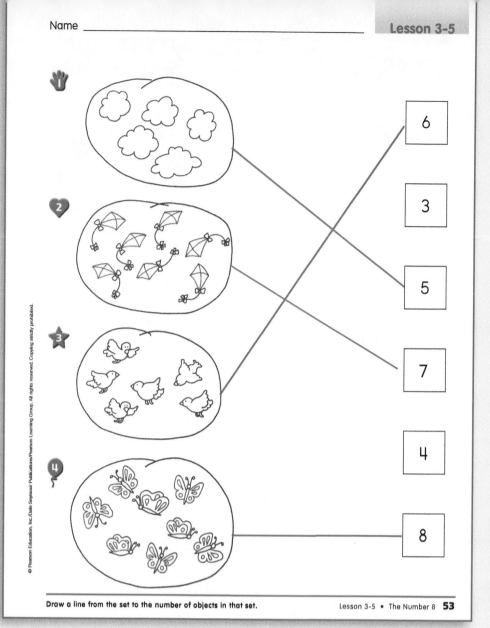

Draw a line from the set to the number of objects in that set.

Lesson 3-5 • The Number 8   **53**

2 Teach

Develop Skills and Concepts Give 8 students each a green crayon. Tell students that you are going to list the names of all the students who are holding a green crayon. Have those students stand. Write **1** on the board or on chart paper and list one name. Have that student sit down. Have students write the successive numbers and even write their names, if possible, to complete the list through the seventh student. Ask how many names are now listed. (7) Have students note that there is 1 more name to be listed. Write **8** on the list and tell students that the number 8 is what we write for 7 and 1 more. Complete the list and then have the 8 students stand again. Tell students to sit down when their number is called out as all students read the numbers in order from the list.

- Display the number cards 1 through 8. Place groups of 1 to 8 counters on a table. Have students select the number that shows how many counters are in each group.

- Have students play a version of the Mother May I game. The leader tells a student to take any number of steps forward, backward, to the left, or to the right. Before the player moves, the leader must be addressed by name as follows: "Nathan, may I?" Nathan then replies, "Yes, you may." All students check to see that the leader's directions were followed exactly.

3 Practice

Using page 53 Have students name the objects in each group and then tell how many groups of objects there are. (4) Have students name the numbers down the right side of the page and then tell how many numbers there are. (6) Ask them if there are more or less numbers than groups of objects. (more) Tell students to count the clouds and tell the number. (5) Ask students to find the number 5 at the right and draw a line from the five clouds to the number 5. Have them count the kites and draw a line from the kites to the number that tells how many are

T53

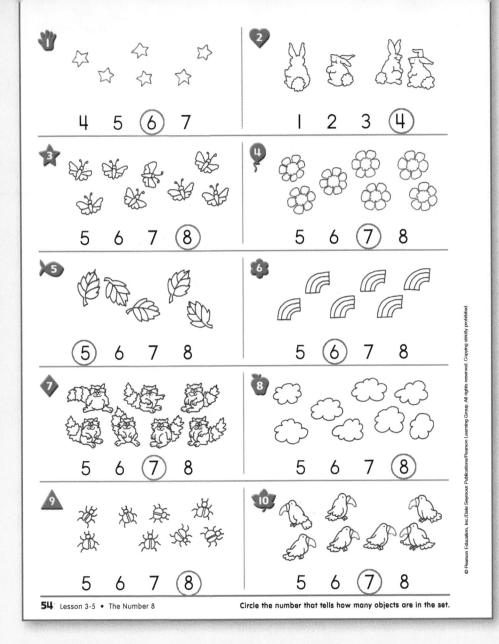

1. 4 5 (6) 7

2. 1 2 3 (4)

3. 5 6 7 (8)

4. 5 6 (7) 8

5. (5) 6 7 8

6. 5 (6) 7 8

7. 5 6 (7) 8

8. 5 6 7 (8)

9. 5 6 7 (8)

10. 5 6 (7) 8

Circle the number that tells how many objects are in the set.

For Mixed Abilities

Common Errors • Intervention

Students who need more work counting from 1 to 8 can practice by setting a table using dishes and silverware. Have students first set a table for four, each student being responsible for a different item, such as spoons, forks, or cups. Then, have them prepare the table for one more at a time as company arrives until they get to eight. As each student collects his or her items, when dismantling the table, the student should count the items aloud.

Enrichment • Number Sense

1. Have students draw the soccer balls that would be needed if eight children were playing and each child had a ball.

2. Have students cut out pictures of eight things that must be kept in a refrigerator or freezer to keep them from spoiling.

shown. (7) Tell students that they are to count the birds and butterflies and draw a line from each group to the number that tells how many there are. Remind them that some of the numbers will not be joined to a group. Have students complete the page independently.

Using page 54 Have students count the stars in Exercise 1 and tell how many there are. (6) Ask students to name the numbers under the stars. (4, 5, 6, 7) Tell them to circle the number that tells how many stars are shown. (6) Help students work through the next several exercises before having them complete the page independently.

4 Assess

Have students build sets of eight using any materials in the classroom. Allow them to select from all over the room and then have students return their eight items to their proper places.

T54

pages 55–56

1 Getting Started

Objectives
- To count one to nine objects and choose the correct number
- To define *nine*

Vocabulary
nine

Materials
*number cards 1 through 9; dominoes; counters

Warm Up • Mental Math
Ask students to answer yes or no.
1. 7 is 6 and 1 more. (yes)
2. 5 comes between 4 and 6. (yes)
3. I am 6 years old.
4. There are 5 people in my family.
5. 8 comes before 7. (no)
6. 5 and 1 more is 6. (yes)
7. 1 less than 8 is 7. (yes)

Warm Up • Number Sense
Have students work in pairs to review counting groups of objects. Have one student name a number from 1 to 8 for the partner to lay out that number of counters. The first student then checks the work. Have partners reverse roles and continue.

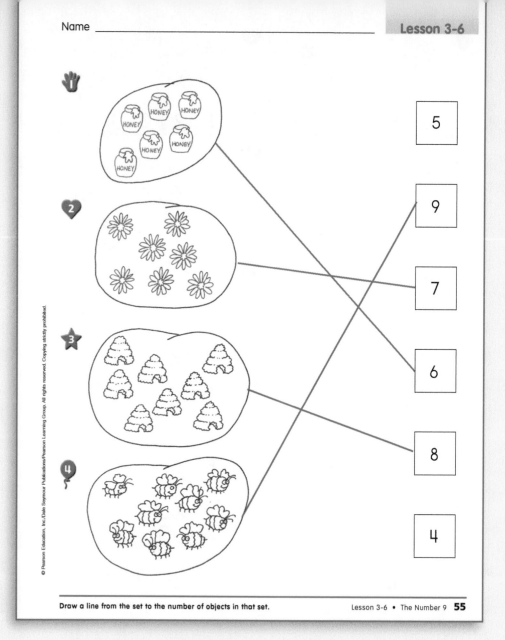

Draw a line from the set to the number of objects in that set.

Lesson 3-6 • The Number 9 **55**

2 Teach

Develop Skills and Concepts Set up 9 chairs in a row and name 8 students to sit on the chairs. Ask students if all the chairs are filled. (no) Ask them to tell why. (There is 1 more chair.) Have students count the chairs one by one as each sitting student goes to the board to write the number. Write **8** for the eighth student and then have students count in unison from 1 to 8. Tell students that 9 is the number for 8 and 1 more as you write **9** on the board. Have a student sit in the ninth chair. Ask how many chairs are filled. (9) Have students count as you point to the numbers 1 through 9 on the board.

- Draw the following on the board:

x	x	x	xx	xx	xx	xxx
x	xx	x	xx		xx	xxx
			x	xx		
x		x		xx	xx	xxx
	x		x	xx		
(3)	(3)	(4)	(4)	(5)	(8)	(9)

Have students count the number of Xs in each group, say the number, and then write the number on the board. Note that you will need to write the number 9 as students are not yet asked to do so.

- Have students count the number of dots on dominoes and match two dominoes end to end if they have the same number of dots.

- Write the number of a student's address on the board. Have students read the number and then ask the student who lives at that address to stand up. Repeat for other addresses but do not use any addresses with zero in them at this time.

3 Practice

Using page 55 Have students name the objects in each group and then tell how many groups of objects there are. (4) Have them name the numbers down the right side of the page and then tell how many numbers there

T55

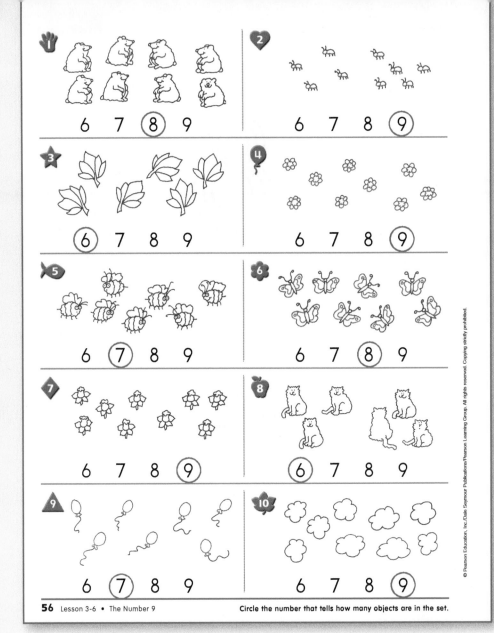

1. 6 7 (8) 9
2. 6 7 8 (9)
3. (6) 7 8 9
4. 6 7 8 (9)
5. 6 (7) 8 9
6. 6 7 (8) 9
7. 6 7 8 (9)
8. (6) 7 8 9
9. 6 (7) 8 9
10. 6 7 8 (9)

56 Lesson 3-6 • The Number 9   Circle the number that tells how many objects are in the set.

For Mixed Abilities

Common Errors • Intervention

Some students may need more work counting from 1 to 9. Have them work with partners to count different groups of objects or counters and then choose the number card that matches the number of objects in the group.

Enrichment • Application

1. List all student addresses by number only. Have students write their name beside their address. As an extension, use telephone numbers.

2. Have students use a toy telephone to practice dialing numbers you write on the board.

are. (6) Ask students if there are more or less numbers than groups of objects. (more) Tell them to count the honey pots in Exercise 1 and tell the number. (6) Ask students to find the number 6 at the right and draw a line from the six honey pots to the number 6. Remind students that they will not need all of the numbers on the right as they draw a line from each group to the number that tells how many objects are in that group. Have them complete the page independently.

Using page 56 Have students count the bears in Exercise 1 and tell how many there are. (8) Ask students to name the numbers under the bears. (6, 7, 8, 9) Tell them to circle the number that tells how many bears are shown. (8) Have students complete the page independently.

4 Assess

Organize the class into pairs and give each pair 9 counters. Have one partner lay out between 1 and 9 counters and have the other partner tell how many counters there are. Partners will switch roles often so that each student has many opportunities to lay out the counters and to tell how many there are.

T56

Trace the paths for writing 8 and 9.

1 Getting Started

Objective
• To write 8 and 9

Materials
*numbers 8 and 9 in relief; Reproducible Master RM7; number cards 8 and 9 in relief; red, green, and black crayons; blank index cards

Warm Up • Mental Math
Have students name the number that is
1. the same as their age
2. first in their phone number
3. first in their address
4. after 8 (9)
5. 1 less than 4 (3)
6. the same as the number of toes on their left foot (5)

Warm Up • Number Sense
Have a student clap hands from 1 to 7 times as classmates count the claps and say the number. Have a student write the number on the board.

2 Teach

Develop Skills and Concepts Have a student count to 8 as you write the numbers on the board. As you begin to write the **8**, sing the following song to the tune of "Mulberry Bush":

Around, around, around, around.
Around, around, around, around.
Around, around, around, around
To make the number 8.

Show students the large 8 in relief and help several students trace it as the song is sung. Give each student a number 8 card in relief and repeat the song as students trace the number.

• Repeat the previous procedure for 9, using the following lyrics:

Around and up and then slant down.
Around and up and then slant down.
Around and up and then slant down
To make the number 9.

• Have students move around the room, locating numbers 8 and 9 to observe their formations. Tell students to trace each number they find. Have them return to their seats to discuss what numbers they found and where they found them.

• Have students practice writing the numbers 8 and 9 on copies of Reproducible Master RM7.

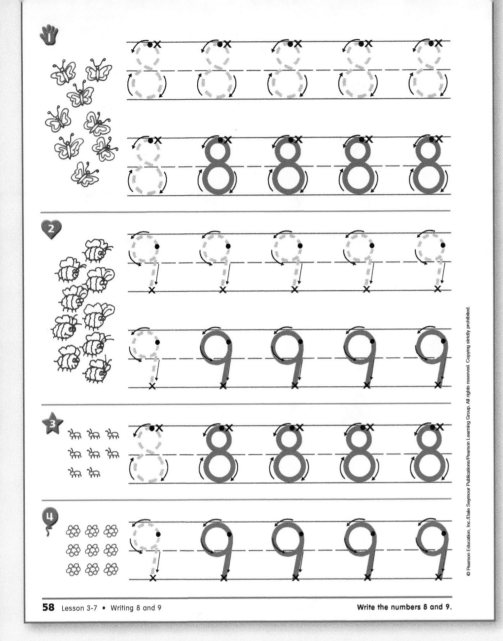

Write the numbers 8 and 9.

For Mixed Abilities

Common Errors • Intervention

Some students may need more practice writing 8 and 9. Make large numbers for 8 and 9 on the floor using masking tape. Use chalk to mark the beginning dot, directional arrows, and ending X for writing the numbers. Have students walk the path as they sing the special words to the tune of "Mulberry Bush." The words can be found in the **Develop Skills and Concepts** activity on page T57.

Enrichment • Number Sense

1. Have students write the numbers 1 to 9 down the left side of a large sheet of ruled paper. Tell students to cut matching numbers from magazines and to paste them by the correct number.

2. Have students draw a large 8 or 9 on a sheet of drawing paper. Tell them to cut out and paste that number of objects on the paper.

3 Practice

Using page 57 Have students count the suns at the top of the page and tell how many there are. (8) Have them count the moons and tell how many there are. (8) Tell students to circle each group of 8 objects. Tell students to find the small circle at the top of the large number 8 and color the circle green. Remind them that green means to go. Tell students to find the X and color it red to remind them to stop at the X. Have students use their index finger to begin at the green circle and trace the path to the red X as they sing the song for writing an 8. Repeat the finger tracing several times before having students trace the 8 with a black crayon. Repeat the procedure to count and circle the groups of 9 before tracing the 9.

Using page 58 Have students count the butterflies and tell the number. (8) Tell them that they will now write the number 8. Have students trace the 8s and then start at

the dots to write more 8s. Repeat the procedure to have students count the 9 bees and trace and write the 9s to tell how many bees are shown. Have them complete the page independently.

4 Assess

Have students create flashcards for the numbers 1 to 9 by drawing objects on one side of a blank index card and writing the number of objects on the other side.

3-8 Problem Solving: Try, Check, and Revise

pages 59–60

1 Getting Started

Objectives
- To use the try, check, and revise strategy to solve problems
- To estimate quantities of nine or fewer

Vocabulary
revise

Materials
tubular pasta; string; 1-lb plastic container with a lid per pair of students; 9 pennies per pair of students; beans; interlocking cubes

Warm Up • Mental Math
Ask students to answer yes or no.
1. I have eight sisters and brothers.
2. There are only nine books in the library. (no)
3. 1 more than 7 is 8. (yes)
4. 9 is less than 8. (no)
5. A spider has eight legs. (yes)
6. 6 is more than 7. (no)

Warm Up • Number Sense
Have students work in pairs to review the numbers 1 to 9. Have students make necklaces with tubular pasta and string. Help them thread one piece of pasta onto a piece of string and then make a knot. Have them repeat with two pieces of pasta and so on, until they have shown the number 9. Help them tie both ends of the string together to form a necklace.

2 Teach

Develop Skills and Concepts Explain the difference between an actual number and an estimate. Make sure students understand that in some situations, an actual number is needed and in other situations an estimate is all that is needed. Have students name some examples of both situations. Ask, *Would you need an estimate or an actual number to pay for a toy?* (actual number)

- Organize the class into pairs and give each pair a 1-lb plastic container with 9 pennies. Have students shake the container to hear what 9 pennies sound like. Repeat with 5 pennies so that students can hear what 5 pennies sound like. Ask, *Can you hear a difference*

All estimates will vary.

1

6   7   8   9

2

2   3   4   5

3

6   7   8   9

4

2   3   4   5

Guess how many marbles are in the jar.
Circle this number. Then count. Write the number.

between the number of pennies? (Yes, 9 pennies sound louder than 5 pennies.) Then, have one student put some pennies into the container and ask his or her partner to guess the number of pennies in the container. After the partner makes a guess, have students pour out the pennies and count to check. After each turn, ask, *Did you guess the correct number? How close was your guess?* Have students take turns and repeat several times.

- Have students estimate quantities using a benchmark of 5. Show what 5 interlocking cubes look like in a pile. Then, show students groups of between 1 and 9 cubes. For each group, students should give a thumbs-up sign if they think the group has more than 5 cubes and a thumbs-down sign if they think the group has fewer than 5 cubes. Have a volunteer count the cubes to check each number. Repeat with other numbers.

T59

All estimates will vary.

1. 4 5 6 7

2. 4 5 6 7

3. 6 7 8 9

4. 6 7 8 9

60 Lesson 3-8 • Problem Solving: Try, Check, and Revise

Guess how many items are shown. Circle
this number. Then count. Write the number.

For Mixed Abilities

Common Errors • Intervention

Some students may have difficulty making estimates that are close to the actual number. For those students, it may be helpful to estimate by comparing to a reference set. Show students a group of 4 interlocking cubes and count them aloud. Then, show students a second group of 8 interlocking cubes. Ask, *Does this group have more cubes than the first group, or does it have fewer cubes?* (more) Continue in this way, asking students to guess whether a group of objects has more cubes, fewer cubes, or the same number of cubes as the group of 4. Repeat the activity using a reference set of a different size.

Enrichment • Estimation

Have students estimate and measure the length of items. For example, have students estimate the length of their math book in interlocking cubes and then make a cube train to count.

• Give each student between 1 and 9 beans to paste onto paper in any arrangement. Call on students, one at a time, to quickly hold up the paper and then turn it upside down. Have the rest of the class guess the number of beans. A volunteer counts the beans to find the actual number. Repeat until each student holds up his or her paper.

3 Practice

Using page 59 Have students look at Exercise 1.

• Now, apply the four-step plan. For SEE, ask, *What are you asked to do first?* (guess how many marbles are in the jar) Then, *What are you asked to do next?* (check my estimate) For PLAN, ask students how they check their estimate. (count the number of marbles in the jar) For DO, have students circle the number of marbles they think are in the jar. For CHECK, have students count the marbles and write the number on the line provided.

• Have students complete the remaining exercises independently.

Using page 60 Have students look at the scene in Exercise 1. Ask a volunteer to describe the scene. (Kittens are playing.) Have students circle the best estimate for the number of kittens playing, count the actual number, and trace the number 6. Ask, *How close was your estimate to the actual number?* Have students complete the page independently. When they have completed the page, ask, *Did your estimates get closer by the end? Why do you think this happened?*

4 Assess

Ask students, *When would you need to estimate a number?* (Possible answer: when you are trying to decide how many items are in a group and you do not need an exact number)

T60

3-9 The Number 0

pages 61–62

pages 61–62

1 Getting Started

Objectives
- To associate 0 with no objects
- To write 0
- To define *zero*, *nothing*, *not any*, and *none*

Vocabulary
zero, nothing, not any, none

Materials
*number 0 in relief; *counters; Reproducible Master RM7; number card 0 in relief; red, green, and black crayons

Warm Up • Mental Math
Ask students how many there are.
1. wheels on a car (4)
2. fingernails on one hand (5)
3. numbers in their phone number (10)
4. numbers in their address
5. thumbs on both hands (2)
6. fingers on both hands if thumbs are not counted (8)

Warm Up • Number Sense
Have students write the numbers 1 to 9 in order on the board. Ask a student to point to each number as the class says the numbers in order and then in reverse order.

2 Teach

Develop Skills and Concepts Have 6 students form a line near the board. Ask the last student to count the number of students in the line and write the number on the board. (6) Have that student sit down. Repeat the instructions for the last student in the line now and ask where the number 5 should be written. (before the 6) Have the student write the 5 in front of the 6 and then sit down. Continue until the numbers 1 to 6 are written on the board and no students are standing. Ask how many students are in the line now. (none) Write **0** in front of the 1 on the board and tell students that a zero is the number we write when there is nothing. Tell students that we also use the words *none* and *not any* to mean "zero." Repeat the activity for 9 students standing to have the numbers 1 through 9 written on the board. Ask what number needs to be written before the 1 to show that no students are standing. (0) Write **0** before the 1.

- Alter the above activity to use counters on a table with one being removed for each new number. Then, start with no counters to build the numbers on the board from 0 to 9.

- Read the poems "Zero" and "None" on page T45a. Have some students act out the poems as classmates say them with you.

- Show the number 0 in relief and trace it as you sing the following song to the tune of "Mulberry Bush":

  Curve down, curve up to make an egg.
  Curve down, curve up to make an egg.
  Curve down, curve up to make an egg
  To make the number zero.

  Then, have each student trace a number 0 card in relief as they sing the song with you.

- Have students practice writing 0s on their copies of Reproducible Master RM7.

Count the objects in each set. Circle the sets of 0 objects.

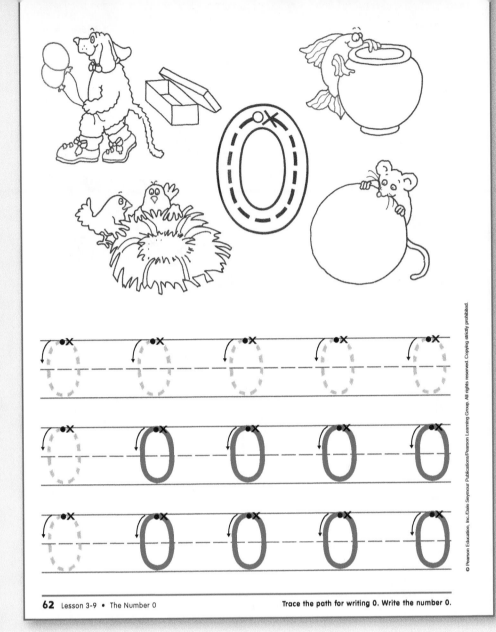

62 Lesson 3-9 • The Number 0   Trace the path for writing 0. Write the number 0.

For Mixed Abilities

Common Errors • Intervention

Some students may have difficulty understanding the concept of zero. Draw 5 sticks on the board. Have a student count the sticks, say the number, and write the number on the board. Repeat for 4, 3, 2, 1, and 0 sticks, erasing 1 stick and the old number each time. Discuss how each time there is 1 less stick so that drawing no sticks for the number 0 will be obvious. Encourage students to use *zero*, *none*, *nothing*, and *not any* to refer to the number 0.

Enrichment • Number Sense

1. Discuss times when having zero of something can present a problem, such as no milk for cereal, no bandages for cuts, and so on. Discuss when having zero of something is a positive situation, such as no chickenpox, cuts, or bruises. Have students draw a picture and a happy or sad face to show an example of each.

2. Have students write the numbers **0** to **9** and draw objects to show each number.

3 Practice

Using page 61 Ask students to name the large number in the example. (0) Ask students how many birds are in the nest. (0) Tell them that they will be circling the nest with zero birds in it in each exercise. Help students complete Exercise 1 if necessary.

Using page 62 Ask students if the box at the top of the page has any balloons in it. (no) Tell students to circle the box to show that it has zero balloons in it. Ask if the fishbowl has any fish in it. (no) Have them circle the bowl to show that it has zero fish in it. Continue similarly for the nest and the large circle the mouse is holding.

• Tell students to use their green crayons to color the dot at the top of the zero and then trace the X with a red crayon. Sing the song for writing a 0 as students trace the large 0 with their index finger several times. Then, have students trace the row of 0s. Tell them to begin at the dot and trace the first 0 on each of the last two lines and then write more 0s.

4 Assess

Have students draw a picture representing the number 0. Have them write the number 0 beneath their picture.

T62

pages 63–64

1 Getting Started

Objectives

• To count zero to nine objects and choose the correct number
• To write 8, 9, and 0

Materials

Reproducible Masters RM7 and RM10; 9 counters per pair of students; number cards 0 through 9; magazines; blank index cards

Warm Up • Mental Math

Have students tell the shape of the following:

1. a clockface (circle)
2. this book (rectangle)
3. the red part of a traffic light (circle)
4. a dinner plate (circle)
5. the board (rectangle)

Warm Up • Number Sense

Have students work in pairs with 9 counters and a sheet of ruled paper. Tell one student to lay out 6 counters. Have his or her partner remove some counters and have the first student write the number of counters left. Have partners reverse roles to begin with 8 counters. Continue until several numbers have been written.

2 Teach

Develop Skills and Concepts Sing "A Whole Lot of Numbers," found on page T45a, to the tune of "He's Got the Whole World in His Hands." Have students lay out the number cards 0 through 9 and show each number as it is heard. You may vary the activity when students know the lyrics well by having a student hold all cards 0 through 9. The student shows a card and students sing that verse. In this case, use the words he, she, his, and her as applicable.

• Stomp your foot 6 times and have students tell the number of stomps heard. Have a volunteer write the number on the board. Continue for students to count the number of finger snaps and hand claps. Be sure to use zero times in the activity.

• Have students write the number **8, 9,** or **0** in the air for

classmates to name the number. Have a student who names the number correctly then write it on the board.

• Tell students that you are thinking of a number that comes between 6 and 8. Have a student write the 6 and 8 on the board and fill in the missing number. (7) Vary the activity by drawing a number of circles on the board and telling students that you are thinking of the number that tells how many circles you have drawn. Have a student write the number.

• Have students write 8s, 9s, and 0s on copies of Reproducible Master RM7.

• Give students copies of Reproducible Master RM10. Write the following on the board: **2 8 9.** Tell students to find a butterfly in the picture and write the number 1 on it. Tell students to look for another butterfly and write the number 2 on it. Have students find the eight pigs one at a time and write the numbers 1 through 8 on them. Repeat the procedure to find and number the nine fenceposts. Have students color the picture.

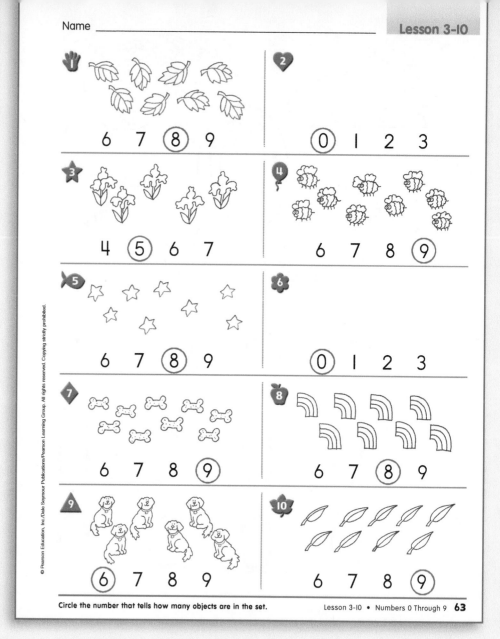

Name _____

Circle the number that tells how many objects are in the set.

Lesson 3-10 • Numbers 0 Through 9 **63**

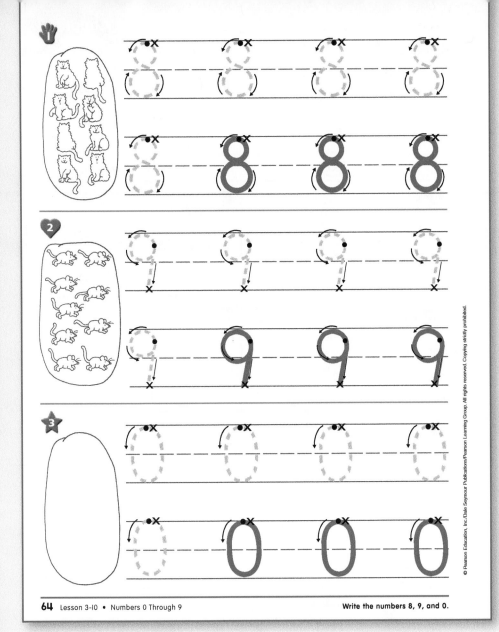

Write the numbers 8, 9, and 0.

For Mixed Abilities

Common Errors • Intervention

Some students may need more practice counting from 0 to 9. Have students work with partners. As you show a number card 0 through 9, have students name the number and draw that many objects. Students should share their pictures with their partners and each should count again to confirm the number.

Enrichment • Application

1. Have students look around their kitchen at home for the numbers 0, 8, or 9. Tell students to write a 0 for each 0 they see and then circle each group of 8 zeros on their papers. For more challenge, have students do all three numbers, each on a separate sheet of paper.

2. Help students explore odd and even numbers. Give pairs of students 50 interlocking cubes and have them count out sets of 1 to 9 cubes. For each set, show students how to connect pairs of cubes. Point out to students that if there are no cubes left over after making pairs, then the number is even. If there is one cube left over, then the number is odd.

3 Practice

Using page 63 Have students count the leaves in Exercise 1 and tell how many there are. (8) Ask them to name the numbers under the leaves. (6, 7, 8, 9) Tell students to circle the number that tells how many leaves are shown. (8) Have students complete the page independently.

Using page 64 Have students count the cats and tell the number. (8) Tell students that they will write the number 8. Have them trace the 8s and then start at the dots to write more 8s. Repeat the procedure to have students count the 9 mice and trace and write the 9s to tell how many mice are shown. Have students notice that there are no objects to the left of the row of 0s. Have them complete the page independently to trace the 0s and write more 0s.

4 Assess

Have students look in magazines to find pictures that show one to nine objects. Have them create flashcards by pasting the picture to one side of an index card and writing the number of objects on the back. Ask how they could show the number 0. (We can leave one side of the index card blank by not pasting any pictures to it.)

T64

Chapter 3 Test and Challenge

pages 65–66

1 Getting Started

Test Objectives

- **Items 1–6:** To count 0 through 9 objects and choose the appropriate number (see pages 45–50, 53–58, 61–62)
- **Items 7–10:** To write numbers 6 through 9 (see pages 51–52, 59–60, 61–62)

Challenge Objective

- To draw one to nine objects

Materials

blank index cards; *objects to be sorted by use in kitchen, bathroom, and bedroom; *pictures of kitchens, bathrooms, and bedrooms; crayons

2 Give the Test

Prepare for the Test Have students create flashcards for numbers 0 to 9. On one side of a blank index card, have students write a number. On the other side, have them draw that number of pictures. Students should then use their flashcards to quiz each other on numbers 0 to 9.

- After students have mastered counting through 9 and the concept of 0, have them complete the Chapter 3 Test. You may wish to review test-taking rules with students. Remind them that they are not to talk to each other and they are to finish the test on their own. Decide ahead of time whether or not to allow use of manipulatives.

Using page 65 Tell students that for Exercises 1 through 6 they are to count the objects in the set and circle the number that tells how many there are. Have them complete the first six exercises independently. Now, have students count the butterflies in Exercise 7 and tell how many there are. (6) Tell students that they are to write the number 6 on the line to tell that there are six butterflies. Have them complete the last three exercises independently.

Name _____

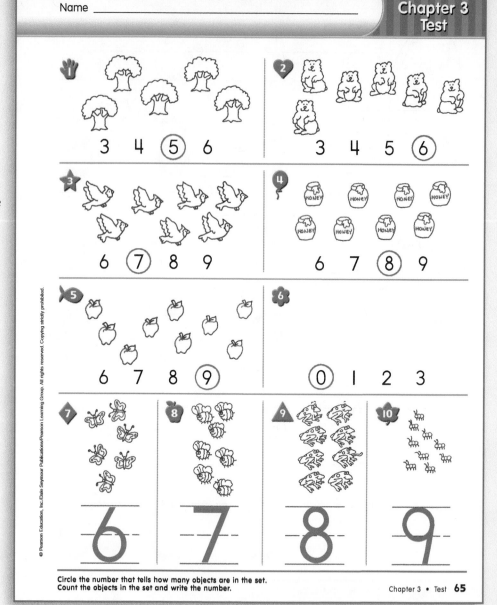

Circle the number that tells how many objects are in the set.
Count the objects in the set and write the number.

Chapter 3 • Test **65**

3 Teach the Challenge

Develop Skills and Concepts for the Challenge Have a student draw 7 objects on the board and draw a circle around them. Now, have a student draw 3 objects on the board and circle them. Have a student write a number under each group to tell how many objects there are. (7, 3) Ask students which group has more. (7) Ask which has fewer. (3) Continue similarly to compare 5 and 9, 8 and 6, 7 and 8, and 6 and 4.

Using page 66 Tell students that they are going to help the baker put rolls into bags. Tell them to trace the rolls that are in the first bag and count them. Then, they should draw more rolls and count them until they have 6 rolls in the bag. Give them the directions to fill each of the other bags one at a time with the designated number of rolls. Be sure students are counting the number of rolls in each bag. Ask which bag has the fewest rolls (the one with 6 rolls) and which has the most (the one with 9 rolls).

Draw 6 rolls in the first sack, 7 in the second, 8 in the third, and 9 in the fourth.

For Mixed Abilities

Common Errors • Intervention

Students may have a difficult time drawing the correct number of objects in a specific set. Have students count aloud as they draw each object in the set. When they reach the correct number of objects that is supposed to be in the set, they should stop drawing.

Enrichment • Number Sense

1. Working in groups, have students cut from magazines 6, 7, 8, or 9 pictures of things that are found below the sea, in a restaurant, in a car, above the ground, and so on. Have students mount their pictures on construction paper and make books for each topic. Have students number the pages in the books.

2. Give students three stapled sheets of lined paper. Have students write six 6s on the first page, eight 8s on the next page, and five 5s on the last page.

4 Assess

Alternate Chapter Test You may wish to use the Alternate Chapter Test on page 207 of this manual for further review and assessment.

Challenge Place 2 objects in a paper bag. Have a volunteer add objects to the bag until there are 6 objects altogether. (add 4) Repeat with different numbers of objects starting in the bag to allow each student a turn.

Numbers 0 Through 12

pages 67–84

Chapter Objectives

- To count 0 to 12 objects and choose the appropriate number
- To define *ten*, *eleven*, and *twelve*
- To write 0 through 12
- To write 0 to 12 in order
- To connect dots numbered 0 to 12 in order
- To write the number before or after a given number and the number between two given numbers
- To order numbers 0 to 12 on a number line
- To use tallies to count objects

Materials

You will need:
*object cards 0 through 12; *counters

Students will need the following:
Reproducible Masters RM7 and RM11; number cards 0 through 12; crayons; 12-cup egg cartons; counters; number cards 0 through 12 in relief; blocks in two colors; blank index cards

Bulletin Board Suggestion

Display stanzas of the "Counting On" poem and have students find pictures to depict each number, 0 through 9. (a tree trunk for 1, knees for 2, and so on) Where no object is mentioned for a number, display that number of pictured items. Add numbers 10, 11, and 12 as they are presented in the lessons.

Verses

Poems and songs may be used any time after the pages noted beside each.

Counting On (pages 69–70)

Zero's the number we say when there's none;
Zero is first; it comes before 1.

1 is a number to count trunks on all trees.

2 is the number to count people's knees.

Next comes a 3 for tricycle wheels;
Breakfast, lunch, dinner—count 3 daily meals.

Then there is 4 to count legs on a bear,
A table, a sofa, or most any ol' chair.

But what do you think we can count with a 5?
How 'bout fingers and toes or 5 bees in a hive?

5 and 1 more, this number comes next.
We use it to count the legs on insects.
'Cause 6 is the number of legs you will find
On insects, no matter what color or kind.

Can 7 be next? Yes, 6 and 1 more
Is 7, my goodness! We're a long way past 4!

We're ready for the number that comes before 9;
Fold in the thumbs and 8 fingers are mine.

Then 9, a good bedtime for children who then
Are fast asleep long before clocks can strike 10.

And 10 is the number that's made when a 1
Is followed by zero. Is our counting done?

No! Counting to 10 doesn't mean that we're done!
11 comes next—1 ten and 1 one.

Then 11 and 1 more—a dozen, no less.
I'll stop now at 12 and take a recess!

One, Two, Buckle My Shoe (anon.)(pages 69–70)

One, two, buckle my shoe.
Three, four, shut the door.
Five, six, pick up sticks.
Seven, eight, lay them straight.
Nine, ten, a good fat hen.

Numbers Song (pages 79–80)

Zero, one, two, three, four, five,
Six, seven, eight, nine, sakes alive!
We have learned our numbers well
Next come ten, eleven, twelve.
We can sing a number song;
Won't you come and sing along?

The Nine-Bee Sneeze

There were six little bees
Sitting in the trees
Trying to sneeze
The leaves off the trees.

The bees in the center
Were really quite little.
They giggled and wiggled,
There, in the middle.

The six bees sneezed
As hard as they could.
But their sneeze didn't do
What they thought it would!

They gave it their all,
But their very best sneeze
Still didn't stir up
Much of a breeze.

The six bees frowned
As they sat in the trees,
Then buzzed off to get
Three more little bees.

Now nine little bees
Sat in three little rows
Each with a little
Sneeze in his nose.

Three bees in front
Three bees between
And three bees behind
To blow the tree clean!

The first row, middle row,
And last row believed
Together they could do it,
And do it with ease!

They started to sneeze,
And just about then
Along came a great big
Gust of wind!

It shook the tree hard,
And blew off the leaves!
But the bees all believed
It was their nine-bee sneeze!

Whenever you see
Zero leaves on the trees,
Remember the sneeze
Of the nine little bees!

Name _____

Count the number of bees in each picture.
Then describe what is happening in each picture.

"The Nine-Bee Sneeze," p. RM11

Activities

Use Reproducible Master RM11, "The Nine-Bee Sneeze," to help students relate the story to the mathematical ideas presented.

- Give each student 9 counters and a sheet of paper. Reread the poem. Have students count out 6 of the counters and lay them on their papers to show the 6 bees. Tell students to place the 3 extra counters on their papers when the poem says that there were 9 bees. Tell them to listen carefully to learn how the bees were arranged as you read on about ". . . three little rows . . ." and "Three bees in front/Three bees between/And three bees behind." Have students arrange their bees accordingly and point to each row as the poem identifies it. Encourage students to identify the first row as the one closest to the top of their page. Now, have 9 students stand to act out the poem as you read it again.

- Place 9 empty chairs in a row. Have 9 students form a line. Write any number 0 through 9 on the board. Have students from the front of the line move to sit in that number of chairs and have the rest of the class count out loud. Then, have them go to the end of the line as you write another number on the board. Continue until all students have participated two or three times. Vary the activity by arranging the 9 chairs in two or three rows.

T67b

pages 67–68

1 Getting Started

Objectives

- To count zero to nine objects and choose the correct number
- To write 0 to 9

Materials

*16 counters; Reproducible Master RM7; number cards 0 through 9; 9 counters

Warm Up • Mental Math

Ask students which is larger.

1. a truck or motorcycle (truck)
2. a plum or grapefruit (grapefruit)
3. a kitten or cat (cat)
4. a boot or sock (boot)
5. a raisin or apple (apple)
6. a schoolbus or car (schoolbus)
7. a purse or suitcase (suitcase)

Warm Up • Number Sense

Shuffle one set of number cards 0 through 9. Have students take turns drawing a card and counting from 0 to the number drawn.

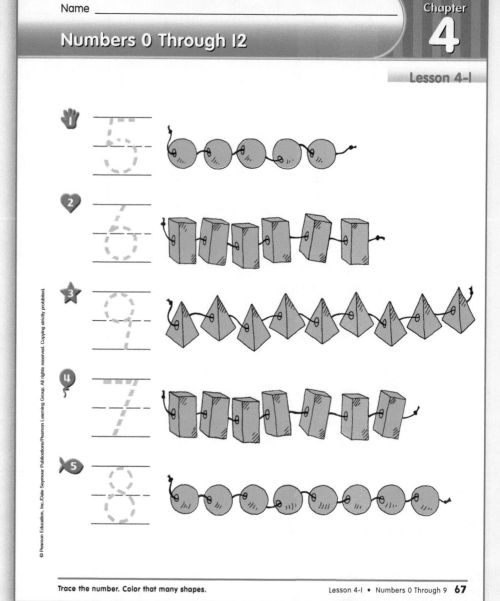

Name _____

Numbers 0 Through 12

Chapter **4**

Lesson 4-1

Trace the number. Color that many shapes.

Lesson 4-1 • Numbers 0 Through 9 **67**

2 Teach

Develop Skills and Concepts With the class, sing "This Old Man" found on page T25a. Stop at the verse for nine. Pause at number words for students to show counters as bones. Sing the song again and have students show their number cards.

- Lay out a group of 7 counters and a group of 9 counters. Lay out the number cards 0 through 9 in random order. Have students find and place the correct number card under each group. Have another student find and place the correct number card between the 7 and 9 and build a group of that number of counters. (8) Repeat with numbers 6 and 8.

- Have students practice writing 0 through 9 on copies of Reproducible Master RM7.

3 Practice

Using page 67 Tell students that they are to trace the number at the beginning of the row and color that many figures on the string. Have them complete the page independently.

Using page 68 Tell students that they are to count the objects in the set and then write the number beside the group. Randomly name objects in some of the groups for students to tell the number they will write. Have students look at Exercise 5 and tell how many objects there are. (none) Ask what number is to be written. (0) Have them now complete the page independently.

68 Lesson 4-1 • Numbers 0 Through 9

Count the objects in the set and write the number.

For Mixed Abilities

Common Errors • Intervention

For students who may still have difficulty writing the numbers 0 through 9, have them form the numbers in the air or in sand as they sing about how to write the numbers to the tune of "Mulberry Bush." The lyrics for the various numbers can be found on pages T35, T37, T41, T51, T57, and T61.

Enrichment • Application

1. Have students write any two different numbers from 0 to 9 on a sheet of paper. Tell them to trade papers with a friend and write all the numbers between the two numbers.

2. Have each student copy his or her phone number from a model. Have students then practice dialing their numbers on a toy telephone.

4 **Assess**

Have students draw a picture with between one and nine objects. Have them switch papers with a partner and tell how many objects there are on the paper.

T68

4-2 The Number 10

pages 69–70

1 Getting Started

Objectives
- To count zero to ten objects and choose the correct number
- To define *ten*

Vocabulary
ten

Materials
number cards 0 through 10

Warm Up • Mental Math
Tell students to count on from
1. 7 to 9 (7, 8, 9)
2. 0 to 6 (0, 1, …, 6)
3. 4 to 9 (4, 5, …, 9)
4. 5 to 8 (5, 6, 7, 8)
5. 3 to 7 (3, 4, 5, 6, 7)
6. 7 backward to 3 (7, 6, 5, 4, 3)
7. 8 backward to 4 (8, 7, 6, 5, 4)

Warm Up • Number Sense
Secretly tell a student to clap hands a number of times from 0 to 9. Choose another student to say the number and write it on the board. Continue for more students to participate.

2 Teach

Develop Skills and Concepts Draw 11 large circles across the board. Show the 0 card and ask students how many sticks should be drawn in the first circle to show the number 0. (none, not any) Adhere the 0 card to the board under the first circle. Continue through the number cards in order through 9, having students draw sticks in the circle above each number card. After each number is completed, ask students how many more sticks were drawn for that number than for the number before it. (1 more) When 9 sticks have been drawn and the numbers 0 to 9 are shown across the board, have a student draw 9 sticks and 1 more in the last circle. Place the 10 card under the last circle and tell students that the word *ten* is said for the number that is 9 and 1 more. Tell students that a ten is written with a 1 and 0 as you write **10** on the board.

- Have students count from 0 to 10 as you remove each respective number card from the board. Now, have students count backward from 10 to 0 as you replace each respective number card.

- Have students lay out their number cards to show each number heard in the poem "One, Two, Buckle My Shoe" on page T67a.
- Read the poem "Counting On" on page T67a, omitting the last two stanzas that refer to numbers 11 and 12.

3 Practice

Using page 69 Have students count the trees in Exercise 1 and tell the number. (7) Tell them to find the 7 down the right side of the page and trace the line from the group of trees to the 7 to show that there are seven trees. Work similarly with students through each of the exercises. Remind them that there will be some numbers that will not match with the pictures.

Using page 70 Tell students that they are to count the objects in the exercise and circle the number that tells how many objects there are. Work through the first two exercises together and then have students complete the page independently.

T69

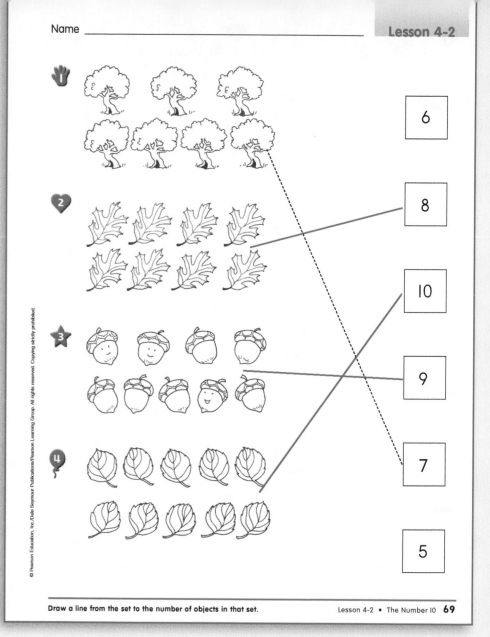

6

8

10

9

7

5

Draw a line from the set to the number of objects in that set.

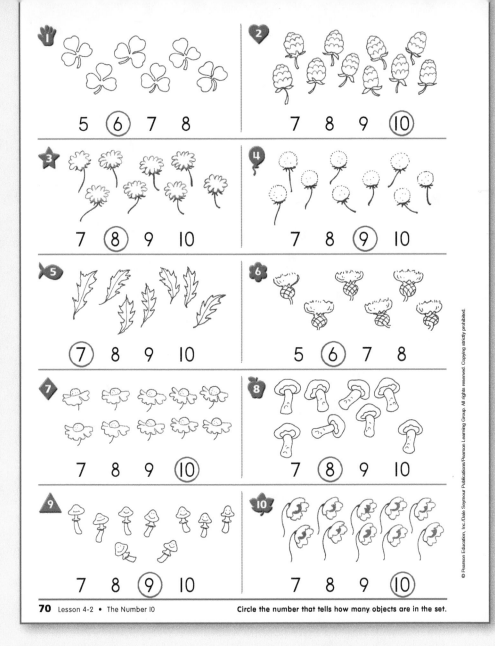

1. 5 (6) 7 8

2. 7 8 9 (10)

3. 7 (8) 9 10

4. 7 8 (9) 10

5. (7) 8 9 10

6. 5 (6) 7 8

7. 7 8 9 (10)

8. 7 (8) 9 10

9. 7 8 (9) 10

10. 7 8 9 (10)

70 Lesson 4-2 • The Number 10   Circle the number that tells how many objects are in the set.

For Mixed Abilities

Common Errors • Intervention

Some students may need more work associating numbers with groups of objects. Have students work with partners with number cards from 6 to 10 and cards in mixed order showing 6, 7, 8, 9, and 10 objects. Have them arrange the two sets of cards to match numerals with the number of objects. Then, have them cooperate in modeling each number using buttons or other counters.

Enrichment • Number Sense

1. Have students write four different numbers on their papers. Then, tell them to rewrite the numbers in order from 0 to 10. Finally, tell students to draw the correct number of sticks or objects beside each number.

2. Have students match the 2s through 10s from a deck of playing cards. Then, have students arrange each of the four suits in order from 2 to 10.

4 Assess

Draw groups of zero to ten triangles on the board. Place the number cards 0 through 10 in random order on a table. Have a student pick a number, say its name, and place it under the group on the board having that number of triangles. Continue until all cards are properly placed.

T70

4-3 The Number 11

pages 71–72

1 Getting Started

Objectives

- To count 0 to 11 objects and choose the correct number
- To define *eleven*

Vocabulary

eleven

Materials

number cards 0 through 11;
11 counters

Warm Up • Mental Math

Ask students which is less.

1. 6 or 7 (6)
2. 3 or 10 (3)
3. 9 or 8 (8)

Ask students which is more.

4. 7 or 10 (10)
5. 9 or 6 (9)
6. 0 or 5 (5)

Warm Up • Number Sense

Write **8** on the board. Give the example of students eating 8 raisins and then 1 more. Ask how many raisins were eaten. (9) Ask students if 1 more than 8 comes before or after 8. (after) Have a student write a 9 on the board after the 8. Continue for the numbers that come after 6, 5, 9, and 7.

Name ____________________ **Lesson 4-3**

11

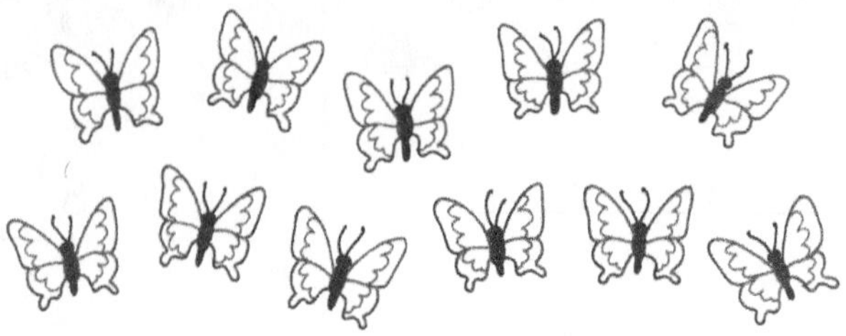

1

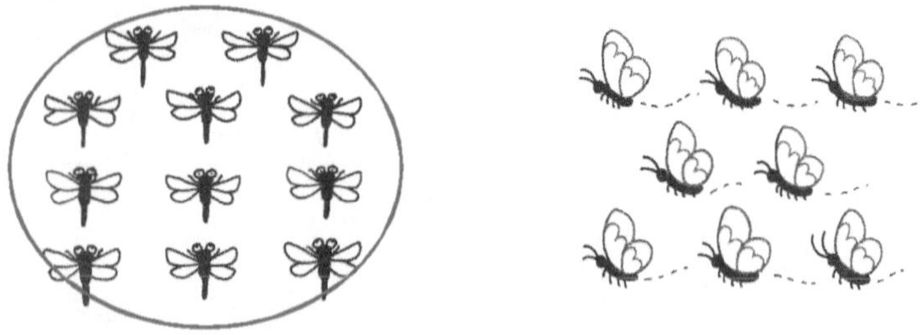

2

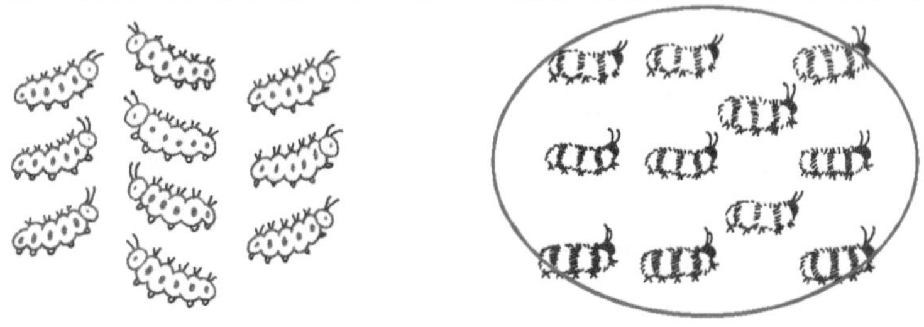

3

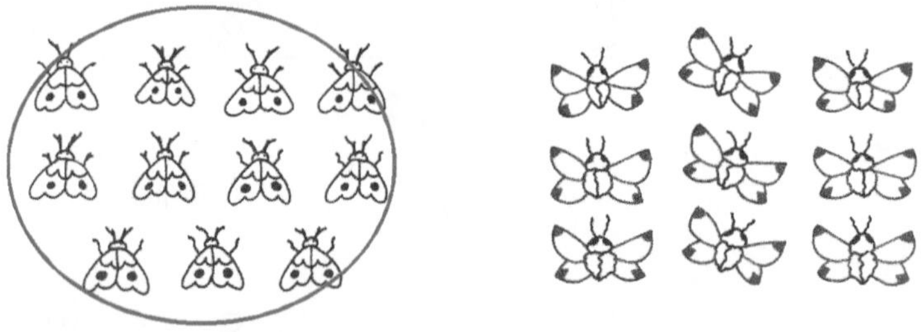

Count the objects in each set. Circle the sets of 11 objects.

2 Teach

Develop Skills and Concepts Have students build a set of 7 counters. Have one student draw 7 circles on the board and number them from 1 to 7. Have students lay out 1 more counter as a student draws another circle on the board and writes its number. (8) Continue through 10. Then, have students lay out 1 more again as you draw 1 more circle on the board and write **11**. Tell students that 10 and 1 more is 11. Have them repeat the statement with you.

- Tell students to lay out their number cards 0 through 11 so that each new number is 1 more than the last number. Have them count with you from 0 to 11.
- Ask a student to count 11 objects in the classroom. Have students count orally to 11 as each new object is pointed out. Have a student write an 11 on the board when 11 objects are noted. Have other students point out 11 objects as classmates count. Having them practice counting out 11 objects and writing the number helps prevent thinking of 11 as 1 and 1 more, or 2.

3 Practice

Using page 71 Ask students to name the number at the top of the page. (11) Have them count the butterflies and tell how many there are. (11) Now, have students count the dragonflies in the first group in Exercise 1 and tell how many there are. (11) Tell students to draw a circle around the 11 dragonflies. Ask them how many butterflies there are in the second group in Exercise 1. (8) Tell students that they are to count the insects in each group and circle the group only if there are 11 in it. Help students decide whether to circle the first group in Exercise 2 (no) and then have students complete the page independently.

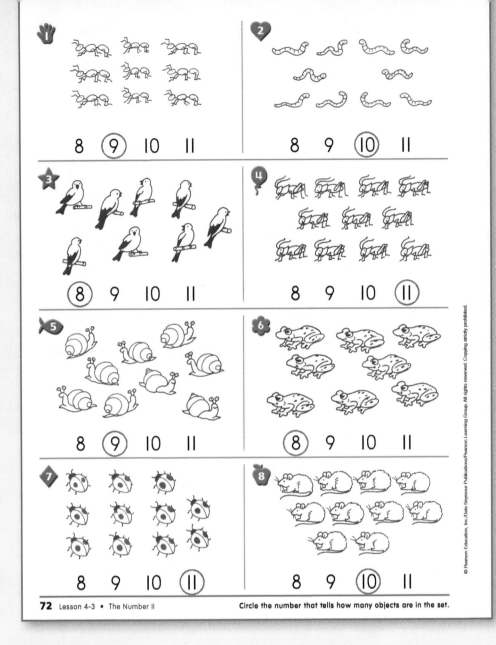

1. 8 (9) 10 11

2. 8 9 (10) 11

3. (8) 9 10 11

4. 8 9 10 (11)

5. 8 (9) 10 11

6. (8) 9 10 11

7. 8 9 10 (11)

8. 8 9 (10) 11

Circle the number that tells how many objects are in the set.

For Mixed Abilities

Common Errors • Intervention

For additional practice with counting, call 7 students forward to stand in a line. Give each student a number card 0 through 6 in mixed order. Have students put themselves in order according to their numbers. Repeat for the numbers 7 through 11. Vary the activity by ordering all numbers 0 through 11 at one time and by having a student direct the ordering.

Enrichment • Number Sense

1. Have students draw 11 objects and color all but 1 or all but 4, and so on. Have them write the numbers that tell how many are colored and how many are not colored.

2. Give students a sheet of large grid paper. Tell them to color 11 squares across the first row, 1 less than 11 on the next row, and so on until the numbers 11 through 0 are shown in order from top to bottom.

Using page 72 Ask how many insects are in Exercise 1. (9) Tell students to find and circle the 9 in the row of numbers below the insects. Help students through the next couple of exercises and then have them complete the page independently.

4 Assess

Show students number cards 0 through 11. For each card you show them, have students draw that many pieces of fruit.

T72

4-4 Writing 10 and 11

pages 73–74

1 Getting Started

Objective
• To write 10 and 11

Materials
Reproducible Master RM7; number cards 0 through 11; green, red, and black crayons

Warm Up • Mental Math
Ask students how many there are.
1. fingers on 1 hand (5)
2. toes on 2 feet (10)
3. windows in this room
4. doors in this room
5. people in their family
6. students in their row or at their table

Warm Up • Number Sense
Write **8** on the board and have students hold up the number card that tells the number after 8. (9) Continue asking for the number after. Then, have students show the number that comes before the number you write on the board.

2 Teach

Develop Skills and Concepts Draw 9 objects on the board and have students tell the number. (9) Have a student write the number. Ask how many there would be if 1 more object was drawn. (10) Write **10** on the board and ask students what two numbers are used to write a 10. (1 and 0) Have several students write the number 10 on the board as classmates sing the songs from earlier lessons for making 1 and 0. The songs are found on pages T35 and T61. Repeat the procedure to show 11 objects and to write the number 11.

• Tell students that you need 10 left hands on the board. Have 10 students trace around their hands on the board. Ask them to help number the hands in order from 1 to 10 to see if there are 10. Have students write the numbers on the hands. Repeat the activity for 11 right hands on the board. It will be difficult for students to trace around their left hand if they are left-handed and vice versa, so another student could do the tracing.

• Have students draw objects on the board to show each number when number cards 0 through 11 are randomly

placed on the board. Now, remove the cards and have them write each number.

3 Practice

Using page 73 Have students count the dogs at the top of the page and tell how many there are. (10) Tell them to use their green crayon to color the circles at the top of the 1 and the 0. Tell students to then use their red crayon to trace the X at the bottom of the 1 and the X at the top of the 0. Have students use their index finger to trace each number from the green dot to the red X. They may enjoy singing the songs for writing 1 and 0 as they trace the numbers. Then, have students use a black crayon to trace the 10. Tell students to first trace the 10s and then to use the dots to write more 10s.

Using page 74 Have students count the cats at the top of the page. (11) Tell them that 10 and 1 more is 11. Have students use a green crayon to color the circle at the top of each 1. Then, have students use a red crayon to trace the X at the bottom of each 1. Tell them to use their index

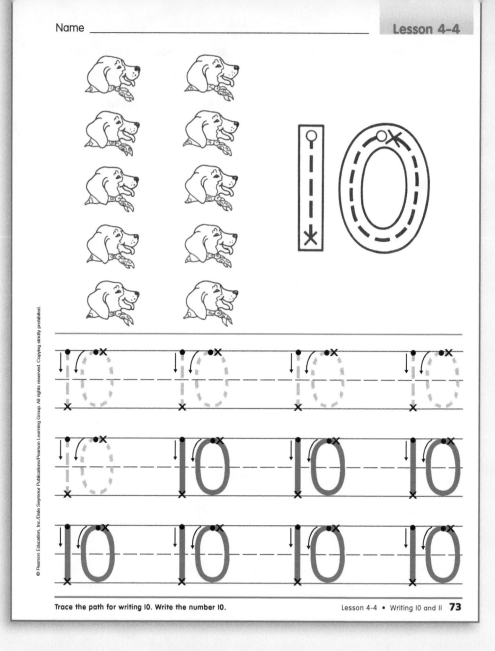

Trace the path for writing 10. Write the number 10.

Lesson 4-4 • Writing 10 and 11   **73**

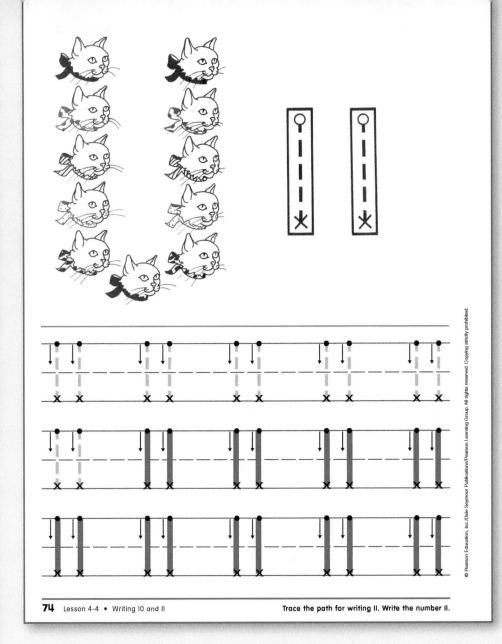

74 Lesson 4-4 • Writing 10 and 11

Trace the path for writing 11. Write the number 11.

Common Errors • Intervention

Some students may have difficulty writing 10 and 11. Have them practice by writing the numbers in the air with their fingers. Tell them to be sure the two digits (1 and 0 in 10 and 1 and 1 in 11) are close enough to each other not to be mistaken for separate numbers. Be sure they transfer the motions from the air to their papers by writing the numbers after each exercise.

Enrichment • Application

1. Have students trace around their hands and number the fingers in order from 1 to 10.

2. Have students write the number 10 on a sheet of paper and cut the paper to form a two-piece puzzle, with one digit on each piece. Repeat for the number 11. Then, have students trade the four puzzle pieces with a classmate in order to reassemble each other's puzzles.

finger to trace the 1s from the green dot to the red X. Then, have students use a black crayon to trace the 11. Have them trace the 11s at the bottom of the page and to use the dots to write more 11s.

4 Assess

Have students write 10 and 11 on copies of Reproducible Master RM7. Then, have students draw a set with ten objects and another set with eleven objects.

4-5 The Number 12

pages 75–76

1 Getting Started

Objectives

- To count 0 to 12 objects and choose the correct number
- To define *twelve*

Vocabulary

twelve

Materials

one 12-cup egg carton per pair of students; number cards 0 through 12; counters

Warm Up • Mental Math

Tell students to count backward from

1. 9 to 6 (9, 8, 7, 6)
2. 10 to 7 (10, 9, 8, 7)
3. 5 to 0 (5, 4, 3, 2, 1, 0)
4. 7 to 3 (7, 6, 5, 4, 3)
5. 11 to 8 (11, 10, 9, 8)
6. 6 to 2 (6, 5, 4, 3, 2)

Warm Up • Number Sense

Distribute the number cards 0 through 11. Write 9 on the board. Have the student who has the number that is 1 less than 9 stand in front of the 9. Repeat for the number that is 1 more than 9. (10; after the 9) Have students say the numbers in order. (8, 9, 10) Continue to form more three-number sequences for numbers 0 through 11.

2 Teach

Develop Skills and Concepts Have students lay out counters to show 1 more than 8 and tell the number. (9) Have a student write a 9 on the board. Tell students to lay out 1 more counter to show 9 and 1 more and to tell the number. (10) Have a student write the number on the board. (10) Repeat for 11 and then have students lay out 1 more counter to show 11 and 1 more as you write **12** on the board. Tell them that a 12 is written with a 1 followed by a 2. Have students write a 12 in the air as they sing the "Mulberry Bush" tune for writing verses 1 and 2. The songs are found on page T35.

- Show students a 12-cup egg carton. Tell students that eggs usually come in a package of 12 and we call a group of 12 things a dozen. Ask students to think of another food that is often purchased a dozen at a time.

(cookies or doughnuts) Have them draw a dozen eggs, cookies, or doughnuts.

- Have students work in pairs to place 1 counter in each cup of an egg carton. Ask students how many counters they used. (12) Then, have them start over and place 1 counter in the first cup, 2 in the second, 3 in the third, and so on to have 12 in the last cup. Vary the activity by having students write a different number from 1 to 12 in each cup and then fill each cup with that number of counters.

- Read the poem "Counting On" on page T67a.

3 Practice

Using page 75 Have students count the watering cans at the top of the page. (12) Have them circle one of the cans and then count the remaining cans to see that 11 and 1 more is 12. Have students count the hand shovels in Exercise 1 and tell the number. (12) Have them circle the group of hand shovels because there are 12. Tell students

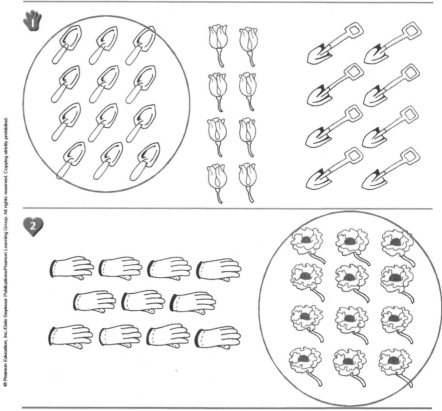

Count the objects in each set. Circle the sets of 12 objects. Lesson 4-5 • The Number 12 **75**

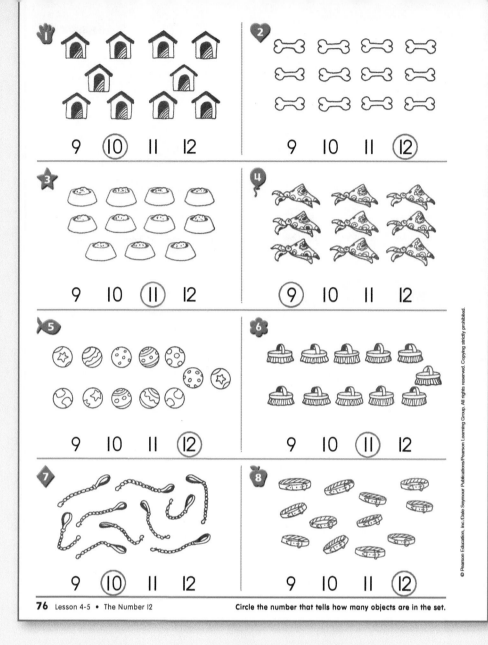

1. 9 ⑩ 11 12

2. 9 10 11 ⑫

3. 9 10 ⑪ 12

4. ⑨ 10 11 12

5. 9 10 11 ⑫

6. 9 10 ⑪ 12

7. 9 ⑩ 11 12

8. 9 10 11 ⑫

Circle the number that tells how many objects are in the set.

For Mixed Abilities

Common Errors • Intervention

Some students may need practice counting 12 objects. Have three students each draw a group of 12 objects on the board. Tell each student to choose a classmate to check the number of objects by counting and then writing the number that tells how many objects there are. (12) Repeat with three different students using the numbers 10 and 11.

Enrichment • Application

1. Have students use counters to show four different ways to lay out a group of 12.

2. Have students draw a dozen eggs. Have students color ten eggs alike and leave two eggs not colored.

ESL/ELL STRATEGIES

Clarify the meaning of *that many* and *how many*. Hold up three fingers and say, *Draw that many circles*. Then, ask, *How many circles did you draw?* Write a number on the board, point to it, and repeat the activity.

that for each exercise, they are to count the objects in each set and circle the set that has 12 objects in it. Have them complete the page independently.

Using page 76 Have students count the dog houses in Exercise 1 and tell how many there are. (10) Tell them to find and circle the number 10 to tell that there are 10 dog houses in the group. Tell students that they are to count the objects in the exercise and circle the number that tells how many there are. Have them complete the page independently.

• Then, ask students to find all the sets on the page that show the same number of objects as the group of bones in Exercise 2. (balls and collars) Point out that all of these groups show 12 objects.

4 Assess

Lay out groups of 10, 11, and 12 counters. Have students tell how many counters are in each group.

1 Getting Started

Objectives
• To write 12
• To practice writing 10 and 11

Materials
Reproducible Master RM7; number cards 0 through 12; green, red, and black crayons

Warm Up • Mental Math
Have students name the following:
1. five objects in the room
2. three members of their family
3. something we buy in a group of 12 (doughnuts, eggs)
4. ten things on their body that are the same (fingers or toes)
5. the number after 10 (11)

Warm Up • Application
Have a student choose a number card 0 through 12 and then put the card on or under something. The student draws that number of objects on the board. The rest of the class then tells what the number is.

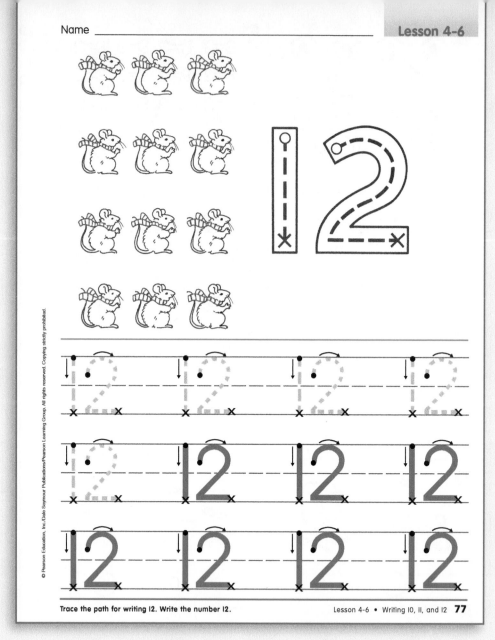

Name _____

Trace the path for writing 12. Write the number 12.

Lesson 4-6 • Writing 10, 11, and 12 **77**

2 Teach

Develop Skills and Concepts Have a student draw 12 objects on the board. Ask students what two numbers are used to write the number that tells how many objects were drawn. (1 and 2) Have a student write a 12 on the board.

• Have several students write the number 12 on the board. Tell them to choose a friend to illustrate their number by drawing 12 objects. Repeat for the numbers 10 and 11.

• Have students practice writing 10, 11, and 12 on copies of Reproducible Master RM7.

3 Practice

Using page 77 Have students count the mice at the top of the page and tell the number. (12) Tell students to use a green crayon to color the circles at the top of the 1 and at the beginning of the 2. Then, have them use a red crayon to trace the X at the bottom of the 1 and again at the end of the 2. Have students use their index finger to trace the 12 and then trace it with a black crayon. Tell students to first trace the 12s and then use the dots and Xs to write more 12s. Have them complete the page independently.

Using page 78 Have students count the insects in the first exercise and tell the number. (10) Tell them to trace the 10 in the first row and then use the dots as starting points to continue writing more 10s. Go through the same procedure with Exercises 2 and 3.

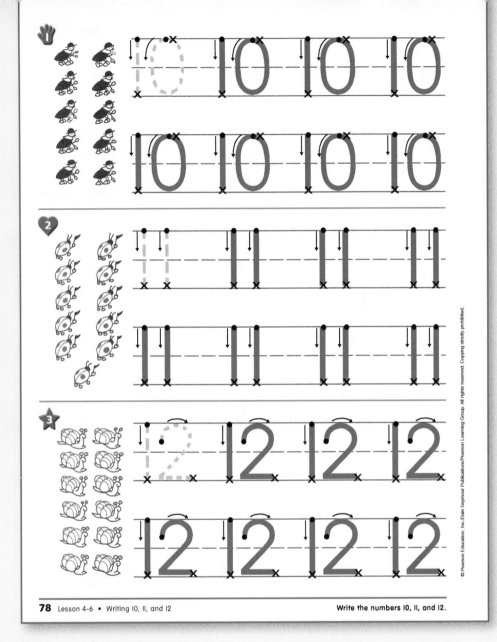

78 Lesson 4-6 • Writing 10, 11, and 12

Write the numbers 10, 11, and 12.

For Mixed Abilities

Common Errors • Intervention

Some students may need practice writing 10, 11, and 12. Have them work with partners with number cards for 10, 11, and 12. Have them use the cards to practice saying, "One followed by zero is ten; one followed by one is 11; one followed by two is 12." Have them write the numbers as they say each description.

Enrichment • Application

1. Have students write the numbers 0, 1, 1, and 2 on four different cards. Tell students to work with a partner who will call for a 10, 11, or 12 to be shown using the cards.

2. Have students look for 10s, 11s, and 12s at home. Tell students to write each number seen. Have students share their numbers and tell where they were seen.

4 Assess

Place the number cards 10 through 12 along the board in random order. Have a student illustrate each number. Remove the number cards and have a different student write the correct number under each group of objects.

T78

4-7 Numbers 0 Through 12 in Order

pages 79–80

1 Getting Started

Objectives

- To write 0 to 12 in order
- To connect dots numbered 0 to 12 in order

Materials

*object cards 0 through 12; number cards 0 through 12

Warm Up • Mental Math

Have students name the numbers that come

1. before 4 (3, 2, 1, 0)
2. between 6 and 10 (7, 8, 9)
3. between 2 and 6 (3, 4, 5)
4. before 12 (11, 10, …, 0)
5. between 8 and 12 (9, 10, 11)
6. between 7 and 11 (8, 9, 10)

Warm Up • Number Sense

Organize the class into pairs. Have one partner draw between 0 and 12 objects on the board for the other partner to count and name the number. Repeat so each partner has a chance to draw and count.

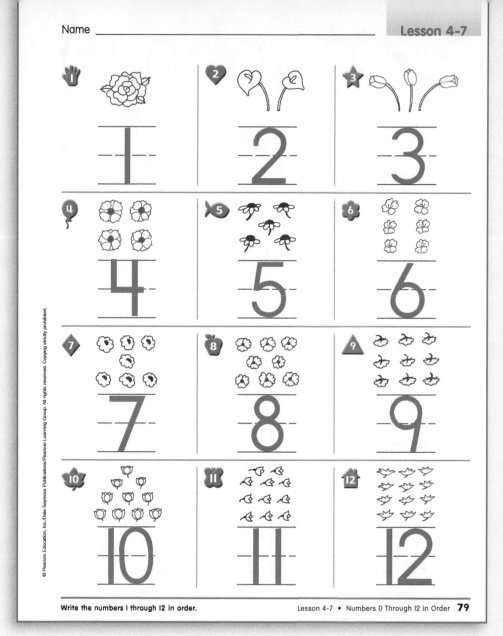

Write the numbers 1 through 12 in order.

Lesson 4-7 • Numbers 0 Through 12 in Order **79**

2 Teach

Develop Skills and Concepts Have a student take any number card from number cards 0 through 12 and lead classmates in counting up to 12 or down to 0 from the drawn number.

- Hold up any object card and have students count up to 12 or down to 0 from the number shown by the objects on the card.
- Give each of 13 students a number from 0 through 12. Tell students to write the number anywhere on the board and draw a circle around it. Have them take turns counting through 12 by pointing to each number in order.
- Give each student a set of number cards 0 through 12. Have one student collect all the 0 cards, another all the 1s, and so on. Place the piles randomly around the room so that all the 1s are together and so on. Have students count as they go from location to location to collect their cards 0 through 12 in order.
- Place the number cards 0 through 12 on a table one at a time as you sing the "Numbers Song," found on page

T67a, to the tune of the "ABC Song." Now, have 13 students stand in a line at the board. Give each student a number card from 0 to 12. Have students place themselves in order and then write their number on the board as they hear it in the song.

3 Practice

Using page 79 Ask students how many flowers are in Exercise 1. (1) Have them write a 1 beneath the one flower. Ask how many flowers are in Exercise 2. (2) Tell students to write 2 under the two flowers. Have students count the flowers in the next exercise and write the number. (3) Have them continue to count the objects and write the number under each. Then, help students notice that they have written the numbers 1 through 12 in order from the least to the greatest.

Using page 80 Have students complete the page independently.

T79

Connect the dots in order from 0 to 12.

For Mixed Abilities

Common Errors • Intervention

Students may need work ordering the numbers 0 to 12. Place the number cards 0 through 5 in random order on the chalk tray and have students place them in order. Repeat for number cards 6 through 12, 0 through 8, 4 through 12, and end with 0 through 12.

Enrichment • Application

1. Have students write the numbers 1 to 12 and then cross out the 1, 3, 5, 7, 9, and 11. Have students name the remaining numbers in order.

2. Have students use chalk to write the numbers 1 to 12 on sidewalk sections or floor tiles. Have students jump over the 1, land on 2, over the 3, land on 4, and so on to count by 2s.

Now Try This! Tell students that they are to join the dots in order from 0 to 12 to complete the picture on this page. Help students find the 0 and its dot on the right side of the page. Tell them to join the dots in order and identify the picture. They may then enjoy coloring the picture after they have joined all the dots.

4 Assess

Have students write the numbers 0 to 12 in order.

T80

© Pearson Education, Inc./Dale Seymour Publications/Pearson Learning Group. All rights reserved. Copying strictly prohibited.

1 Getting Started

Objectives
- To write the number between two given numbers
- To write the number before or after a given number
- To order numbers 0 to 12 on a number line

Materials
number cards 0 through 12 in relief

Vocabulary
number line

Warm Up • Mental Math
Have students name
1. three blue things
2. four things larger than they are
3. a body part above their waist
4. 1 less than 6 children (5 children)
5. a number smaller than 8 (7, 6, …, 0)

Warm Up • Number Sense
Have students do the following activities in the order given:
1. count from 0 to 12
2. count forward to 12 from a number
3. count backward to 0 from 12 or less
4. tell what number comes between two numbers that you name

2 Teach

Develop Skills and Concepts Have students write the numbers 0 to 12 in the air or trace on cards having the numbers in relief. Although this lesson begins having students write numbers without directional clues, allow students to use models if needed.

- Write 8 ____ ____ ____ 12 on the board. Have students write the missing numbers in their proper places. Repeat for other sequences having three or fewer missing numbers.

- Write descending sequences such as 12 ____ 10 ____ 8 ____ on the board for students to fill in the missing numbers. Encourage students to use the terms *before* and *after* as they find the missing numbers. Ask, *What number comes before 12?* (11) *Before 10?* (9) *After 6?* (7) *After 3?* (4)

Name _____

Write the missing number before, after, or between the given numbers.

Lesson 4-8 • Sequencing **81**

It's Algebra! The concepts in this lesson prepare students for algebra.

3 Practice

Using page 81 Have students count the peppers in the first set in Exercise 1 and tell how many there are. (7) Ask students to tell if the number under the peppers tells how many peppers there are. (yes) Repeat for the pears, the last set. (9; yes) Ask students what number is missing under the apples. (8) Tell them to write the number that tells how many apples there are. (8) Help students complete Exercise 2 by writing 11 as the missing number. Students will probably realize that they need only to write the missing number in each sequence, but some may need the aid of counting the objects to fill in the missing numbers. Have students complete the page as independently as possible.

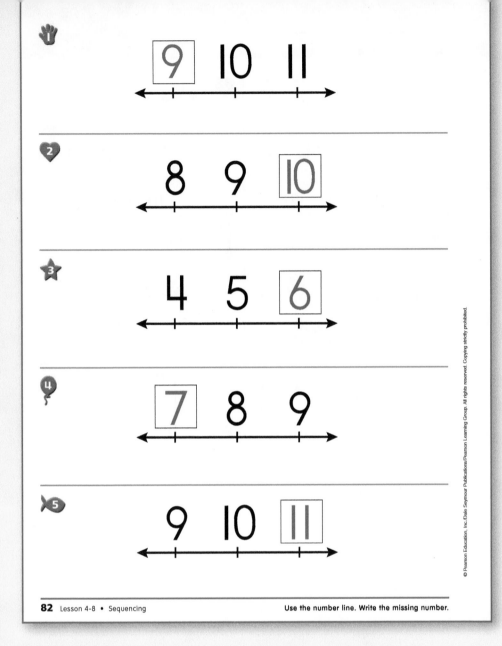

82 Lesson 4-8 • Sequencing   Use the number line. Write the missing number.

For Mixed Abilities

Common Errors • Intervention

For students who still have difficulty recognizing numbers before, between, and after numbers from 0 to 12, display a number line of these numbers. Then, play "What Number Is It?" For example, say, *I am thinking of the number between 9 and 11.* The student can respond by asking "Are you thinking of 10?" Encourage students to make up the "I am thinking" sentences and ask their classmates to respond.

Enrichment • Number Sense

1. Give students a deck of playing cards with face cards removed. Have students place the cards of all four suits in order.

2. Have students make number books with a page for each number, 1 to 12, in order. Have students draw the correct number of objects on each page and then number the pages.

Using page 82 Point out that a number line shows numbers in order. Have students look at the first number line and ask them to say the numbers aloud. When they come to a missing number, they should say "blank." For example, for Exercise 1, they should say, "Blank, 10, 11." Ask, *What number comes before 10?* (9) Encourage students to use the terms *before* and *after* as they describe each number line and complete the page.

4 Assess

Draw a number line for 0 to 12 on the board. Write the numbers **0, 3, 6, 9,** and **12** on the line. Then, have students fill in the missing numbers.

T82

Chapter 4 Test and Challenge

pages 83–84

© Pearson Education, Inc./Dale Seymour Publications/Pearson Learning Group. All rights reserved. Copying strictly prohibited.

1 Getting Started

Test Objectives
- **Items 1–4:** To count 0 to 12 objects and choose the correct number (see pages 67–72; 75–76)
- **Items 5–7:** To write 10, 11, and 12 (see pages 73–74; 77–78)

Challenge Objective
- To draw one to nine objects

Vocabulary
tally

Materials
blank index cards; number cards 0 through 12 in relief; blocks in two colors

---

Name _____

1 9   **(10)**   11   12

2 9   10   **(11)**   12

3 9   10   11   **(12)**

4 **(9)**   10   11   12

5 10

6 11

7 12

Circle the number that tells how many objects are in the set.
Count the objects in the set and write the number.

Chapter 4 • Test **83**

---

2 Give the Test

Prepare for the Test Have students create flashcards for numbers 0 through 12. On one side of a blank index card, have students write a number. On the other side, have them draw that number of pictures. Students should then use their flashcards to quiz each other on numbers through 12.

- After students have mastered counting and writing numbers through 12, have them complete the Chapter 4 Test. You may wish to review test-taking rules with students. Remind them that they are not to talk to each other and they are to finish the test on their own. Decide ahead of time whether or not to allow the use of manipulatives.

Using page 83 Students should be able to complete this page independently. For Exercises 1 to 4, students will count the number of items in the set and circle the correct number. For Exercises 5 to 7, students will count the number of items in the set and write the correct number. Allow use of number cards as needed for help in writing the numbers in these exercises.

3 Teach the Challenge

Develop Skills and Concepts for the Challenge Draw six Xs and four Os in a mixed group on the board. Write the following on the board:

X _____ _____

O _____ _____

Tell students that we want to know how many Xs and Os are in the group. Point to an X and say, *Here is an X, so I will make a tally mark on the line beside the X to show that I have counted one X. Here is another X, so I will make another tally mark.* Continue until all the Xs have been tallied. Then, tell students to count the tally marks aloud with you. Tell students that you will write the number of tally marks on the line to show that there are that number of Xs in the group. Be sure students understand that the tally mark for 5 is four sticks with a fifth written across them.

For Mixed Abilities

Common Errors • Intervention

If students are having difficulty tallying numbers greater than 5 because they are losing count, encourage them to use different-colored crayons to make their tallies. For example, for the number 10, students may use a blue crayon to make the tally marks for 1 through 5, and a red crayon to make the marks for 6 through 10.

Enrichment • Number Sense

Give students a grid paper with blocks of ten squares outlined in different patterns, including some with unequal sides. Have students place counters on each square to show different ways of laying out a group of ten.

ESL/ELL STRATEGIES

Point to the tally marks in the example on page 84 and say, *These are tallies. You make tally marks when you count things.* Then, point to and count the mice in Exercise 1, having students trace each of the tally marks as you say each number.

Tally the number of the animal in the picture and then write the number.

• Now, have students help tally the Os and record the number. Repeat for another group of two different kinds of objects on the board. Lay out 12 or fewer blocks in each of two colors and have students help count them by making tally marks on the board.

Using page 84 Have students name all the different groups of animals in the picture at the top of the page. Tell them to find the first exercise and tell what animal is shown. (mouse) Then, have students look at the picture at the top of the page and trace a tally mark on the line as each mouse is found. Tell them to count the tally marks and trace the 3 to show that there are three tally marks for three mice in the picture. Work similarly through each exercise with students.

4 Assess

Alternate Chapter Test You may wish to use the Alternate Chapter Test on page 208 of this manual for further review and assessment.

Challenge To assess students' understanding of tallies, lay out four pencils and have students create a tally showing the number. Repeat with five and six pencils.

5

Time and Money

pages 85–102

Chapter Objectives

- To arrange pictures in order of time sequence
- To compare the duration of two activities, using more time or less time
- To write numbers 1 to 12 on a clock face
- To match a time with its clock face
- To tell time to the hour
- To match time on a digital clock to time on an analog clock
- To count a dime, nickels, and pennies through 12¢ and write the amount
- To identify coins needed to buy items priced through 12¢
- To define *minute hand, hour hand, o'clock, penny, cent, coin, heads, tails, nickel, value, price,* and *dime*
- To act out buying items priced through 12¢ using coins
- To buy items using dimes, nickels, and pennies through 10¢ and find the change

Materials

You will need the following:
*pictures showing sequences of events; *pictures of breakfast, lunch, and dinner foods; *pictures of babies, older children, and adults; *number cards 0 through 12; *demonstration analog clock and digital clock; *compass for drawing circles; *12 cans with tape labels; *cards showing 1 through 12 pennies; *cards with amounts 1¢ through 12¢; *items priced through 12¢; *cards with items priced through 12¢; *cards with coins for amounts through 12¢; *colored chalk; *counters

Students will need the following:
Reproducible Masters RM12–RM14; wind-up toys; clock faces; number cards 1 through 12; real or play pennies, nickels, and dimes; crayons

Bulletin Board Suggestion

Help students locate or create pictures to show a store scene in which a clock face shows the time and priced items are displayed. Have students change the time and items each day and tell a story about the new scene.

Verses

Poems and songs may be used any time after the pages noted beside each.

Numbers Song (pages 85–86)

Zero, one, two, three, four, five,
Six, seven, eight, nine, sakes alive!
We have learned our numbers well
Next come ten, eleven, twelve.
We can sing a number song;
Won't you come and sing along?

Wee Willie Winkie (anon.) (pages 91–92)

Wee Willie Winkie runs through the town,
Upstairs and downstairs in his nightgown.
Rapping at the window,
Crying through the lock,
"Are the children all in bed?
For now it's eight o'clock."

What Time Is It? (pages 91–92)

"What time is it?" I asked my mom,
"Is it almost time to play with Tom?"
"No, no," said Mom, "it's not yet two;
Two o'clock is when he's due.
You haven't had your breakfast yet,
I've started to cook your omelet.
It's eight o'clock and then comes nine,
Ten, eleven . . . then it's time
At twelve o'clock to have your lunch
And read your book about Judy and Punch.
Then one o'clock is time for calm,
We'll prepare a snack for you and Tom.
By then it will be nearly two,
And two o'clock is when Tom's due!"

What Coin Do I Have? (pages 101–102)

I have a mystery coin in my hand
I have a mystery coin in my hand
I have a mystery coin in my hand
Now what coin do I have?

The coin I am hiding in my hand
Could be traded for ten pennies, in my hand
Could be traded for two nickels, in my hand
Now what coin do I have? (dime)

The coin I am hiding in my hand
Could be traded for five pennies, in my hand
Could buy a five-cent item, in my hand
Now what coin do I have? (nickel)

The coin I am hiding in my hand
Has the very smallest value, in my hand
Shows the head of Abraham Lincoln, in my hand
Now what coin do I have? (penny)

The Flea Race

One day, 12 little fleas decided to have a jumping contest on Rover's back. Rover, the farmer's dog, did not know the fleas were planning to do this. If he had known, he might have started scratching to try to knock the fleas off. However, since he didn't, they went ahead with their plans.

Each little flea put a number on its back. "The race starts as soon as Rover lies down to take a nap," said flea number 3. Every flea agreed.

"Get in line," said number 9. "We're going to have a jumping race from Rover's tail to the top of his head!"

Then, all the little fleas formed a straight line at Rover's tail. Number 2 was between number 1 and number 3. Number 4 was between number 3 and number 5. Number 6 was before number 7 and after number 5. Number 8 was between number 7 and number 9. Number 10 was between number 9 and number 11. And number 12 was on the very end.

The instant Rover laid down and closed his eyes, flea number 1 yelled, "On your mark! Get set! Go!" And all the fleas began to jump! They sprang up Rover's back as fast as they could.

Just then, a cat dashed into the yard chasing a mouse. Rover sprang up and raced for the cat! The fleas still raced up Rover's back. Everybody raced everybody! But since Rover was racing the same way as the fleas were racing, the fleas finished much sooner than they expected.

The mouse got away from the cat. The cat got away from Rover. And poor Rover couldn't stop soon enough to get away from the fence. He crashed—KERSPLAT—right into it! And the fleas went flying.

The cat didn't catch the mouse. Rover didn't catch the cat. And the fleas didn't know who won the race . . . so everyone ended up with nothing, a big fat 0!

Name _____

Draw a line from flea 1 to flea 12 in order.   Chapter 5 • "The Flea Race" **RM12**

"The Flea Race," p. RM12

Activities

Use Reproducible Master RM12, "The Flea Race," to help students relate the story to the mathematical ideas presented.

- Have 12 students stand. Give each student a different number card from 1 to 12. Have students arrange themselves as fleas in order from 1 to 12 as you reread the story. Have students exchange number cards to repeat the activity. Allow time for all students to participate with at least two different number cards. Have them arrange themselves in order from 1 to 12 and then from 12 to 1.

- Help students understand why the fleas finished their race sooner than they expected. Ask if any of them had ever visited an airport where there was a moving walkway or a department store with an escalator. Discuss with them the difference between standing still on the walkway or moving stairs and walking along with the movement.

1 Getting Started

Objective
- To arrange pictures in order of time sequence

Materials
*groups of pictures showing a sequence of events; *pictures of breakfast, lunch, and dinner foods; *pictures of babies, older children, and adults

Warm Up • Mental Math
Ask students which is less.

1. 12 or 11 (11)
2. 9 or 10 (9)
3. 0 or 10 (0)
4. 11 or 2 (2)
5. 10 or 7 (7)
6. 1 or 11 (1)
7. 7 or 1 more than 7 (7)

Warm Up • Number Sense
Sing the "Numbers Song" on page T85a to the tune of the "ABC Song" as students show each number heard in the song. Help students realize that having the cards in order makes it easier to find the correct number.

Name _____

Time and Money

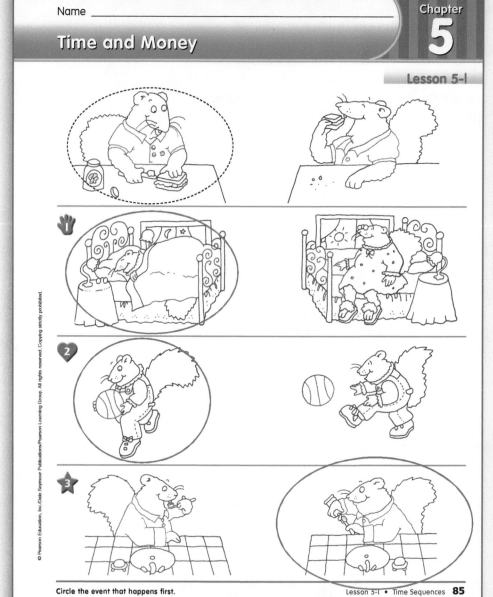

Circle the event that happens first.

2 Teach

Develop Skills and Concepts Ask students what meal they eat when they get up in the morning. (breakfast) Show a breakfast picture. Talk with them about many activities they do after eating breakfast and then ask what meal they eat after completing these activities. (lunch) Place a lunch picture to the right of the breakfast picture. Continue to discuss afternoon activities to show the time passing and then ask students what meal they eat in the evening. (dinner or supper) Place a dinner picture to the right of the lunch picture. Now, orally review the sequence of a day's events and ask students to name each meal in the sequence. Mix up the pictures and have a student place them in order by telling about each one. Then, ask students which meal happens first. (breakfast) Continue for the next meal (lunch) and the last meal (dinner or supper). Ask which meal is eaten in the middle of the day. (lunch)

- Use the collection of pictures to discuss other events the students do in sequence, such as the following: go

to school, work at school, go home from school; get up, eat breakfast, go to school; get up, eat breakfast, brush teeth. Give students three such events to tell about or show in order with pictures. Ask them to tell which event happens first and which happens last for each sequence. Make the activity fun by telling stories of a child who brushes his or her teeth, then gets up and eats breakfast, and so on for students to see the humor and impossibility of such events being out of order. Have volunteers retell each story to place the events in correct order.

3 Practice

Using page 85 Ask students what the squirrel is doing in the first picture in the example. (making a sandwich) Ask what is happening in the second picture in the example. (The squirrel is eating the sandwich.) Ask students if the squirrel would make the sandwich first or eat it first. (make it) Tell them to trace the circle around the picture

Circle the event that happens first.
Put a check on the one that happens second.

For Mixed Abilities

Common Errors • Intervention

Some students may have difficulty ordering pictures. Show these students a picture of a child and a picture of that child as a baby. Ask, *Which came first?* (baby) Help students tell in their own words how first a person is a baby and then the person gets older. Have students arrange in order more sets of two pictures of a baby and an adult, a child and an adult, a baby and a child. Then, show students sets of two pictures of other related events and have students place them in the order in which they happen.

Enrichment • Time Activity

1. Have students make a time book by drawing or cutting and pasting six pictures of events in the order in which they happen.

2. Have students bring in three or more pictures of themselves at different ages. Have students exchange pictures and place the pictures in order from the most recent to the oldest.

that tells which event happened first. (making the sandwich)

• Similarly, discuss the first exercise and have a student tell the story with the events in order. Have students circle the event that happened first. (sleeping) Tell them to complete the next two exercises by looking at each row of pictures and circling the event that happened first.

Using page 86 Discuss the pictures in the first exercise with students and ask which would happen first. (getting the cereal) Have a student tell the story with the events in order. Tell students to circle the picture that shows what happened first. Tell them to put a check mark on the picture that shows what happened second. (eating the cereal) Work through the second exercise before having students complete the last two exercises independently.

4 Assess

Have students make a time book by drawing or cutting and pasting two related pictures in the order in which they occur.

5-2 More Time, Less Time

pages 87–88

1 Getting Started

Objectives

- To compare the duration of two activities
- To identify which activity takes more time or less time

Materials

*counters; wind-up toys

Warm Up • Mental Math

Ask students which happens first.

1. wake up or get dressed (wake up)
2. get in your car or drive home (get in your car)
3. tie your shoe or put on your shoe (put on shoe)
4. play a computer game or turn on a computer game (turn on a computer game)
5. cook dinner or eat dinner (cook dinner)

Warm Up • Number Sense

Show students one set of 12 counters and one set of 5 counters. Have students compare the two sets to determine which has more counters. Repeat several times, comparing sets of different quantities.

Name _____

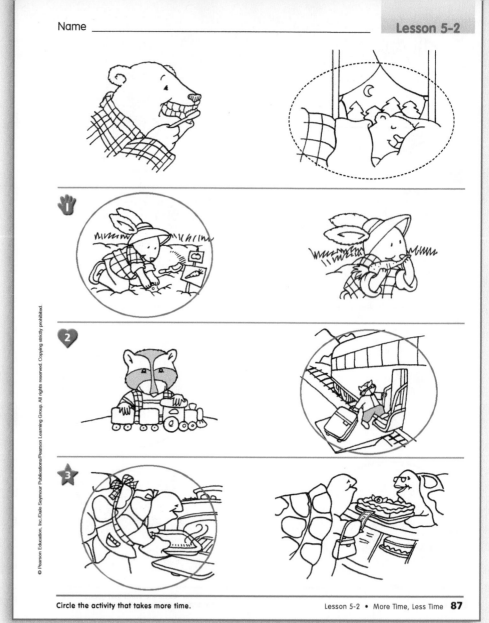

Circle the activity that takes more time.

Lesson 5-2 • More Time, Less Time **87**

2 Teach

Develop Skills and Concepts In order to effectively compare the duration of activities, it is important to identify whether an activity takes a long time or a short time. Name one example of a short activity and one of a long activity, such as tying your shoes (short) and going on vacation (long). Then, have each student draw a picture of one short and one long activity. Ask each student to pick one of his or her pictures to compare with a classmate's. Have the pair decide which pictured activity takes more time.

- Have students work in pairs to act out and time two activities. For example, have one student tie his or her shoe while the partner evenly claps and counts to keep time. Then, have students reverse roles for a different activity, using the same method of timing. Ask, Which activity took more time? Encourage students to look at the number of claps to compare the length of the

activities. Point out that the greater number is the longer activity.

- Have pairs of students compare the length of time that two wind-up toys will run. Have students predict which toy will run for a longer time and then have them wind up and release the toys at the same time. Make sure students understand that the toy that keeps going after the other one stops is running for a longer time.

- Have two students perform two different activities. Make sure they both start their activities at the same time. The rest of the class will observe and tell who finished first and who finished second. Point out that the student who finishes first has a shorter activity. Have students compare activities such as building a tower of ten blocks and writing the first ten letters of the alphabet, or doing ten jumping jacks and singing "The Itsy Bitsy Spider."

T87

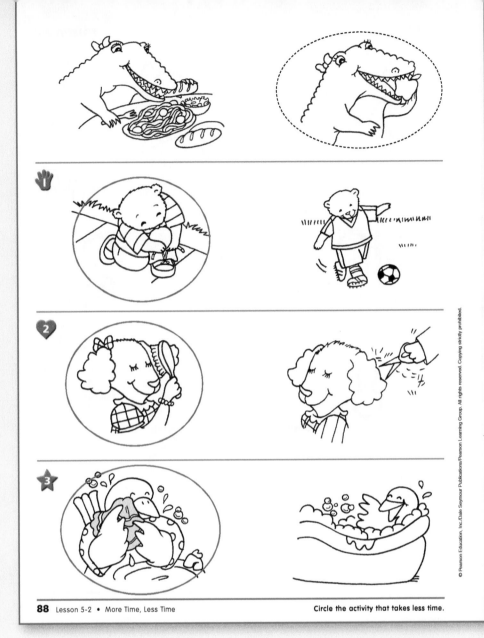

88 Lesson 5-2 • More Time, Less Time

Circle the activity that takes less time.

For Mixed Abilities

Common Errors • Intervention

Some students may have difficulty understanding that if one activity takes more time to complete, then the other activity takes less time. Have them look back at page 88. For each exercise, have them circle again the activity that takes less time but also mark with an X each activity that takes more time.

Enrichment • Time Activity

1. Give students magazine pictures of three different activities that take different amounts of time. Have students compare the duration of the activities and put them in order from shortest to longest.

2. Have students measure the duration of an activity using an egg timer. Repeat for a second activity and then have students discuss how to compare the duration of the activities using the measurements from the timer.

3 Practice

Using page 87 Have students look at the example and describe the activity in each picture. (a bear brushing his teeth and a bear sleeping) Ask, *Which activity takes more time?* (sleeping) Have students trace the circle around the picture of the sleeping bear. Ask, *Why do you think this activity takes more time?* (It takes a few minutes to brush your teeth, but you sleep all night.) Help students complete Exercise 1 before assigning the page to be completed independently. After students have completed the page, have them explain how they decided which activities to circle.

Using page 88 Have students look at the example and describe the activity in each picture. (an alligator eating dinner and an alligator eating an apple) Point out to students that on this page, they need to circle the activity that takes less time. Have students trace the circle around the alligator eating the apple. Ask, *Why do you think this*

activity takes less time? (It takes much longer to eat an entire dinner than to eat an apple.) Have students complete the page independently. Then, have students explain how they decided which activities to circle.

4 Assess

Ask students, *Which takes less time, getting to school or being in school?* (getting to school)

T88

1 Getting Started

Objectives
- To write numbers 1 to 12 on a clock face
- To match a time with its clock face
- To define *minute hand, hour hand,* and *o'clock*

Vocabulary
minute hand, hour hand, o'clock

Materials
*demonstration clock; number cards 1 through 12; clock faces; crayons

Warm Up • Mental Math
Have students complete the following:

1. _____ comes before 11. (10)
2. 0 comes before _____. (1)
3. 8 comes before _____. (9)
4. _____ comes before 12. (11)
5. 9 comes before _____. (10)
6. _____ comes before 8. (7)

Warm Up • Number Sense
Have 12 students hold one of each number card 1 through 12 and form a line in order. Help them form a circle and sit in place. Have a student start at 1 and move around the circle counting each number. Have other students repeat this, each starting at another number.

Write the numbers on the clock face.
Color the minute hand red and the hour hand blue.

2 Teach

Develop Skills and Concepts Ask questions that use the word *time*. Include questions about bedtime, time of favorite television shows, and so on. Help students answer the questions in sentences using the word *time*. (For example, "My bedtime is 8:00.")

- Discuss the need for clocks as a way to tell time to know when things happened or will happen. Tell students a story about Christopher and Colleen, who wanted to go to a movie but they had no clock. Ask why a clock would be needed. (to know when it is time to go to the movie theater) Have students tell of other times when clocks are needed.

- Have students tell the numbers on a clock. (1 through 12) Have them compare the sizes of the hands on the clock. Tell students that the longer hand is called the minute hand and the shorter hand is called the hour hand. Have students identify each hand. Show them how the minute hand goes faster around the clock from 12 and back to 12, whereas the hour hand moves slowly from one number to the next number. Have a student move the long hand around clockwise as classmates tell about the short hand's movement. (slowly moves from one number to the next)

- Have students show 7:00 on their clock faces, say the time, and tell where each hand is. Tell them that 7 o'clock is written **7:00** as you write the time on the board. Have students tell where each hand points. Write several times on the board to have students show each time on their clock faces.

- Tell students that there are different times of the day that correspond to times on the clock. The term *morning* is usually used to describe times from when we wake up until right before lunch. Ask students to name morning activities and times. (7:00 to 11:00; eat breakfast, go to school) *Afternoon* is usually used to

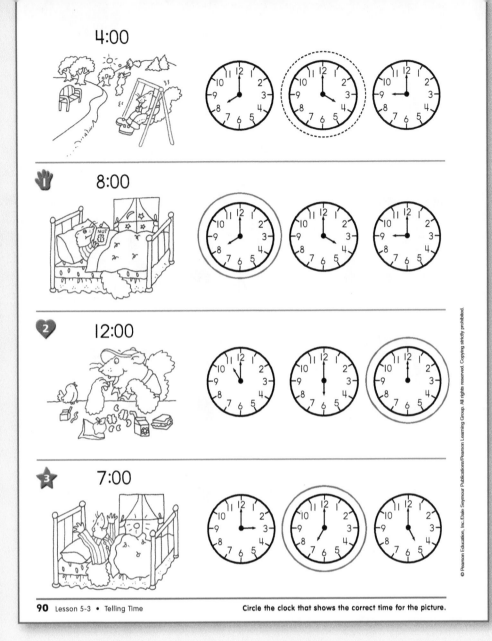

4:00

8:00

12:00

7:00

Circle the clock that shows the correct time for the picture.

For Mixed Abilities

Common Errors • Intervention

Students may need additional work reading a clock face. Lay out the number cards 1 through 12 on a large taped circle on the floor. Choose two students to be the hour and minute hands, pointing with outstretched arms. Have them point to different o'clock times as other students tell the time.

Enrichment • Time Activity

1. Have students act out what they are doing at a particular hour for classmates to guess the time.

2. Have students cut out pictures from newspapers or magazines that show people using time in their lives. (people waiting for a bus, woman looking at her watch, people going into a movie)

ESL/ELL STRATEGIES

Draw a clock face on the board, including the hands. Point to and say *hour hand* and *minute hand* and have students repeat these phrases. Then, draw the hands in various positions and have students practice saying times, such as 8 o'clock.

---

describe times from lunch until right before dinner. Ask students to name afternoon activities and times. (12:00 to 5:00; eat lunch, go home) *Evening* is the term used to describe times from dinner until bedtime. Have students name evening activities and times. (5:00 to 8:00; dinner, homework, reading books)

3 Practice

Using page 89 Tell students that each new day starts at 12 o'clock at night while they are sleeping. Remind them that 1 hour later than 12 o'clock is 1 o'clock. Tell students to trace the 1 on the clock face and to trace or write the other numbers in order around the clock face. Remind students that they will be writing the numbers in order from 1 to 12. Supervise as they write the numbers.

• Have students find the longer hand. Remind students that this is called the minute hand. Tell them to color the minute hand red. Remind students that the shorter hand is called the hour hand. Tell students to color the

hour hand blue. Ask what time is shown on the clock face. (8 o'clock) Write **8:00** on the board and tell students that this is the way we write 8 o'clock.

Using page 90 Help students discuss the picture in the example as you note that playing in the park usually happens after school. Ask a student to read the time over the picture. (4 o'clock) Tell students that they are to look at the three clock faces beside the picture and find the clock face that shows 4 o'clock. Tell them to trace the circle around the clock that shows 4 o'clock. Work through Exercise 1 before assigning the page to be completed independently.

4 Assess

Show 1:00, 6:00, and 10:00 on three clock faces. Name each time and have students identify the correct clock face. Now, write each time on the board and have students identify the correct clock face.

T90

1 Getting Started

Objective

- To match time on a digital clock to time on an analog clock

Vocabulary

digital clock

Materials

*demonstration analog clock; *demonstration digital clock; *compass; Reproducible Masters RM13–RM14; clock faces

Warm Up • Mental Math

Ask students which happens first.

1. go to school or eat breakfast (breakfast)
2. button shirt or put a shirt on (put it on)
3. lunch or bedtime (lunch)
4. eat a sandwich or make a sandwich (make it)
5. put on a helmet or ride a bike (put on a helmet)

Warm Up • Time Activity

Use a compass to draw a circle on the board. Mark where each clock number goes and have students write the numbers in order from 1 to 12. Leave the clock face on the board for later use.

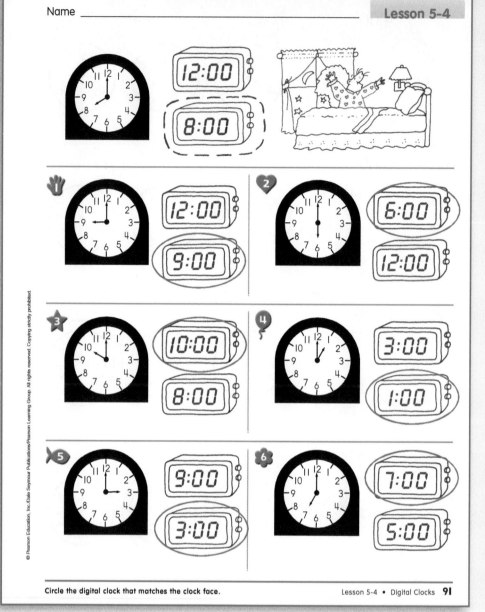

Circle the digital clock that matches the clock face.

2 Teach

Develop Skills and Concepts Set the demonstration analog clock at 6:00. Have a student tell the time as you write **6:00** on the board. Move the minute hand around the clock one time and have students tell what the hour hand does. (moves to the next number) Have a student tell the time as you write **7:00** on the board. Continue for more times.

- Have a student draw the minute and hour hands on the clock on the board to show 7:00. Have students tell what they are doing at 7:00 in the morning as you write **7:00** on the board. Have a student change the hour hand to show 8:00 and tell an activity done at that time in the morning.

- Read the poem "Wee Willie Winkie" on page T85a and have students show the time. Change the time in the poem for other practice.

- Show the demonstration analog clock at 12:00. Write **12:00** on the board and have students read the time.

Now, show a digital clock set at 12:00 and have students read the time. Continue showing more times on both clocks. Write each time on the board.

- Read the poem "What Time Is It?" on page T85a and have students show the times on their clock faces. Write each time on the board.

- Show 2:00 on the demonstration analog clock and write **10:00** and **2:00** on the board. Have students tell which time is shown on the clock. Continue for practice with more times.

- Give each student copies of Reproducible Master RM13. Have students write the numbers on a clock, draw the minute and hour hands to show any time, and draw a picture to tell what they are doing at that hour of the day.

- Give each student copies of Reproducible Master RM14. Have them draw a line from the clock face to the digital clock showing the same time.

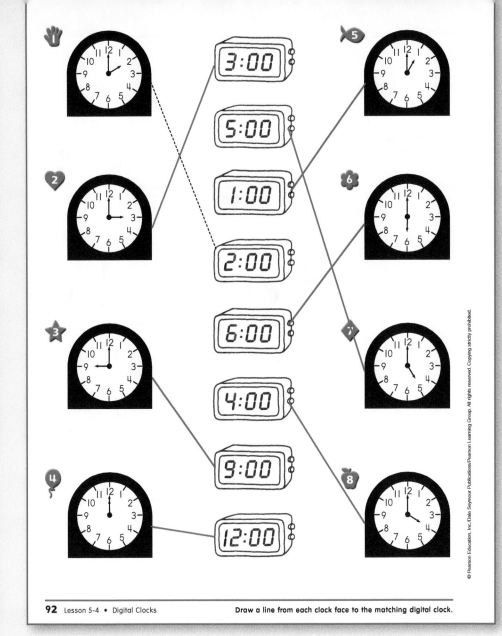

92 Lesson 5-4 • Digital Clocks   Draw a line from each clock face to the matching digital clock.

For Mixed Abilities

Common Errors • Intervention

Some students may need more practice reading clocks. Give students a time to show on their clock faces. Have them tell the time as they place their clocks in order on the chalk tray. Repeat this activity several times with different times for each student.

Enrichment • Time Activity

1. Have students place clock faces showing 5:00, 9:00, and 12:00 in order.

2. Have three students show different times on clock faces and place themselves in order. If students show 8:00, 2:00, and 4:00, have them explain the sequence as 8:00 in the morning, and 2:00 and 4:00 in the afternoon.

3. Have students draw pictures to show their activities at 7:00 in the morning and 7:00 in the evening.

3 Practice

Using page 91 Tell students that the squirrel in the example is getting ready to go to bed at night and her clock shows the time. Ask what time the squirrel gets ready to go to bed. **(8 o'clock)** Tell students that they are to find the digital clock that tells the same time as the squirrel's clock. Tell them to trace the circle around the correct digital clock. Work through Exercise 1 with students and then have them complete the page independently.

Using page 92 Tell students that they are to match the round clock faces to their digital clocks. Ask the time on the clock face in Exercise 1. **(2:00)** Tell them to trace the line to the digital clock that shows 2:00. Work through the next exercise with students. Have them complete the page independently.

4 Assess

Show 3:00 on a demonstration digital clock. Have students show this time on their analog clocks.

T92

5-5 Counting Pennies

pages 93–94

1 Getting Started

Objectives

- To count pennies through 12¢ and write the amount
- To define *penny, cent, coin, heads,* and *tails*

Vocabulary

penny, cent, coin, heads, tails

Materials

*demonstration clock; *12 cans with tape labels; *cards showing 1 through 12 pennies; *cards showing 1¢ through 12¢; clock faces; real or play pennies

Warm Up • Mental Math

Tell students to clap as they count to show the following:

1. 1 less than 11 (10)
2. 1 more than 8 (9)
3. 12
4. their age
5. the first number in their phone number
6. the number of people in their family
7. 1 more than 11 (12)

Warm Up • Time Activity

Have students tell the time as you move the hour hand on the demonstration clock from 1:00 to 12:00. Read the poem "What Time Is It?" on page T85a and have students show each time on their clock faces.

2 Teach

Develop Skills and Concepts Give each student 12 pennies. Have students turn each penny so that the picture of Abraham Lincoln shows. Tell students that Lincoln was once the president of the United States. Have them count the coins and tell how many there are. Write **12¢** on the board and tell students that the cent sign is written after the number to tell that we mean 12 pennies or 12 cents. Have students form a group of 7 pennies as you write **7¢** on the board. Have them read with you to say "seven cents." Write **9¢** on the board and have students form a group of 9 pennies and read the amount as nine cents. Continue for 10¢, 6¢, and 11¢.

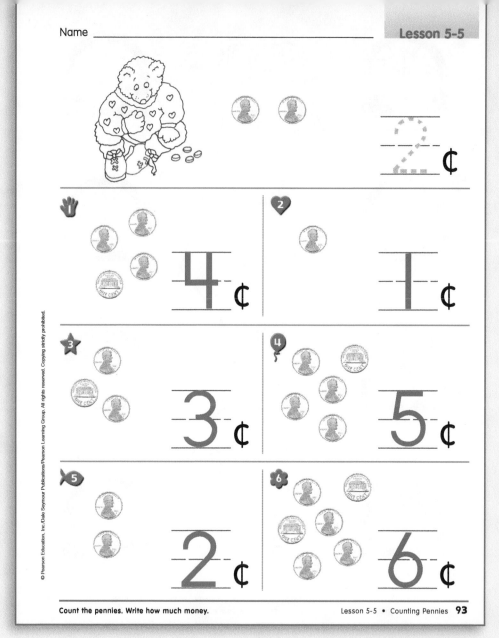

Count the pennies. Write how much money.

Lesson 5-5 • Counting Pennies **93**

- Tell students that the side of the penny with the head of Abraham Lincoln is called heads. Have students turn each penny over to the tails side to see the picture of a building. Tell them that the Lincoln Memorial is a building in Washington, D.C. Write **8¢** on the board and have students show 8¢ with the tails side up. Continue for more amounts. Now, lay out 6 pennies, some heads up and some tails up, and have students tell the amount. (6 cents) Repeat for more amounts through 12¢.

- Have a student drop 3 pennies into a can as students count the clinking sounds. Have the student write the number 3 on the tape label. Print the cent sign after the 3 and have students read the amount of money. Continue to have them fill and mark cans until there is one can for each amount 1¢ through 12¢. Distribute the cans for students to check the work. Then, have students place the cans in order from 1¢ through 12¢.

- Lay out the penny and amount cards. Have students draw an amount card and match it to the correct penny card. Continue for all the cards.

T93

Common Errors • Intervention
Students may need more practice counting a group of pennies. Have them work with partners to play a matching game in which they match penny cards and amount cards. Mix the cards and lay them facedown for students to turn over two at a time. A match wins the cards and another turn, whereas unmatched cards are turned facedown again.

Enrichment • Probability

1. Have students see how many different ways they can lay 5 pennies heads or tails up as you record the results.
   (4 heads, 1 tail; 3 heads, 2 tails; and so on)

2. Have students see how many different arrangements of 12 pennies they can make.
   (one row of 12, two rows of 6, three rows of 4, and so on)

94 Lesson 5-5 • Counting Pennies   **Count the pennies. Write how much money.**

3 Practice

Using page 93 Discuss the heads and tails sides of the penny. Help students remember that the heads side shows the head of Abraham Lincoln and the tails side shows the Lincoln Memorial, a building in Washington, D.C. Remind students that we write the value of 1 penny as 1¢ as you write **1¢** on the board. Tell students that the bear is counting pennies and they are to help by counting the pennies in each exercise and then writing the number. Ask how many pennies are in the example. (2) Tell students to trace the 2 in front of the cent sign. Have students read the amount in unison. (2 cents) Work through Exercise 1 with them before assigning the page to be completed independently.

Using page 94 Tell students that they are to count the pennies in each exercise and write the number that tells how many cents there are. Have them complete the page independently.

4 Assess

Lay out 5 pennies. Have students tell how many cents there are. (5 cents) Repeat with other amounts.

T94

pages 95–96

© Pearson Education, Inc./Dale Seymour Publications/Pearson Learning Group. All rights reserved. Copying strictly prohibited.

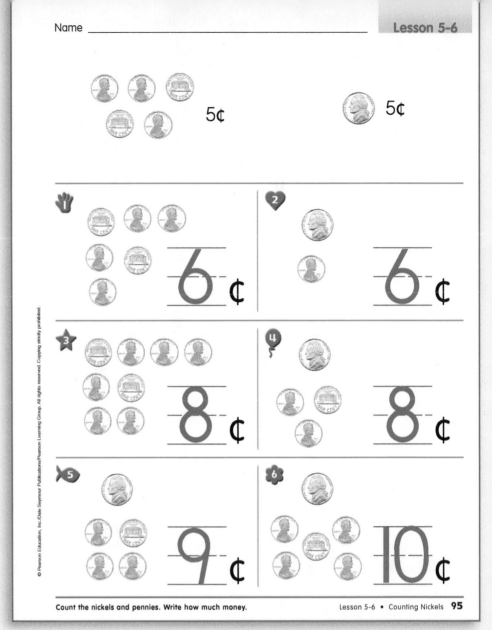

1 Getting Started

Objectives

- To count a nickel and pennies for amounts through 12¢
- To choose nickels or pennies to buy an item priced through 12¢
- To define *nickel, value,* and *price*

Vocabulary

nickel, price

Materials

*cards with items priced through 12¢; *cards with coins for amounts through 12¢; 2 real or play nickels and 12 real or play pennies

Warm Up • Mental Math

Ask students to fill in *before* or *after*.

1. 9:00 comes (before) 10:00
2. 2:00 comes (after) 1:00
3. 12:00 comes (after) 11:00
4. 3:00 comes (before) 4:00
5. 10:00 comes (after) 9:00
6. 6:00 comes (before) 7:00

Warm Up • Money Activity

Have students draw 7 circles on the board and write **1¢** in each to show 7¢. Repeat for other amounts through 12¢.

2 Teach

Develop Skills and Concepts Help students compare a penny and a nickel for shape, size, and color. Tell students that the larger coin is a nickel and has a picture of Thomas Jefferson, another president, on it. Tell students that the building on the tails side is called Monticello, which was the home of Thomas Jefferson. Write **5¢** on the board and tell students that a nickel is the same as five pennies or 5 cents.

- Draw a nickel and 3 pennies on the board and point to the nickel and say, *5 cents.* Point to each penny in turn as you say, *6 cents, 7 cents, 8 cents.* Write **8¢** on the board and have students read the amount and count the coins aloud with you. Draw another penny on the board and have students count. (5¢, 6¢, . . . , 9¢) Write ¢ on the board and have a student write the number in front. (9) Give students one nickel and three pennies one at a time to begin at 5¢ and count the totals.

- Have students lay out 12 pennies as you write **12¢** on the board. Have a student trade 5 pennies for a nickel. Ask how many pennies there are now. (7) Ask if there are enough pennies to trade for another nickel. (yes) Make the trade and ask what coins make up the 12¢ now. (2 nickels, 2 pennies) Have students work in pairs to repeat this activity.

- Hold up an item card for 11¢. Ask what coins could be used to buy the item. (11 pennies; 2 nickels, 1 penny; or 1 nickel, 6 pennies) Repeat for more item cards. Have students then match the item cards to the amount cards. More than 1 amount card may match an item card.

- Have students help set up a classroom store with items priced through 12¢.

T95

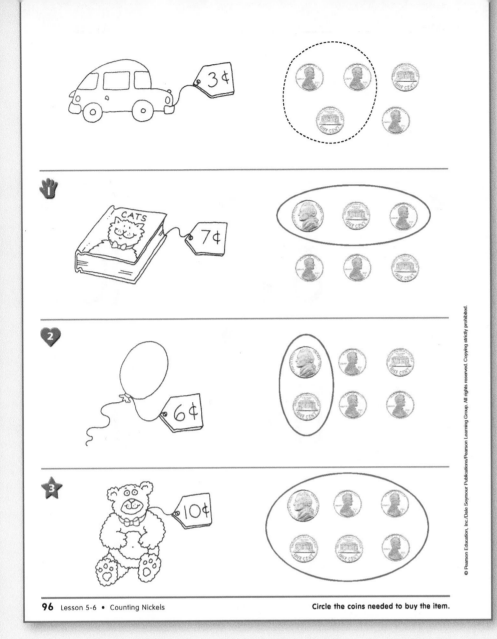

Circle the coins needed to buy the item.

For Mixed Abilities

Common Errors • Intervention

Some students may have difficulty counting nickels with pennies. Use real or play coins for students to practice amounts up to 10¢. Encourage them to always start with 5 for a nickel and then count on for each penny. If they pause after each nickel, they will better appreciate the different values of the coins.

Enrichment • Money Activity

1. Have students draw coins to show three different ways to pay for an item costing 12¢.

2. Have students place 1 nickel and 7 pennies in a bag, draw out a handful of coins, and tell the amount.

3. Ask students to compare the values of two groups of coins and tell whether one group of coins is worth more, less, or the same as the other group. Have students look at page 95 and pick any two groups to compare. Encourage them to use the terms *more, less,* and *the same as* to describe the groups they are comparing.

3 Practice

Using page 95 Have students look at the example at the top of the page. Ask students the name of the coins in the first group. (pennies) Ask how many pennies are shown. (5) Write 5¢ on the board for students to read in unison. Ask the name of the other coin in the example. (nickel) Have them read the amount next to the nickel. (5 cents) Remind students that the nickel is larger in size than a penny but a nickel has the same value as 5 pennies. Tell students to count and then write the amount of money in the first exercise. (6¢) Remind them to start at 5 and count on to find the value of a nickel and any number of pennies. Help students complete the next three exercises and then assign the remainder of the page to be completed independently.

Using page 96 Ask students the cost of the toy car in the example. (3¢) Ask students what coins are shown. (5 pennies) Ask how many of the pennies would be needed to buy the car. (3) Tell students that they are to trace the circle around the coins that are needed to buy the car. Have them count the pennies that are left. (2) Note that students are only identifying the number of coins spent and those left. Help students work through Exercise 1 where the nickel and 2 pennies are needed to buy the book. Have them complete the page independently by circling the coins needed to buy the balloon and the bear.

4 Assess

Give students a nickel and 3 pennies. Show a price tag for an object at 6¢. Ask, *Do you have enough money to buy this?* (yes) *How much money will be left over?* (2¢)

5-7 Counting Dimes

1 Getting Started

Objectives

- To count a dime, nickels, and pennies and write the amount through 12¢
- To define *dime*

Vocabulary

dime

Materials

real or play dimes, nickels, and pennies

Warm Up • Mental Math

Have students count to 12 from the following numbers:

1. 7 (7, 8, …, 12)
2. the first number in their phone number
3. their age
4. 1 more than 3 (4, 5, …, 12)
5. 1 less than 6 (5, 6, …, 12)
6. none (0, 1, …, 12)

Warm Up • Money Activity

Have students draw nickels or pennies on the board, writing 5¢ or 1¢ in each to show 10¢. Continue for other amounts through 12¢.

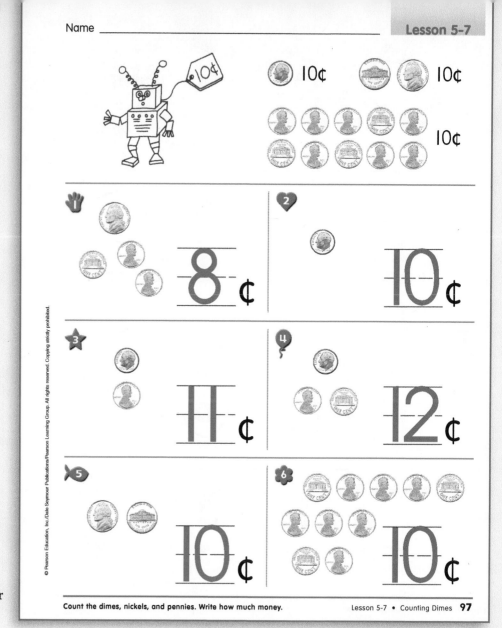

Name _____

Lesson 5-7

Count the dimes, nickels, and pennies. Write how much money.   Lesson 5-7 • Counting Dimes **97**

2 Teach

Develop Skills and Concepts Draw 10 pennies on the board. Ask students how 10¢ can be made with nickels. Have a student draw 2 nickels on the board and write the value. (10¢) Ask students to count the coins in each amount on the board. (10 pennies, 2 nickels) Tell students that there is another way to make 10¢ using only one coin as you show a dime. Draw a dime on the board and write **10¢**. Tell students that a dime has the same value as 10 pennies or 2 nickels. Give each student a dime, 10 pennies, and 2 nickels, and ask students to lay out coins to show three ways to make 10¢. (10 pennies; 2 nickels; 1 dime) Remind them that each coin has a heads and a tails side. Have students look at the heads side of the dime as you tell them that this picture is of another president whose name was Franklin Roosevelt. Review the heads side of the penny and nickel. Have students find the torch and the olive and oak branches on the dime's tails side. Tell students that the torch is to show that people are free in America and the olive and oak branches are to show that people are strong and want peace.

- Have students compare the three coins for size, shape, and color by asking which coin is silver, thinnest, a circle, the same as five pennies, and so on.

- Show a dime and have a student trade you two coins so you will still have 10¢. (2 nickels) Continue for more trading to have the same values.

- Draw 1 dime and 1 penny on the board and have students count with you from 10 to 11. Write **11¢** on the board. Repeat for 12¢.

- Have one student be a banker who makes trades so that students can shop at the classroom store with the correct coins to buy any item.

3 Practice

Using page 97 Ask students the cost of the robot in the example. (10¢) Tell students that 1 dime is 10 cents or 10 pennies and that 2 nickels is also 10 cents. Have students

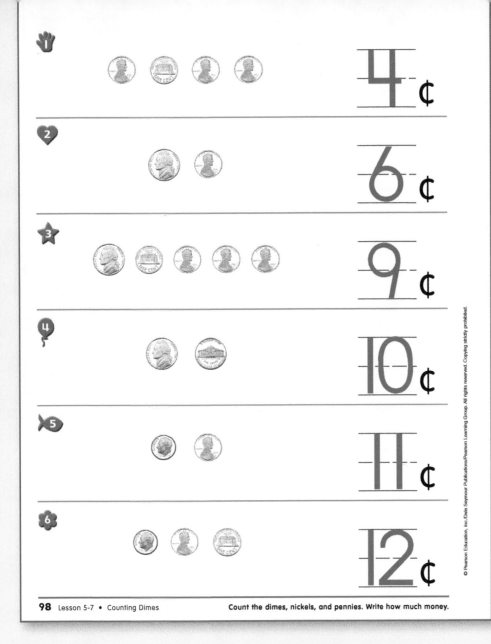

1. 4 ¢

2. 6 ¢

3. 9 ¢

4. 10 ¢

5. 11 ¢

6. 12 ¢

98 Lesson 5-7 • Counting Dimes

Count the dimes, nickels, and pennies. Write how much money.

For Mixed Abilities

Common Errors • Intervention

Some students may have difficulty counting dimes and pennies. Draw a dime and 2 pennies on the board. Have students count orally with you as you draw 10 sticks under the dime and 1 under each penny. Repeat for other coin combinations for amounts through 12¢.

Enrichment • Money Activity

1. Have students draw and identify the coins that have Abraham Lincoln's, Franklin Roosevelt's, or Thomas Jefferson's picture on them.

2. Tell students to cut pictures of two items from newspapers. Help students price each item at 12¢ or less. Then, have them draw coins to show two different ways to purchase the items.

ESL/ELL STRATEGIES

Give each student a different number of pennies, nickels, and dimes. Elicit the name of each coin. Repeat each name and ask students to repeat the name. Then, ask students to tell how many of each coin they have. For example, "I have 3 nickels."

count the money in the first exercise and write the amount. (8¢) Remind students that they should start at 5 and count on when there is a nickel and pennies and start at 10 and count on when there is a dime and pennies. Have them complete the page independently.

Using page 98 Tell students that they are to find the value of the coins in each exercise and write the amount. Have them complete the page independently.

4 Assess

Give students dimes, nickels, and pennies. Ask them to lay out 12¢ using any combination of coins. (Possible answers: 1 dime, 2 pennies; 2 nickels, 2 pennies; 1 nickel, 7 pennies; 12 pennies)

T98

5-8 Problem Solving: Act It Out

pages 99–100

1 Getting Started

Objective
• To act out buying items priced through 12¢ using coins

Materials
*items priced 1¢ through 12¢; *colored chalk; real or play dimes, nickels, and pennies

Warm Up • Mental Math
Ask students how much money there is.

1. 2 nickels (10¢)
2. 1 nickel, 1 penny (6¢)
3. 10 pennies (10¢)
4. 1 penny more than 6 (7¢)
5. 2 nickels, 2 pennies (12¢)
6. 1 nickel, 5 pennies (10¢)

Warm Up • Money Activity
Tell a student to draw coins on the board to show 12¢. Have another student check the amount. Repeat for other amounts through 12¢.

2 Teach

Develop Skills and Concepts Give one student 10 pennies to hold, another student 2 nickels, and a third student 1 dime. Ask students to draw their coins on the board and write the amount of money they have. (All have 10¢.) Ask how they could all have 10¢ when each one has different coins. (same value) Discuss who has more coins, who has less, and how 1 dime or 2 nickels are easier to carry and not as heavy in a purse or pocket. Hold up an item priced at 10¢ and ask if any of these three students could buy it. (All three could buy it.) Ask if any of them would need to trade any coins to buy the item. (no) Show other priced items and ask who could buy the item without trading coins. Encourage students to talk about the need to have the right coins in order to buy a drink at a lemonade stand, to get a gumball or a toy from a machine, and so on.

• Draw a dime and 2 pennies on the board. Have a student count the money on the board and write the amount in front of the cent sign you provide. Show an item priced at 11¢ and ask students the cost of the item and if there is enough money to buy the item. (yes) Ask a student to use colored chalk to color the coins needed. Ask if any coins are left. (yes) Ask how much

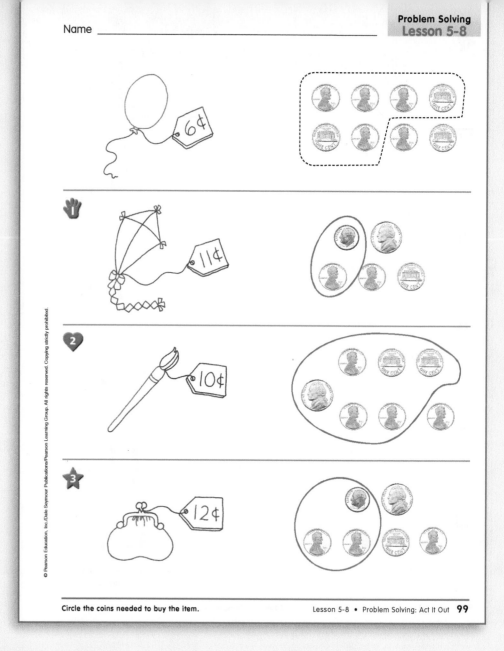

Circle the coins needed to buy the item.

money is left. (1¢) Repeat for other priced items and draw coins through 12¢.

3 Practice

Using page 99 Have students look at the example at the top of the page.

• Now, apply the four-step plan. Tell students that they are to use real or play coins to act out buying each item. Then, they will circle the same number of coins. For SEE, ask, *What information are you given?* (A balloon costs 6¢. There are 8 pennies.) For PLAN, ask students how they will figure out how many pennies to circle. (I will circle 6 pennies because I counted out 6¢ with my coins.) For DO, have students trace the circle around the six pennies. For CHECK, have students check the price of the balloon and then count the number of pennies in the circle to make sure they have circled 6 pennies.

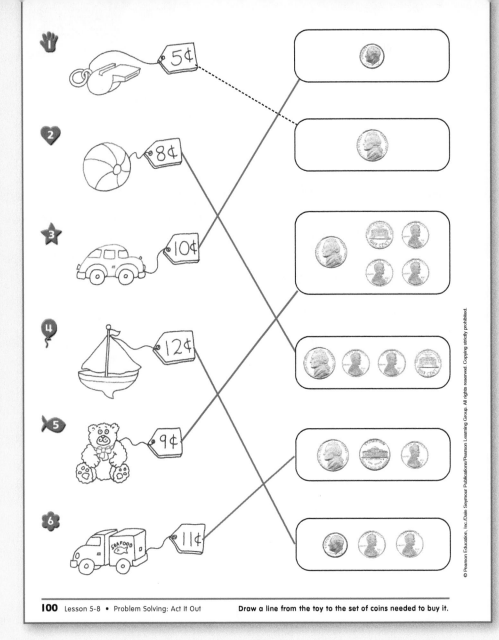

Draw a line from the toy to the set of coins needed to buy it.

For Mixed Abilities

Common Errors • Intervention

Some students may have difficulty matching coins with a printed amount. Draw a dime and a penny on the board and have a student draw sticks under each to show the total value. Because there are 10 + 1 or 11 sticks, they should look for a price tag that shows 11¢ from a group of price tags showing 1¢ to 12¢. Repeat for other amounts.

Enrichment • Money Activity

1. Have students start with 12 pennies to buy two items at the classroom store and have 2¢ left.

2. Have students identify all items that could be purchased from the classroom store with 1 nickel and 4 pennies or less.

3. Have a group of students cut pictures from magazines or catalogs to make their own catalog. Have them price each item up to 12¢ and take orders from friends.

• Have students complete the remaining exercises independently.

Using page 100 Be sure students understand that the items to be bought are down the left side of the page and groups of coins are down the right side of the page. Tell them that they need to find the correct group of coins to buy each item. Ask the cost of the whistle in Exercise 1. (5¢) Have students trace the line to the nickel to show that the whistle, which costs 5¢, can be bought with the nickel, which is worth 5¢. Help students complete the next exercise to draw a line from the ball to the group of coins showing a nickel and 3 pennies. Have them complete the page independently.

4 Assess

Draw 2 nickels and 1 penny on the board. In another area of the board, draw 1 nickel and 3 pennies. Show students an item priced at 8¢ and ask which group of coins is needed to buy the item. (1 nickel, 3 pennies) Repeat for more practice in choosing the correct group of coins for items priced 12¢ or less.

T100

Chapter 5 Test and Challenge

pages 101–102

1 Getting Started

Test Objectives
- **Item 1:** To arrange pictures in order of time in sequence (see pages 85–86)
- **Items 2–3:** To match time on a digital clock to time on an analog clock (see pages 91–92)
- **Items 4–5:** To count coins for amounts through 12¢ (see pages 95–98)
- **Items 6–7:** To choose coins to buy an item priced through 12¢ (see pages 96–100)

Challenge Objective
- To buy items using dimes, nickels, and pennies through 10¢ and find the change

Materials
*pictures showing sequences of events; *demonstration analog clock and 2 demonstration digital clocks; real or play dimes, nickels, and pennies; *items priced up to 12¢; colored chalk

Name _____

Circle the event that happens first. Circle the digital clock that matches the clock face. Count the dimes, nickels, and pennies. Write how much money. Circle the coins needed to buy the item.

Chapter 5 • Test **101**

2 Give the Test

Prepare for the Test Help students review the concepts presented in this chapter by showing them pictures of sequences of events and having them determine which event occurred first. Set a demonstration analog clock and one demonstration digital clock to 5:00 and the second demonstration digital clock to 2:00 and have students determine which digital clock time matches the time on the analog clock. Then, give each student a different amount of real or play coins and an item priced up to 12¢ to determine which coins they will need to buy the item. Vary and repeat these activities until students have mastered the concepts presented in this chapter.

- After students have mastered telling time and counting pennies, nickels, and dimes, have them complete the Chapter 5 Test. You may wish to review test-taking rules with students. Remind them that they are not to talk to each other and they are to finish the test on their own. Decide ahead of time whether or not to allow the use of manipulatives.

Using page 101 Have students look at the two skiing pictures in Exercise 1. Tell them that they are to circle the picture that shows the action that happened first. (walking up the slope) For Exercises 2 and 3, tell them to look at the round clock face and circle the digital clock that shows the same time. For Exercises 4 and 5, tell students to count the two groups of coins and write the amounts. In the last two exercises, have students draw a circle around the coins needed to buy the bear and the dump truck.

3 Teach the Challenge

Develop Skills and Concepts for the Challenge Give students an item priced at 7¢ and then give them 2 nickels and 2 pennies. Ask students which coins they would need to buy the item. (1 nickel and 2 pennies) Ask, *Which coins are left over?* (1 nickel) Explain that if they had given a clerk in a store 12¢ for an item that cost 7¢, they would get a nickel back as change. Repeat this activity with other items priced up to 10¢ and different amounts up to 12¢.

For Mixed Abilities

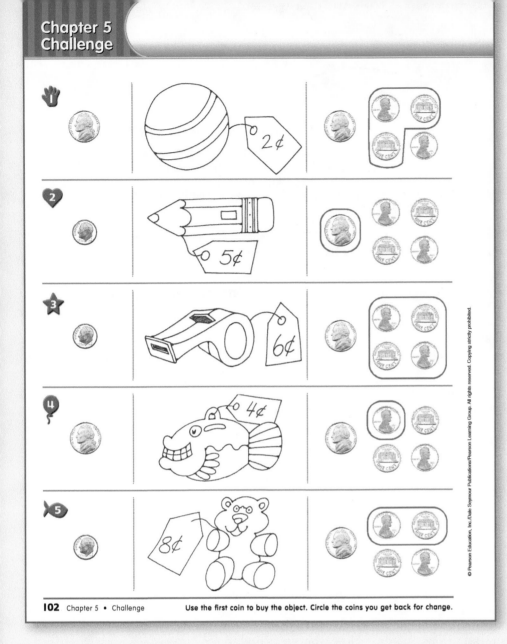

Use the first coin to buy the object. Circle the coins you get back for change.

Common Errors • Intervention

Some students may count the number of coins rather than consider the value of each. For example, they may count a nickel and two pennies as 3¢ because it is three coins. Have them work in cooperative teams to review counting on by 10s, 5s, and 1s. Then, give each group some coins and have the groups practice counting on by all dimes first, then all nickels, and finally all pennies.

Enrichment

1. Tell students that they bought an item costing 4¢ with a dime. Have them draw the coins they would have left. Have them repeat the purchase with a nickel and draw the change.

2. Have students draw dimes, nickels, and pennies to show different ways to make 12¢.

3. Have students write their phone number and the numbers in their address.

Using page 102 If students have experienced the store and bank activities, this page should be familiar to them. Have students point to the nickel in Exercise 1. Tell them they have this nickel to shop with and they are to buy the ball. Ask the cost of the ball. **(2¢)** Ask how many pennies can be traded for 1 nickel. **(5)** Ask how many pennies will be left after the ball is bought. **(3)** Tell students to circle the 3 pennies to show that 3 pennies will be left. Work similarly with them through the next two exercises before assigning the page to be completed independently. When they have completed the page, have some students show the work with their coins as one student draws the coins on the board. Have students take turns describing the transaction in sentences. **(I have a dime. The pencil costs 5¢. I will have a nickel left.)**

4 Assess

Alternate Chapter Test You may wish to use the Alternate Chapter Test on page 209 of this manual for further review.

Challenge Place a dime on an overhead projector. Tell students that you want to buy an item that costs 6¢ with your dime. Have students draw the coins that you would receive for change. **(4 pennies)**

pages 103–122

Chapter Objectives

- To review counting objects through 12
- To write 0 through 20
- To count 0 through 20 objects and choose the correct number
- To define *thirteen, fourteen, fifteen, sixteen, seventeen, eighteen, nineteen, twenty,* and *missing*
- To construct and interpret a picture graph
- To compare information from a picture graph
- To count combinations of dimes, nickels, and pennies through 20¢ and choose or write that amount
- To count and choose the correct coins for items priced through 20¢
- To write missing numbers in a sequence through 20
- To put numbers 0 through 20 in order
- To read a bar graph

Materials

You will need the following:
*"Spotted Worm" display; *tape; *large number cards 0 through 20; *number cards 0 through 20; *items priced through 20¢; *money amount cards through 20¢; *floor number line through 20; *20 objects to count

Students will need the following:
Reproducible Masters RM7, RM15, RM16, and RM17; number cards 0 through 20; object cards 0 through 20; 20 counters; clock faces; number cards 0 through 20 in relief; yellow paper; 1 dime; 2 nickels; 10 pennies; crayons; markers; picture graph titled "Markers and Crayons"

Bulletin Board Suggestion

Have students help make Spotted Worm, who has a head and then a body segment for each number. Connect the segments with paper clasps. Each segment shows its number by having that number of spots on it. For numbers greater than 10, draw a group of 10 spots and then single spots. You may also want to write the corresponding number on each segment. Add segments to the body as the numbers through 20 are learned. Once students are comfortable with the order of the numbers, mix up the segments and have students place them in order again.

Verses

Poems and songs may be used any time after the pages noted beside each.

We Have a Whole Lot of Numbers (pages 103–104)

We have a whole lot of numbers in our hands.
We have a whole lot of numbers in our hands.
We have a whole lot of numbers in our hands.
Let's see what we have.

We've got the number 1 in our hands.
We have 1 more than zero in our hands.
We have 1 less than 2 in our hands.
We have a 1 in our hands.

You may wish to extend this song as follows:

2: 1 more than 1;   1 less than 3

3: 1 more than 2;   1 less than 4
⋮   ⋮   ⋮
19: 1 more than 18;   1 less than 20

Mystery Version (pages 113–114)

I have a mystery number in my hand.
I have a mystery number in my hand.
I have a mystery number in my hand.
Let's see what I have.
I have the number ? in my hand.
I have 1 more than 15 in my hand.
I have 1 less than 17 in my hand.
What number do I have? (16)

What Coins Do I Have? (pages 117–118)

I have some mystery coins in my hand.
I have two mystery coins in my hand.
I have fifteen cents in my hand.
Now what coins do I have? (1 dime and 1 nickel)

Spotted Worm (pages 119–120)

Worm, Worm, Spotted Worm,
What number's been left out?
You're missing one of your segments!
To find it, we will count!

Zero, one, two, three, four, five,
Six, seven, eight, nine, ten,
Eleven, twelve, thirteen, oh, whoops!
Fourteen must be put in!

Eight, nine, ten, eleven,
Twelve, thirteen, fourteen,
Fifteen, sixteen, seventeen, whoops!
Eighteen must be put in!

Time for 12-Nut Pie

Mrs. Chipmunk got up at seven o'clock in the morning, which was very early. She had a big day ahead of her. Her grandson, Chippy, was coming to visit at seven o'clock that evening. She decided she would surprise him and bake a 12-nut pie.

"I'd better hurry," said Mrs. Chipmunk. "Chippy will be here in 12 hours, and I have a lot to do!"

By eight o'clock, Mrs. Chipmunk had made her bed and cleaned her house. By nine o'clock, she had done the wash. At ten o'clock, she went out to look for 12 perfect nuts to bake Chippy a 12-nut pie.

By eleven o'clock, she had found two good nuts. By twelve o'clock, she had found three more good nuts. But at one o'clock, she was tired and had to rest. Then one hour later, at two o'clock, she began looking for nuts again. She found three more good ones. By three o'clock, she had found four more. At last, she had 12 perfect nuts to bake a 12-nut pie!

She took the nuts home and hurried to the store. There, she talked to Mr. Toad who ran the store.

"I need 12 pinches of sugar to bake a 12-nut pie," she said.

"Sugar is a penny a pinch," said Mr. Toad. "Here you are. That will be 12 cents, please."

"Would you like one dime and two pennies or two nickels and two pennies?" asked Mrs. Chipmunk.

"It doesn't matter," said Mr. Toad. So Mrs. Chipmunk gave Mr. Toad two nickels and two pennies.

"Oh, dear!" said Mrs. Chipmunk, "It's five o'clock! I have to hurry!"

By six o'clock, the pie was baked. And at seven o'clock, Chippy burst through the door. "Grandmother!" he cried. "I'm so glad to see you! I could smell your delicious 12-nut pie baking long before I got here!"

"Hello, Chippy!" said Mrs. Chipmunk, as she hugged her grandson and gave him a large piece of his favorite pie.

RM15 Chapter 6 • "Time for 12-Nut Pie"   Count various objects in the picture such as the nuts or the hearts on the napkin.

"Time for 12-Nut Pie," p. RM15

Activities

Use Reproducible Master RM15, "Time for 12-Nut Pie," to help students relate the story to the mathematical ideas presented.

- Reread the story as students move the hands of their clock faces from seven o'clock to show the passing of 12 hours.

- Read the story while one student moves the hands of the large demonstration clock and other students act out the roles of Mrs. Chipmunk, Mr. Toad, and Chippy. Assign objects to represent the 12 nuts and the 12 pinches of sugar. Provide the coins mentioned in the story. Repeat for students to play different roles.

- Show seven o'clock on the demonstration clock and tell students that Mrs. Chipmunk had finished the wash one hour later than the time shown. Have students show and tell the new time on their clock faces. (eight o'clock) Continue telling students of Mrs. Chipmunk's accomplishments for each hour and have students show and tell the new time. Now, begin at 7:00 in the evening, when Chippy arrives, and work backward to 7:00 in the morning to help students see the events in reverse order.

1 Getting Started

Objectives
- To review counting objects through 12
- To write 0 through 12

Materials
Reproducible Masters RM7 and RM16; number cards 0 through 12; object cards 0 through 12; 12 counters; clock faces

Warm Up • Mental Math
Ask students to answer yes or no.
1. 1 dime is 10¢. (yes)
2. 5 apples and 1 more apple is 6 apples. (yes)
3. 12 comes before 11. (no)
4. 1 less than 6 is 5. (yes)
5. Zero is not any. (yes)
6. 2 pennies is 5¢. (no)

Warm Up • Time Activity
Have students show 12:00 on their clock faces. Discuss how the minute hand goes around the clock one time for each hour and is always pointing to the 12 for an o'clock. Have students show 6:00, 9:00, 2:00, and then 12:00 again.

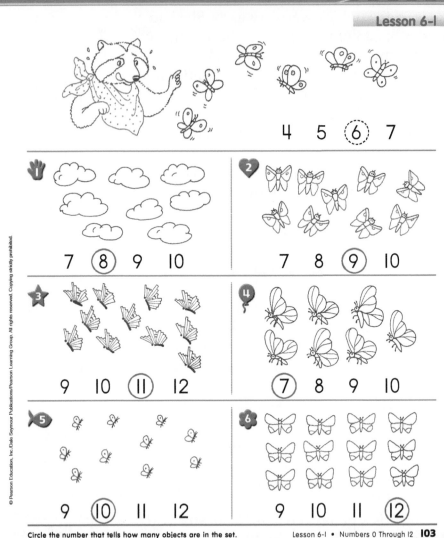

Name _____

Numbers Through 20

Chapter **6**

Lesson 6-1

Circle the number that tells how many objects are in the set.

Lesson 6-1 • Numbers 0 Through 12 **103**

2 Teach

Develop Skills and Concepts Show the number cards 0 through 12 in random order while students show the corresponding object cards.

- Have students place their object cards in order from 0 to 12 and then from 12 to 0. Now, have them count aloud in unison from 0 to 12 and then backward from 12 to 0.
- Play "I'm Thinking of a Number." Tell students that you are thinking of a number that is 1 more than 11. Have students show both the number card and the object card for this number. (12) Repeat with more number clues for the numbers 0 through 12 and then have students lead the activity.
- Place different sets of counters on a table in random order. Each set should have no more than 12 counters. As you point to each set, have students hold up the corresponding object card. Then, repeat the activity and have students hold up the corresponding number card.

- Give students copies of Reproducible Master RM16. Have them find the numbers 0 through 12 in sequence to help in drawing a line for the best route through the maze.
- Give students copies of Reproducible Master RM7. Have them write numbers from 0 to 12 in random order as you say the numbers.

3 Practice

Using page 103 Have students silently count the butterflies at the top of the page and tell the number. (6) Then, have students count them aloud in unison. Have a student use a sentence to tell how many butterflies there are. (There are 6 butterflies in the group.) Have students read the four numbers aloud in unison and tell which of these numbers shows how many butterflies there are. (6) Tell them to trace the circle around the 6. Work through Exercise 1 with students to have them circle the 8 to show there are 8 clouds. Then, have them complete the page independently.

104 Lesson 6-1 • Numbers 0 Through 12   Count the objects in the set and write the number.

For Mixed Abilities

Common Errors • Intervention

Some students may need more review with the numbers 1 through 12. Make sure each student has a set of number cards. Then, have students sing "We Have a Whole Lot of Numbers," found on page T103a, through the number 12 and have them show the number card for each number they sing.

Enrichment • Number Sense

1. Have students make number books by cutting many versions of the numbers 1 through 12 from newspapers or magazines. Have them number the pages and put all number 1s on page 1 and so on.

2. Have students work with a partner to number the cups of an egg carton 1 through 12. Have one partner show a number card for the other partner to place that number of counters, seeds, or beans into its proper cup.

3. Have students practice writing their phone number from a model.

ESL/ELL STRATEGIES

Have students discuss the meaning of the word *order*. To demonstrate, hand each student a large number card, 0 through 12. As the class counts from 0 to 12 aloud, have the student with that number card stand in the front of the room. Explain to students that the way we say the numbers is the correct *order* of the numbers.

Using page 104 Have students look at the examples at the top of the page. Tell students that they should count the objects in the group and write the number that tells how many there are. Have students tell how many they see in the first example. (none) Tell them to trace the zero to show that there are no objects in the group. Work through the next two examples with students before having them complete the page independently. Once students have completed the exercises, ask them to say the numbers in unison to see that they have counted the groups of objects and written the numbers from 0 to 12 in order.

4 Assess

Show the number cards 0 through 12 in random order. As you hold up each card, have one student say the number, another write the number on the board, and all students hold up the corresponding object card. Repeat for all numbers 0 through 12.

T104

1 Getting Started

Objectives

- To count 0 through 14 objects and choose the correct number
- To write 13 and 14
- To define *thirteen* and *fourteen*

Vocabulary

thirteen, fourteen

Materials

*tape; *large number cards 0 through 14; *floor number line marked for numbers 0 to 14, but has space for numbers through 20; Reproducible Master RM7; number cards 0 through 14 in relief; object cards 0 through 14

Warm Up • Mental Math

Ask students how many cents there are in the following:

1. 1 nickel (5¢)
2. 1 dime (10¢)
3. 7 pennies (7¢)
4. 1 nickel and 3 pennies (8¢)
5. 1 dime and 2 pennies (12¢)

Warm Up • Number Sense

Have students write the numbers 0 through 12 in order across the board and draw under each number a group with the corresponding number of objects. Stress that each new number is one more than the last number.

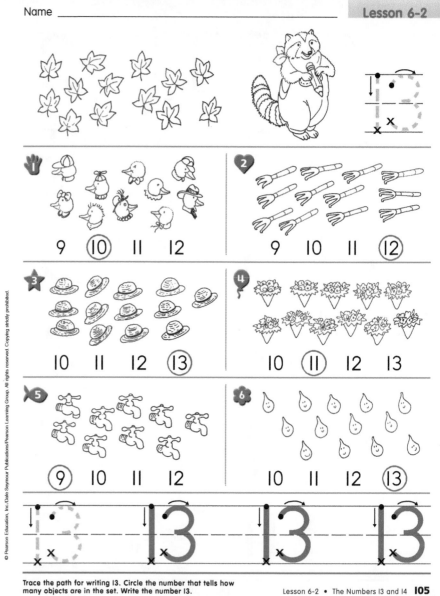

Trace the path for writing 13. Circle the number that tells how many objects are in the set. Write the number 13.

Lesson 6-2 • The Numbers 13 and 14 **105**

2 Teach

Develop Skills and Concepts Have a student draw 12 objects and then draw 1 more. Write **13** on the board and tell students that 12 and 1 more is 13. Have them count the 13 objects in unison. Ask what two numbers are used to write 13. (1 and 3) Have several students write 13 on the board. Repeat for 14.

- Say a number from 0 to 14. Have students show the correct number and object cards.
- Tape a number line on the floor to accommodate the numbers 0 to 20 for use throughout this chapter. Mark large dots at each point on the number line, evenly spaced apart, and then write each number 0 through 14. Give one large number card to each of 15 students. Have the student with the number 0 stand on the dot

above zero. Repeat for 1 through 14. Have students count to 14 in unison.

- Shuffle the large number cards and place them facedown on a table. Have a student draw a card, stand on zero on the floor number line, and take the number of steps called for on the card to stand on that dot. Continue until all cards through 14 are in order. Have students count orally through 14.

- On the board, randomly draw groups with 13 and 14 objects. Have students silently count the objects and write the correct number, 13 or 14, on Reproducible Master RM7 to show the number of objects you have drawn on the board. Be sure to draw each group in two sets, one ten and the other ones.

- Distribute the following number cards to students: 0, 1, 2, 3, and 4. Tell students that you want to make a 13 and you need two number cards. Ask which numbers are needed. (1 and 3) Have students with those numbers stand to correctly show the number 13. Then, have one student write 13 on the board and another draw 13 objects under it. Repeat for 11, 14, 10, and 12.

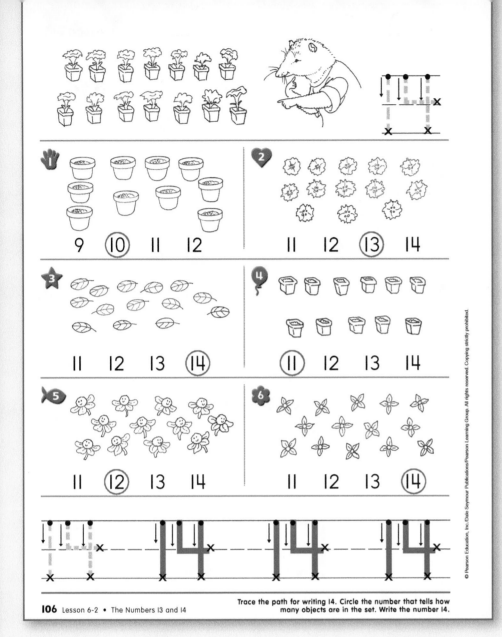

1) 9 (10) 11 12

2) 11 12 (13) 14

3) 11 12 13 (14)

4) (11) 12 13 14

5) 11 (12) 13 14

6) 11 12 13 (14)

Trace the path for writing 14. Circle the number that tells how
many objects are in the set. Write the number 14.

For Mixed Abilities

Common Errors • Intervention

Some students may have difficulty associating a number with the set containing that number of objects. Place large number cards in their proper place on a floor number line made with masking tape. Have a student draw a number of objects on the board and choose a classmate to stand on the number that tells how many objects were drawn. Repeat with other students and numbers.

Enrichment • Number Sense

1. Have students cut out 13 or 14 objects. Have students see how many different ways the objects can be arranged into two groups. (Possible answers: for 13: 12 and 1, 11 and 2; for 14: 13 and 1, 12 and 2)

2. Have students count out 14 counters and place 1 on each of two sheets of paper until all are placed. Have students tell how many are in each group. (7) Repeat for other numbers, 2 through 13, and have students note any numbers that end up with unequal groups. (odd numbers)

3. Have students extend their number books to 14.

3 Practice

Using page 105 Have students look at the example at the top of the page. Then, have them count and circle a group of 12 leaves. Ask how many leaves are not in the circle. (1) Have students say with you, *Twelve leaves and one more is thirteen leaves.* Tell them to trace the 13.

• For Exercises 1 to 6, tell students that they are to count the objects in the group and circle the number that tells how many objects there are. Have students complete the six exercises independently. Now, tell students that they are to trace the 13 at the bottom of the page and then write three 13s. Be sure students begin writing at the dots and end at the Xs.

Using page 106 Have students look at the example at the top of the page. Then, have students count and circle a group of 13 potted plants. Ask, *How many plants are not in the circle?* (1) Have students say with you, *Thirteen plants and one more is fourteen plants.* Tell students to trace the 14.

• For Exercises 1 to 6, tell students that they are to count the objects in the group and circle the number that tells how many objects there are. Have them complete the six exercises independently. Now, tell students that they are to trace the 14 at the bottom of the page and then begin at the dots to write three 14s.

4 Assess

Draw a set of 14 objects on the board. Have students write the number that tells how many are in this group. (14) Now, have students draw a group with 13 objects.

pages 107-108

1 Getting Started

Objectives

- To count 0 through 16 objects and choose the correct number
- To write 15 and 16
- To define *fifteen* and *sixteen*

Vocabulary

fifteen, sixteen

Materials

*16 objects to count; *floor number line through 16; Reproducible Master RM7; number cards 0 through 16; object cards 0 through 16; 16 counters; yellow paper

Warm Up • Mental Math

Tell students to count from
1. 7 to 14 (7, 8, . . . , 14)
2. 8 to 13 (8, 9, . . . , 13)
3. 0 to 10 (0, 1, . . . , 10)
4. their age to 14 (Starting number will vary.)
5. 5 to 12 (5, 6, . . . , 12)

Warm Up • Number Sense

Have one student stand up for each number as classmates count from 1 to 14. Then, have students count backward from 14 to 1 as one student at a time sits down.

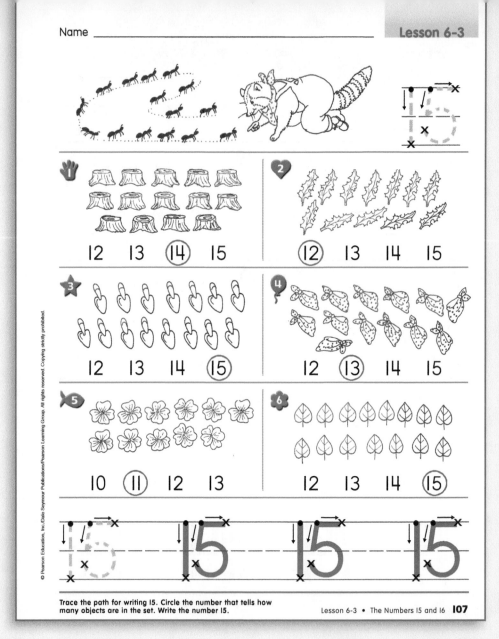

Trace the path for writing 15. Circle the number that tells how many objects are in the set. Write the number 15.

Lesson 6-3 • The Numbers 15 and 16 **107**

2 Teach

Develop Skills and Concepts Show a group of 15 objects. Tell students that they are going to count the objects to see how many there are. Have them count aloud as you remove one object at a time from the group. When 14 is reached, tell students that 1 more than 14 is 15 as you write **15** on the board. Have students count the objects from 1 to 15 again. Repeat for counting 16 objects.

- Extend the floor number line to 16. Have students choose an object card through 16, show it to the class, stand on zero, and walk the number of dots shown on their card. Continue until all object cards are in place for 0 through 16.
- Have students work in pairs with 16 counters. Tell one student to write a number on a sheet of paper for the partner to lay out that number of counters on the paper. Tell students to reverse roles for more practice.
- Have students work in pairs to match number cards to object cards.

- Have students practice writing 15 and 16 on copies of Reproducible Master RM7.
- Tell students a story about a cheese hole counter who always counted each hole in a slice of Swiss cheese before eating it and again as he ate it. Tell students that the cheese hole counter is about to eat a slice of cheese with 16 holes in it. Have them use yellow paper to draw the 16 holes the cheese hole counter counted. Have students tear off one hole at a time as they count from 1 to 16.

3 Practice

Using page 107 Have students count the ants in the example at the top of the page and tell the number. (15) Tell them to trace the 15.

- Tell students that for Exercises 1 to 6, they are to count the objects in the set and circle the number that tells

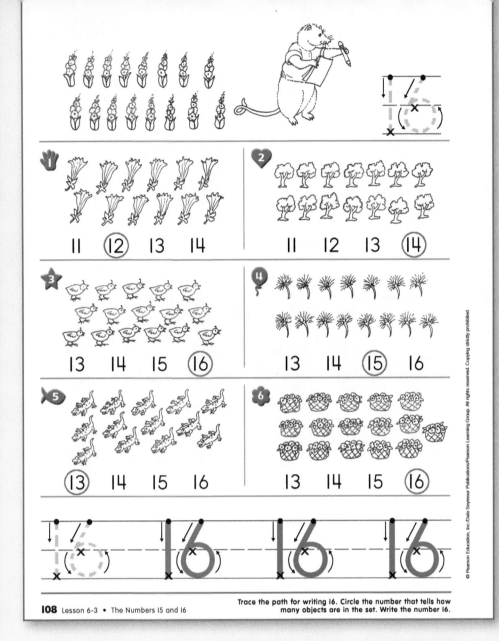

Trace the path for writing 16. Circle the number that tells how many objects are in the set. Write the number 16.

108 Lesson 6-3 • The Numbers 15 and 16

For Mixed Abilities

Common Errors • Intervention

Students may need additional practice with the numbers 13 through 16. Have students work in pairs. One student draws 13, 14, 15, or 16 sticks and the other student writes the number. Both check to make sure the answer is correct. Then, they should reverse roles and repeat the activity.

Enrichment • Number Sense

1. Have students write the numbers 1 through 16 that tell the ages of some teenagers. (13 to 16) Have students use magazines to find pictures of teenagers.

2. Have students make a collection of 16 animals. Tell them to write the numbers 1 to 16 on separate cards and cut out and paste a magazine picture of an animal on each card. Have students arrange their cards in order by number and show their animals to the class.

how many objects there are. Have students complete the six exercises independently. Then, have students trace the 15 and write three 15s.

Using page 108 Ask students to count the flowers in the example at the top of the page (16) and then say with you, *There are 16 flowers.* Tell students to trace the 16.

• Tell students that for Exercises 1 to 6, they are to count the objects in the set and circle the number that tells how many objects there are. Have students complete the six exercises independently. Then, have students trace the 16 and write three 16s.

4 Assess

Have students work in pairs. Have them lay their stack of number cards and any number of counters from 7 to 16 in front of them. Now, tell students to move to their partner's work and find the correct number card to match the given number of counters. Have students return to their own work to check the partner's counting. Repeat for more numbers.

T108

6-4 Problem Solving: Make and Use a Picture Graph

pages 109–110

1 Getting Started

Objectives
- To construct and interpret a picture graph
- To compare information from a picture graph
- To count 0 through 6 objects

Vocabulary
picture graph

Materials
16 counters; markers; crayons; picture graph titled "Markers and Crayons"

Warm Up • Mental Math
Ask students which number comes after the given number.
1. 11 (12)
2. 15 (16)
3. 2 (3)
4. 13 (14)
5. 5 (6)
6. 0 (1)
7. 9 (10)

Warm Up • Number Sense
Have students compare the number of counters in two sets by lining the counters up one-to-one. The set with all matched items has less and the other set has more. Repeat several times with other sets of counters. Make sure to include two sets that have the same number of items.

2 Teach

Develop Skills and Concepts Tell students that a picture graph shows information by using pictures. On the board, create a picture graph titled "Boys and Girls," with a column for boys and another for girls and a key with the letter *B* representing 1 boy from the class and the letter *G* representing 1 girl. Have each student place the appropriate letter in the graph in the correct column to represent himself or herself. Mention that this graph has the information listed up and down, or vertically.

Cats and Dogs

There are more dogs.

Color a picture on the graph for each cat and dog counted. Then decide if there are more cats or dogs.

3 Practice

Using page 109 Have students look at the picture at the top of the page. Ask students what types of animals are shown. (cats and dogs) Then, have students look at the picture graph. Ask them for the title. ("Cats and Dogs") Ask students, *What does the cat face in the picture graph represent?* (1 cat in the picture) Then ask, *What represents one dog from the picture?* (1 dog face in the picture graph) Mention that in this picture graph, the information is shown across, or horizontally.

- Now, apply the four-step plan to making a picture graph. For SEE, ask, *What are you asked to do?* (make a picture graph) Then, *What information is given?* (a picture with cats and dogs; a picture graph with one row for cats and one row for dogs; 1 face in the graph equals 1 animal; 4 cats; 6 dogs) For PLAN, have students count the number of cats and the number of dogs in the picture to find out how many faces they will color in each row of

T109

Things on the Shelf

There are more books.

110 Lesson 6-4 • Problem Solving: Make and Use a Picture Graph

Color a picture on the graph for each book, bear, and block counted. Then decide if there are more books, bears, or blocks.

Common Errors • Intervention

Some students may have difficulty understanding that each picture on the graph represents one item in the picture. Have students create a human picture graph and then show the information using pictures. Ask students which they like better, apples or oranges, and have students choose by standing in the correct row. Create a horizontal picture graph titled "Apples and Oranges" on the board with an apple and orange at the left of the designated row. Then, have each student in the apple row draw a red circle on the graph and every student in the orange row draw an orange circle on the graph.

Enrichment • Application

Ask each student what his or her favorite season is: summer, fall, winter, or spring. Have students draw and cut out a picture of a sun to represent summer, a leaf for fall, a snowflake for winter, or a flower for spring. Then, draw a picture graph on the board for students to display their choices. Now, ask students to use the graph to decide which season is liked the most and which is liked the least.

---

the picture graph. (cat, 4; dog, 6) For SOLVE, have students color in 4 cats and 6 dogs to complete the graph. For CHECK, have them count how many faces are used in their completed graphs. (10) Then, have students count the number of animals shown in the picture altogether. (10) Ask, *Does this match the number of animals in the picture?* (yes)

• Have students match each colored cat with each colored dog in the picture graph until no more pairs can be made. Then, ask students which group has more animals in the picture graph. (dogs) Then, ask students if there are more dogs or cats in the picture. (dogs)

Using page 110 Ask students to look at the picture at the top of the page. Have students describe the three different types of objects they see in the graph. (teddy bears, blocks, and books)

• Now, ask students to color in 1 object on the picture graph to represent 1 object in the picture. When students are finished, review how many objects should be colored in each row. (6 books, 3 teddy bears, and

4 blocks) Now, ask students to decide if there are more books than teddy bears in the picture graph using one-to-one correspondence. (books) Then, ask students to decide if there are more books than blocks in the picture graph using one-to-one correspondence. (books) Ask, *Are there more books, teddy bears, or blocks in the picture graph?* (books)

4 Assess

Have students work in small groups. Give each group a mixed set of markers and crayons so that there are 10 altogether. Give each group a copy of a picture graph titled "Markers and Crayons" with markers in one row and crayons in another. Tell students that they are to complete the graph so that it represents their combination of markers and crayons. Then, have groups show the mixed set of markers and crayons to the class while they share their completed picture graph.

6-5 Counting Money Through 16¢

pages 111–112

1 Getting Started

Objectives
- To count and write the amount for combinations of pennies, nickels, and dimes through 16¢
- To match items priced through 16¢ to the correct set of coins

Materials
*items priced through 16¢; *money amount cards through 16¢; 1 dime, 2 nickels, and 10 pennies

Warm Up • Mental Math
Ask students which type of coin is needed to complete the trade.
1. 2 coins for 1 dime (nickel)
2. 5 coins for 1 nickel (penny)
3. 1 coin for 10 pennies (dime)
4. 1 coin for 5 pennies (nickel)
5. 10 coins for 1 dime (penny)

Warm Up • Number Sense
Have students compare the dime, nickel, and penny for color, size, and value. Discuss the heads and tails sides of each coin. Have students show two ways to trade nickels or pennies for a dime. (Possible answers: 2 nickels; 10 pennies; 1 nickel and 5 pennies)

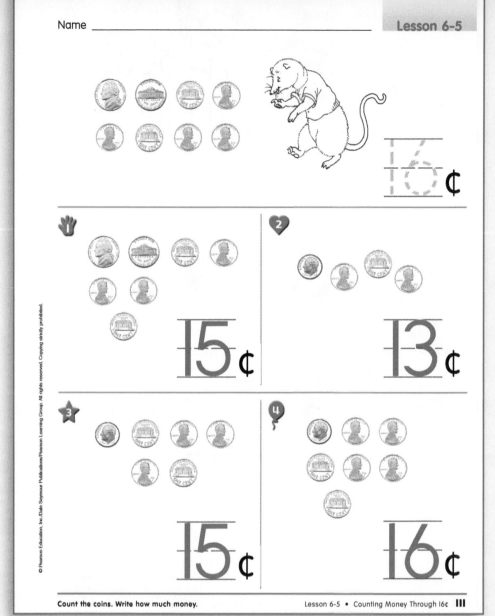

Name _____

16¢

1. 15¢

2. 13¢

3. 15¢

4. 16¢

Count the coins. Write how much money.

Lesson 6-5 • Counting Money Through 16¢ **111**

2 Teach

Develop Skills and Concepts Have students lay out 1 nickel and 10 pennies. Ask how many pennies are in 1 nickel. (5) Tell students to begin counting at 5 with you and count on each of the 10 pennies to see how much money there is in all. (5, 6, . . . , 15¢) Remind them to use the word *cents* to mean money as you write **15¢** on the board. Now, draw 1 dime and 1 penny on the board and help students begin at 10 and count on to tell the total. (11¢) Write **11¢** on the board. Continue to draw coins to have students begin at 5 or 10 and count on for amounts through 16¢. Point out that counting always starts with the coin that has the greater value.

- Show students a card with 1 dime and 4 pennies and have students count from 10 for each of the 4 pennies. (10, 11, . . . , 14¢) Have a student write the number on the board. (14) Write the cent sign and have students read the amount. (14 cents) Show a card with 2 nickels

and 4 pennies and say, *Two nickels are the same as 10 pennies, so I count 10, 11, 12, 13, 14 cents.* Have students count with you. Now, lay out an item that costs 15¢ and coin cards showing several ways to make 15¢. Have students count to see if each amount of money could buy the item. (yes) Repeat and add a card with less or more money to encourage careful counting.

- Have students take 1 dime, 2 nickels, or 10 pennies to the classroom store to purchase items. Have a banker make trades so that correct coins are used for purchases.

- Play "I'm Thinking of Some Coins." Tell students that you are thinking of 3 coins that make 12¢. (1 dime and 2 pennies) Have a student draw the 3 coins on the board and write the amount. Repeat for more amounts through 16¢.

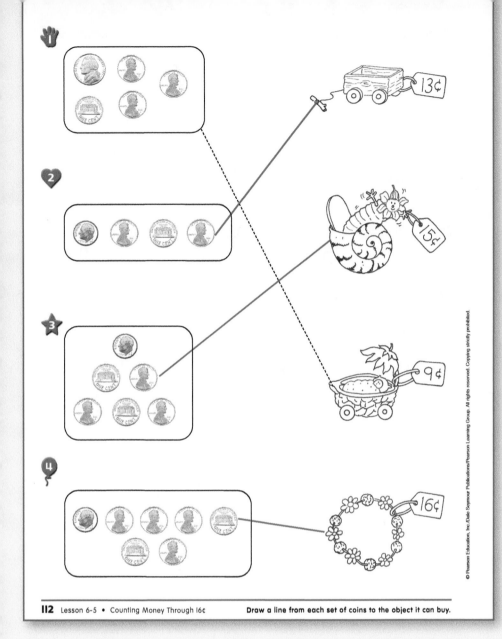

1

2

3

4

112 Lesson 6-5 • Counting Money Through 16¢   Draw a line from each set of coins to the object it can buy.

For Mixed Abilities

Common Errors • Intervention

Some students may count coins incorrectly. Have two students work together with 1 dime, 1 nickel, and 5 pennies to create as many different combinations as they can for 15¢. Remind them to start counting with 10 for a dime and 5 for a nickel. Encourage them to pause when counting from one kind of coin to another to remind themselves of the changing values. Also, students can move each coin to the side as it is counted. Continue with more coin combinations for amounts through 16¢.

Enrichment • Number Sense

Have students sort a collection of dimes, nickels, and pennies into stacks totaling 10¢, 14¢, 15¢, or 16¢.

ESL/ELL STRATEGIES

Some students may have difficulty with the final -ts blend at the end of the word *cents*. Make the -ts sound in isolation and have students repeat it. Then, say several amounts, such as *three cents*, and have students repeat them.

3 Practice

Using page 111 Have students count the money in the example at the top of the page. **(16¢)** If necessary, have them place a real or play coin on top of each pictured coin to show the 2 nickels and 6 pennies. Have students then trade the 2 nickels for 1 dime and count 10, 11, 12, . . . , 16¢. Tell students to trace the 16 in front of the cent sign. Tell them that the cent sign means we are talking about money as you read the amount as *16 cents.* Work similarly with students through Exercise 1 and then have students complete the page independently.

Using page 112 Ask students to count the group of coins in Exercise 1 and tell the amount of money. **(9¢)** Ask students which toy on the page costs 9¢. **(the baby carriage)** Tell students that they are to trace the line from the set of coins worth 9¢ to the baby carriage that costs 9¢. Help students complete Exercises 2 through 4 so that each group of coins is joined to the correct toy that costs that amount.

4 Assess

Show the class a priced item that costs no more than 16¢. Have students draw coins to show the money needed to buy the item. Have them write the amount of money and then draw a different group of coins for the same amount.

T112

1 Getting Started

Objectives

- To count 0 through 18 objects and choose the correct number
- To write 17 and 18
- To define *seventeen* and *eighteen*

Vocabulary

seventeen, eighteen

Materials

*number cards 0 through 18; *floor number line through 18; *large number cards 0 through 18; *18 items to count; 18 counters; 1 dime, 2 nickels, and 10 pennies; Reproducible Master RM7; object cards 0 through 18

Warm Up • Mental Math

Ask students who is older.

1. Ben is 15 and Eric is 12. (Ben)
2. Sara is 16 and Tara is 16. (same)
3. Bret is 13 and Alex is 14. (Alex)

Ask students who is younger.

4. Kregg is 15 and Ali is 5. (Ali)
5. Cristi is 16 and Burt is 15. (Burt)
6. Kelly is 16 and Rod is 16. (same)

Warm Up • Number Sense

Have students lay out coin combinations worth 16¢ and compare the different ways they chose to show the amount. Repeat for 14¢ and 15¢.

2 Teach

Develop Skills and Concepts Have students count out 16 counters and then put 1 more counter in the group as you write **17** on the board. Tell students that 16 and 1 more is 17. Ask what is the number that comes before 17. (16) Ask what is the number that is 1 less than 17. (16) Repeat the activity for developing the number 18.

- Have students practice writing 17 and 18 on copies of Reproducible Master RM7.

- Have students place object cards in order from 0 to 18 at their seats. Then, show the number card for 17. Have students place their object card for 17 facedown in front of them while you place number card 17 on a table. Continue for other numbers in random order so that all students' object cards are in a stack facedown and your number cards are set across the table in the order called for. Now, have students turn their stack over and count the number of objects on the first card. (17) Have a student find the correct number card on the table and write that number on the board. (17) Continue through the stack of object cards.

- Have a student count 15 items and write the number on the board. (15) Have students count the objects backward from 15 to 0. Repeat for 16, 17, and 18.

- Write ? on the board and hold up a number card so that students cannot see the number. Sing the mystery version of "We Have a Whole Lot of Numbers" on page T103a and point to the question mark on the board instead of singing the number. Repeat for more numbers and then have students lead the activity. For the number 18, you will need to repeat the line about 1 more than 17 because 19 has not yet been taught.

- Extend the floor number line to 18 and have students put large number cards in place by walking the necessary number of spaces.

(13)	14	15	16
14	15	16	(17)
13	(14)	15	16
13	14	15	(16)
14	15	16	(17)
14	(15)	16	17

Trace the path for writing 17. Circle the number that tells how many objects are in the set. Write the number 17.

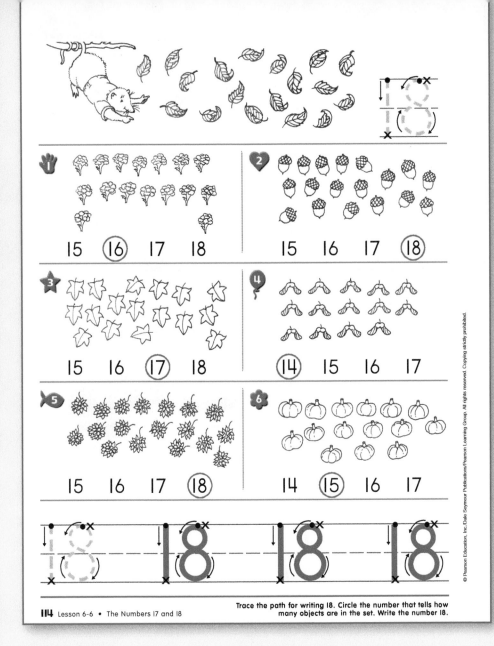

15 (16) 17 18   15 16 17 (18)

15 16 (17) 18   (14) 15 16 17

15 16 17 (18)   14 (15) 16 17

18 18 18 18

114 Lesson 6-6 • The Numbers 17 and 18

Trace the path for writing 18. Circle the number that tells how many objects are in the set. Write the number 18.

For Mixed Abilities

Common Errors • Intervention

Students may need more practice counting to 18. Have students arrange number cards from 0 to 18 in order. Remove two cards and have a student count aloud, filling in the missing numbers. Replace the two cards and remove two others. Continue the activity, removing three and four cards.

Enrichment • Number Sense

1. Have students lay out the number cards in order from 1 to 18. Have them begin at 2 and count by skipping every other number. Then, have students write the numbers in the order said.

2. Have students work in pairs with a book. One student says a number 1 through 18 and the other finds that page as quickly as possible. Reverse roles and repeat.

3 Practice

Using page 113 Have students count the plants in the example and tell the number. (17) Have a student use a sentence to tell the number of plants. (Possible answer: The raccoon has planted 17 plants.) Tell students to begin at the dots to trace the 1 and the 7 to write 17 and to show that 17 is the number that tells how many plants are shown.

• Tell students that for Exercises 1 to 6 they are to count the objects in the group and then circle the number that tells how many objects there are. Tell students to then trace the 17 at the bottom of the page and write three 17s. Have them complete the page independently.

Using page 114 Ask students how many leaves there are in the example for the opossum to count. (18) Have a student use a sentence to tell about the opossum and the number of leaves. (Possible answer: The opossum counted 18 leaves.) Tell students to trace the 1 and the 8 to write 18.

• For Exercises 1 to 6, tell students that they are to count the objects in the group and circle the number that tells how many objects there are. Tell students to then trace the 18 at the bottom of the page and write three 18s. Have them complete the page independently.

4 Assess

Have students work in pairs. Have each partner draw a group of 18 circles. Then, have partners switch papers and check each other's work.

T114

6-7 The Numbers 19 and 20

pages 115–116

1 Getting Started

Objectives
- To count 0 through 20 objects and choose the correct number
- To write 19 and 20
- To define *nineteen* and *twenty*

Vocabulary
nineteen, twenty

Materials
*20 objects to count; *large number cards 1 through 20; *floor number line through 20; Reproducible Master RM7; object cards and number cards 0 through 20

Warm Up • Mental Math
Ask students who has more money.
1. José has 10¢ and Maurice has 15¢. (Maurice)
2. Janice has 17¢ and Nikkia has 7¢. (Janice)
3. Ryland has 15¢ and Abbie has 16¢. (Abbie)
4. Gina has 16¢ and Rob has 14¢. (Gina)
5. Liam has 13¢ and Tara has 15¢. (Tara)

Warm Up • Number Sense
Draw an array of objects on the board as follows:

```
xxxxx      xxxx
xxxxx      xxxx
```

Have students count the Xs and write the number. Erase one X at a time to have students write and name the number that is 1 less.

2 Teach

Develop Skills and Concepts Lay out 18 objects as students count them. Place 1 more object in the group and tell students that 18 and 1 more is 19. Write **19** on the board. Repeat for developing the number 20.

- Extend the floor number line to include 19 and 20. Have students put the large number cards in place by walking the necessary number of spaces.

- Shuffle the large number cards 11 through 20. Have a student draw a card, say the number name, and tell the

number before it. Then, have students count backward to zero from this number.

- Point to a dot on the floor number line for a student to tell the number by counting from the starting point.

- Have students practice writing 19 and 20 on copies of Reproducible Master RM7.

- Distribute one large number card, 0 through 9, to individual students. Call a number between 11 and 20 and have the two students having the necessary number cards form the number. Have a third student write the number on the board and a fourth student draw that number of objects on the board.

- Draw 20 objects on the board in two groups of 10. Have students tell the number. Erase 1 object to have students tell the number that is 1 less. (19) Continue until one group of 10 objects remains. Starting with 19 objects, repeat the activity for students to gain experience in naming 1 more than each number.

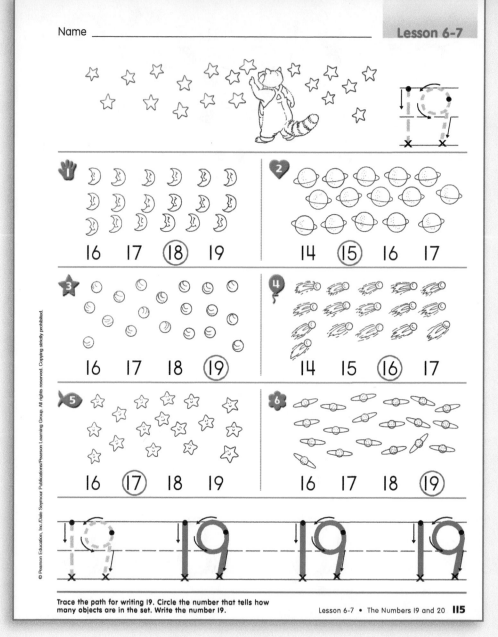

Name _____

Trace the path for writing 19. Circle the number that tells how many objects are in the set. Write the number 19.

Lesson 6-7 • The Numbers 19 and 20 **115**

T115

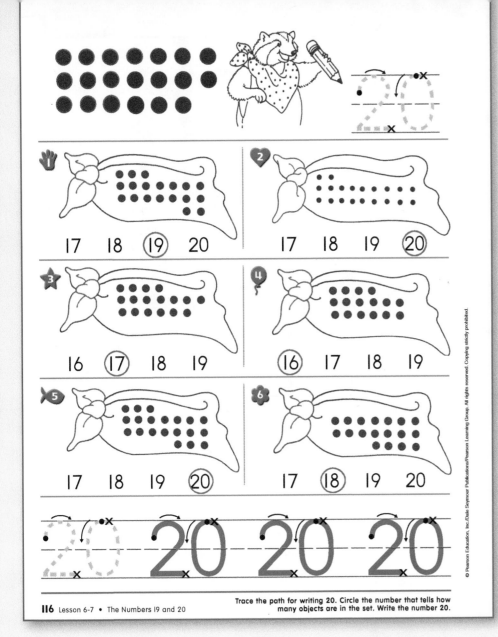

17 18 ⑲ 20

17 18 19 ⑳

16 ⑰ 18 19

⑯ 17 18 19

17 18 19 ⑳

17 ⑱ 19 20

20 20 20 20

Trace the path for writing 20. Circle the number that tells how
many objects are in the set. Write the number 20.

116 Lesson 6-7 • The Numbers 19 and 20

For Mixed Abilities

Common Errors • Intervention

Some students may need more practice counting to 20. Write the following across the board:

<u>11</u> <u>12</u> ___ <u>14</u> <u>15</u>

<u>16</u> ___ <u>18</u> ___ <u>20</u>

Have a volunteer fill in the missing numbers. (13, 17, 19) Erase three different numbers and repeat the activity.

Enrichment • Number Sense

1. Have students work in pairs. Have each partner choose a number between 11 and 20. Have partners give clues of more than or less than to guess each other's number.

2. Have students extend their number books through 20 by writing each number on a page and drawing that number of objects. Have them number the pages.

3. Have students lay out the number cards 0 through 20 in order. Have them begin at 3 and say the name of every other number card. Tell students to write the numbers said.

---

3 Practice

Using page 115 Have students look at the example and count the stars with the raccoon. (19) Have a student use a sentence to tell the number of stars. (Possible answer: The raccoon sees 19 stars.) Tell students to trace the 19 to show that 19 is the number that tells how many stars the raccoon sees.

• For Exercises 1 to 6, tell students that they are to count the objects in the group and circle the number that tells how many objects there are. Then, have students trace the 19 at the bottom of the page and write three 19s. Have them complete the page independently.

Using page 116 Ask students how many dots the raccoon has drawn altogether in the example at the top of the page. (20) Have a student use a sentence to tell about the raccoon and the number of dots. (Possible answer: The raccoon has drawn 20 dots.) Tell students to trace the 2 and the 0 to write 20. Ask students what the number 20 means to this example. (There are 20 dots.)

• For Exercises 1 to 6, tell students that they are to count the objects in the group and circle the number that tells how many objects there are. Then, have students trace the 20 at the bottom of the page and write three 20s. Have them complete the page independently.

4 Assess

Draw 19 circles on the board. Ask students to hold up the object card with the same number of objects. (19 card) Then, ask students to hold up the number card that represents this number of objects. (19)

T116

6-8 Counting Money Through 20¢

pages 117–118

1 Getting Started

Objectives

- To circle or write the amount for combinations of pennies, nickels, and dimes through 20¢
- To count coins and choose the correct amount of money for items priced through 20¢

Materials

*items priced through 20¢; Reproducible Master RM17; number cards 0 through 20; 1 dime, 2 nickels, and 10 pennies

Warm Up • Mental Math

Tell students to name the number.

1. the o'clock before 10:00 (9 o'clock)
2. pennies in a dime (10)
3. all their fingers and toes (20)
4. girls in this class (Answers will vary.)
5. the number the minute hand is pointing to at 3:00 (12)
6. 18 and 1 more (19)

Warm Up • Number Sense

Have students find and place in order shuffled number cards 0 through 20 as you give a clue for each number, such as 1 less than 1 (0), 1 more than zero (1), the number before 3 (2), and so on.

2 Teach

Develop Skills and Concepts Draw 1 dime, 1 nickel, and 1 penny on the board and have students write the value above each. Have a student draw 10 sticks under the dime, 5 under the nickel, and 1 under the penny. Tell students, *We are going to count by 1s to find out how much money there is.* Point to the dime. Count aloud from 1 to 10, saying the numbers softly as you point to each of the 10 sticks and stressing the last number, 10. Then, point to the nickel, count from 10 by 1s softly, and stress the last number, 15. Now, point to the penny, count from 15, and stress 16. Have students repeat the counting with you. As you point to each coin, have students count with you, saying *10, 15, 16¢.*

- Draw another penny and repeat for counting through 17¢. Continue through 20¢. Repeat for 1 dime and

2 nickels. Then, draw 2 dimes and repeat the procedure, stressing 10 and 20.

- Draw 1 nickel and 4 pennies on the board and have one student write the value of each coin under it. Have students count aloud as you write **5, 6, 7, 8, 9¢** under the respective coins. Now, have a student draw 1 dime and 1 penny and write the numbers that should be stressed. Repeat for other coin combinations through 20¢.

- Have students draw on the board and count orally the coins needed to buy items priced through 20¢.

- Have students lay out 20¢ in coins and then draw on the board the various coin combinations they chose. Have them count each group orally, counting softly and stressing the correct numbers. As students become able to do so, encourage them to say only those numbers that are stressed. Repeat for 19¢, 17¢, and 18¢.

- Give students copies of Reproducible Master RM17. Ask, *What coins do you see at the top of the page?* (1 dime, 1 nickel, and 4 pennies) Tell students to help the

T117

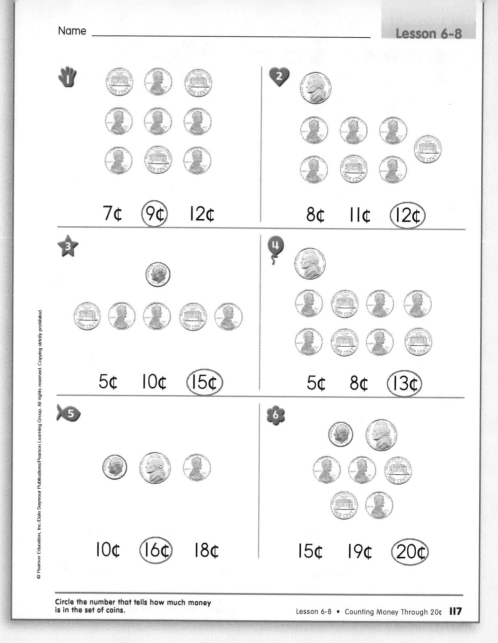

1. 7¢ (9¢) 12¢
2. 8¢ 11¢ (12¢)
3. 5¢ 10¢ (15¢)
4. 5¢ 8¢ (13¢)
5. 10¢ (16¢) 18¢
6. 15¢ 19¢ (20¢)

Circle the number that tells how much money is in the set of coins.

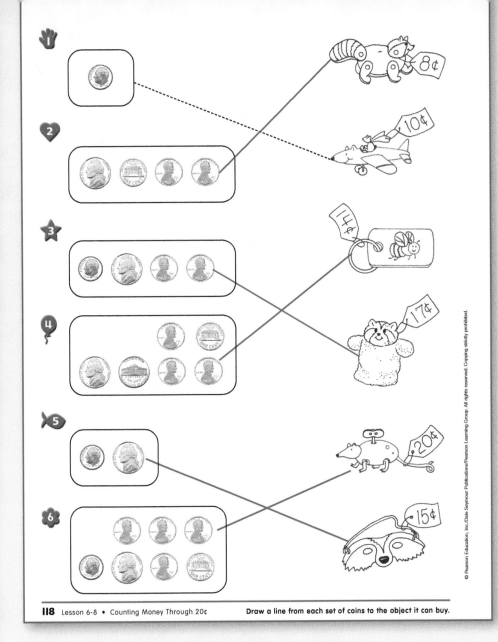

1

2

3

4

5

6

118 Lesson 6-8 • Counting Money Through 20¢   Draw a line from each set of coins to the object it can buy.

Common Errors • Intervention

Students may need more practice counting money. Have them work in pairs with 2 dimes, 4 nickels, and 20 pennies and the number cards 1 through 20. Have one partner flip a card and the other place the correct number of pennies on each number card. Then, have the first partner, where possible, replace sets of pennies with nickels and dimes. Students should then switch roles and continue.

Enrichment • Number Sense

1. Have students lay out an amount of money through 20¢, buy an item, and then count to tell the amount of money left.

2. Have students count on to solve problems such as how many more pennies they would need to buy a 19¢ item if they only had 1 dime and 1 nickel.

3. Display 1 quarter and 1 dollar bill. Ask students to name each and its value. Write **25¢, $0.25,** and **$1.00** on the board, and point out the dollar sign and the decimal point. Explain that the dollar sign is written before the number of dollars and the decimal point is placed between the dollars and the cents.

animals count these coins. Have students count aloud. Then, have them complete the page, counting each set of coins to find its total.

• Continue classroom store experiences for items through 20¢.

3 Practice

Using page 117 Ask students what coins are being counted in Exercise 1. **(9 pennies)** Have students find the value of the 9 pennies and circle the answer. **(9¢)** Now, have students look at Exercise 2 and remind them that when we count money, we start with the coin with the greatest value. Ask, *Which coin in this exercise has the greatest value?* **(nickel)** Tell students that they will start with 5 to count the value of the coins. Have students count the money aloud. **(5, 6, 7, . . . , 12¢)** Ask students how much money is shown. **(12¢)** Have a student read the three amounts of money under the coins and tell which one tells the amount of money pictured. **(12¢)** Tell students to circle 12¢. Work through each exercise

similarly and have students choose and circle each correct amount.

Using page 118 Ask students to tell the amount of money shown in Exercise 1. **(10¢)** Tell them to find the toy that costs 10¢ **(airplane)** and then to trace the line from the dime to that toy.

• Tell students to begin at 5 in the group of coins in Exercise 2 and count the money aloud. **(5, 6, 7, 8¢)** Tell students to find the toy that costs 8¢ **(raccoon)** and draw a line from that group of coins to that toy. If necessary, continue to work through each of the remaining exercises with students or have them complete the page independently.

4 Assess

Write **18¢** on the board. Give each student 1 dime, 2 nickels, and 10 pennies. Have students lay out 18¢ in any combination. Review the different combinations with the class.

pages 119–120

1 Getting Started

Objectives

- To write missing numbers in a sequence through 20
- To put numbers 0 through 20 in order
- To define *missing*

Vocabulary

missing

Materials

*object cards 0 through 20; *number cards 0 through 20; *"Spotted Worm" display; number cards 0 through 20

Warm Up • Mental Math

Have students name the time if the minute hand is on 12 and the hour hand is on

1. 6 (6:00)
2. 12 (12:00)
3. 3 (3:00)
4. 1 (1:00)
5. 11 (11:00)

Warm Up • Number Sense

Sing "What Coins Do I Have?" from page T103a to the tune of "He's Got the Whole World." Substitute different numbers of coins and total amounts from 15¢ to 20¢.

Name _____

Write the numbers 1 through 20 in order.

---

2 Teach

Develop Skills and Concepts Have students count in unison from 0 to 20 and then backward from 20 to zero.

- Place the number cards 0 through 20 in a pile on a table. Remove one number card from this set. Have a student arrange the cards in order on a table and write the missing number on the board. Repeat with the object cards and have a student draw the number of objects for the missing card. Continue both activities, removing two or three cards.

- Hold a set of number cards that has the number 19 card removed. Sing the following altered version of "We Have a Whole Lot of Numbers" from page T103a:

  The number that is missing in my hand
  Is 1 more than 18, in my hand

My number's 1 less than 20, in my hand
What number don't I have?

Repeat for other numbers less than 19.

- Arrange every other number card in order across a table, leaving spaces for the missing numbers. Have students put in the missing number cards. Now, remove three to five numbers in sequence and arrange the remaining number cards across the table, leaving spaces for missing numbers. Have students fill in the numbers.

- Have students shuffle their sets of number cards and remove one card secretly. Have students work in pairs and exchange their incomplete sets of cards. Each student is to place the number cards in order and write the missing number on paper. Partners check each other's work. Repeat for removing two or three numbers.

- Remove segment 14 of Spotted Worm on the bulletin board and sing "Spotted Worm" from page T103a to the tune of "Row, Row, Row Your Boat." Sing the verse for the missing number 18 and then remove other segments for continued verses you and the students create.

Count from 0 to 20. Write the missing numbers.

For Mixed Abilities

Common Errors • Intervention

Some students may need practice numbering from 0 to 20 in order. Draw a zigzag row of 21 dots across the board. Write 0 under the first dot. Have a student write 1 under the next dot, another student write 2 under the next dot, and so on until a student writes 20 under the last dot. Repeat this activity with different zigzag patterns.

Enrichment • Number Sense

1. Explain the "License Plate Numbers Game." Combine this game with a trip to the school bus garage or some other parking area. Have students look for a license plate with a number 1 and then for a plate with a 2, 3, 4, or other number sequence.

2. Shuffle a set of number cards and give half to each of two students to place all numbers in order.

---

3 Practice

Using page 119 Tell students to find the butterfly that is flying along from left to right across the top of the page. Have students find and follow the butterfly across the middle of the page and then across the bottom picture to the raccoon. Tell them that they should follow the butterfly to write the numbers from 1 to 20 in order. Have students return to the top of the page, trace the 1 and the 2, and tell what number they will write next. (3) Tell students to write 3. Ask what number is 3 and 1 more. (4) Tell them to write 4 and then trace the 5 and the 6. Continue to ask students which number comes next as they write or trace each number in order through 20.

Using page 120 Have students complete the page independently.

Now Try This! Tell students that some of the numbers are missing from the path. Encourage them to start at zero and count each square until they get to a blank square. They should then write the missing number. If necessary, students can start over at zero and repeat the process until they reach the next missing number. Have students complete the page independently.

4 Assess

Write **11, 12, ___, ___, 15, ___** on the board. Have students fill in the missing numbers. **(13, 14; 16)**

T120

Chapter 6 Test and Challenge

pages 121–122

1 Getting Started

Test Objective

• **Items 1–8:** To count 13 through 20 objects and choose or write the correct number (see pages 105–108, 113–116)

Challenge Objective

• To make and read a bar graph

Vocabulary

bar graph

Materials

number cards 1 through 20; object cards 1 through 20; crayons

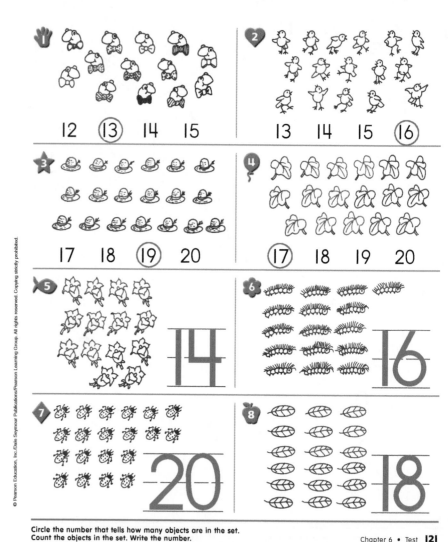

Circle the number that tells how many objects are in the set.
Count the objects in the set. Write the number.

2 Give the Test

Prepare for the Test Say the following rhyme and have students show the correct object card and number card each time you begin a new stanza. Have students show the number and object cards for 20 as you say, *Twenty little monkeys hanging from the trees. One jumped off to scratch his fleas.* Continue repeating the rhyme with a new number through 1. When you reach the last stanza for 1, say, *One little monkey hanging from a tree. He won't jump off! He has no fleas!*

• After students have mastered counting through 20, have them complete the Chapter 6 Test. You may wish to review test-taking rules with students. Remind them that they are not to talk to each other and they are to finish the test on their own. Decide ahead of time whether or not to allow the use of manipulatives such as counters.

Using page 121 Students should be able to complete this page independently. For Exercises 1 to 4, students will count the number of objects in the set and circle the correct number that tells how many objects there are. For Exercises 5 to 8, students will count the objects in the set and then write the correct number that tells how many objects there are.

3 Teach the Challenge

Develop Skills and Concepts for the Challenge Draw two circles toward the bottom of the board and shade one entirely as you tell students that the shaded circle is to show a chocolate cookie, whereas the unshaded circle shows a vanilla cookie. Draw ten squares above each cookie to make a bar graph. Name ten students to stand and ask each student in turn to shade one box above the kind of cookie he or she likes more. When all students have responded, ask students which kind of cookie has more votes, which has less votes, are there more votes for vanilla cookies or for chocolate cookies, how many more votes are for one kind of cookie than the other, and so on. Encourage students to explain how they got their answers.

Using page 122 Have students count the number of bananas on the left side of the page. (4) Then, have them count up the bar to show how many boxes they would color to show this fruit on the bar graph. (4)

For Mixed Abilities

Common Errors • Intervention

Some students may not clearly write their numbers, making it difficult for them to be read. Have students practice tracing the numbers printed on the number cards 0 through 20 with their index finger. Then, have students immediately practice writing that number on Reproducible Master RM7.

Enrichment • Application

1. Have students work as a group using magazines and newspapers to cut out one 1, two 2s, three 3s, and so on through twenty 20s. Provide 20 strips of paper in various lengths for students to paste the one 1 on one strip, all the 2s on another strip, 3s on a third, and so on. Have students arrange the strips in order on a bulletin board to create a bar graph.

2. Have students make a bar graph to compare the number of sides on a square, a rectangle, and a triangle.

Count the number of each kind of fruit. Write that number.
Then color that many boxes.

• Now, have students do the same for the apples. (5) Ask students how many boxes should be colored. (5) Then, have students count the apples in the picture to be sure there are 5.

• Be sure that students begin to count from the bottom up in the bar graph and that they are counting spaces, not lines.

4 Assess

Alternate Chapter Test You may wish to use the Alternate Chapter Test on page 210 of this manual for further review and assessment.

Challenge To assess students' understanding of bar graphs, lay out 4 pencils and 4 crayons. On the board, draw a bar graph for students to fill in. Have different students go to the board to color in different squares of the bar graph representing the 4 pencils and the 4 crayons. Have other students in the class explain if they feel the completed graph is correct and why.

pages 123–142

Chapter Objectives

- To count a set of 10 and count on through 20
- To write 0 through 20
- To identify 0 through 20 in sequence
- To review counting and writing the numbers for 10 or less
- To circle a group having fewer or more objects than another group
- To define *greater* and *less*
- To use ordinal numbers first through tenth
- To define ordinal number words *first* through *tenth*
- To count 0 through 31 objects
- To write 11 through 31
- To say the days of the week in order
- To use the correct words for the days of the week
- To associate days of the week with dates of a month
- To complete a monthly calendar
- To define *calendar, month, week,* and *day*
- To use groups of 10 to identify and write numbers through 31
- To put 0 through 30 in order
- To count from 1 to 100 using a hundred board
- To find patterns on a hundred board
- To create a bar graph; to read a bar graph

Materials

You will need the following:
*floor number line through 31; *large number cards 0 through 31; *number cards 0 through 31; *1 paper bag; *"Spotted Worm" display; *ordinal word cards first through tenth; *color word cards for red, green, blue, yellow, brown, and purple; *colored paper in red, green, blue, yellow, brown, and purple; *current monthly calendar showing full names of days of the week; *word cards for Sunday through Saturday; *small paper squares; *strip of paper with months written in order top to bottom; *cards showing bundle and/or single-stick combinations; *items priced through 9¢; *31 ice-cream sticks; *3 rubber bands; *31 counters; *yarn; *31 items to count; *2 paper circles and 1 paper triangle

Students will need the following:
Reproducible Master RM7 and RM19; 31 counters; 1 paper bag; object cards 0 through 31; number cards 0 through 31; 31 objects to count; yarn; 31 ice-cream sticks; 3 rubber bands; string and beads; word cards for Sunday through Saturday; copy of a hundred board; 10 cubes; at least 10 crayons

Bulletin Board Suggestion

Continue to build on the "Spotted Worm" display for a total of 31 segments. Show the spots on Segments 11 through 31 as groups of 10 spots and groups of ones.

Verses

Poems and songs may be used any time after the pages noted beside each.

We Have a Whole Lot of Numbers (pages 131–132)

We have a whole lot of numbers in our hands.
We have a whole lot of numbers in our hands.
We have a whole lot of numbers in our hands.
Let's see what we have.

We've got the number 1 in our hands.
We have 1 more than zero in our hands.
We have 1 less than 2 in our hands.
Let's write it on the board.

You may wish to extend this song as follows:

2: 1 more than 1;   1 less than 3

3: 1 more than 2;   1 less than 4

⋮   ⋮   ⋮

30: 1 more than 29;   1 less than 31

Tommy and Bessy (anon.) (pages 133–134)

As Tommy Snooks and Bessy Brooks
Were walking out one Sunday,
Says Tommy Snooks to Bessy Brooks,
Tomorrow will be Monday.

Penelope Lundy (pages 133–134)

Born on Sunday
Weaned on Monday
Toothed on Tuesday
Walked on Wednesday
Talked on Thursday
Trained on Friday
Schooled on Saturday
This is the growth
Of Penelope Lundy

Lining Up (pages 127–128)

If I am first, then you can't be,
Unless you stand in front of me.
Then I am second. I'm behind
The one who is the first in line.
If someone else now joins our line,
He will be third and stand behind.
Then fourth is next and fifth is last,
Unless we all turn around real fast.
Now who is first and who is third?
Your number changed! Where is your word?

Ronnie Raccoon's Petunia Patch

Ronnie Raccoon lived in a big hollow tree stump. One day he decided to plant flowers to make his home more beautiful. He carefully studied the ground around the stump. He figured he needed at least thirteen petunia plants to put a garden all the way around his house. So he went to Peter Opossum's Plant Store.

"Good morning, Peter," said Ronnie. "I need thirteen petunia plants to make my home more beautiful."

"Very well," said Peter Opossum. "The petunia plants are one penny each. That will be thirteen cents, please."

Ronnie gave Peter one dime and three pennies. Then he hurried home to plant his petunias. He started on one side of the front door and was working his way around when he ran out of plants. "Oh dear," said Ronnie, "I didn't buy enough!"

He figured again and realized he needed seven more plants. That would be twenty plants all together. Quickly, he ran back to Peter Opossum's store.

"I didn't need thirteen plants!" said Ronnie. "I needed twenty plants to go all the way around my house."

"Then you need seven more," said Peter. "That will cost seven cents."

Ronnie gave Peter one nickel and two pennies.

That evening as Peter Opossum walked home, he passed Ronnie Raccoon's house. Together they admired the beautiful petunia plants as Ronnie counted them out loud: "One, two, three, four, five, six, seven, eight, nine, ten, eleven, twelve, thirteen, fourteen, fifteen, sixteen, seventeen, eighteen, nineteen, twenty!"

"For twenty cents, you got a lot of pretty petunias," said Peter.

"Yes," laughed Ronnie, "and for twenty petunias, you got one dime, one nickel, and five pretty pennies!"

They laughed together and waved goodbye.

Name _____

Count the petunias around the stump. Draw and color enough petunias to make 20 in all.

Chapter 7 • "Ronnie Raccoon's Petunia Patch" **RM18**

"Ronnie Raccoon's Petunia Patch," p. RM18

Activities

Use the Reproducible Master RM18, "Ronnie Raccoon's Petunia Patch," to help students relate the story to the mathematical ideas presented.

• Reread the story as students act it out, using a desk or table as the tree stump. Have students play the parts of Ronnie Raccoon, Peter Opossum, and the petunias. Provide the appropriate coins for the purchase of the petunias.

• Draw a tree stump with a flower beside it on the board. Have a student write the number 1 on the flower. Draw another flower and have a student write the number 2 on it. Continue until 13 flowers are shown and numbered in order, but leave room for more flowers around the stump. Have a student draw another flower and write 14 on it. Continue until 20 numbered flowers completely surround the stump. Have students read the flower numbers in order from 1 to 20.

1 Getting Started

Objectives

- To count a set of 10 and count on through 20
- To write 0 through 20
- To identify 0 through 20 in sequence

Materials

*large number cards 0 through 20; *floor number line through 20; *20 ice-cream sticks; *2 rubber bands; 20 counters; 1 paper bag; number cards 0 through 20; 20 ice-cream sticks; 2 rubber bands; Reproducible Master RM7; object cards 0 through 20

Warm Up • Mental Math

Have students tell the numbers that come between those given.

1. 16 to 20 (17, 18, 19)
2. 0 to 5 (1, 2, 3, 4)
3. 10 to 14 (11, 12, 13)
4. 13 to 17 (14, 15, 16)
5. 19 to 15 (18, 17, 16)

Warm Up • Number Sense

Have a student secretly choose a large number card from 0 to 20 and move that number of dots from 0, on the floor number line. Have classmates name the number.

Circle each set of 10. Write the total number of objects in the set.

---

2 Teach

Develop Skills and Concepts Have a student count out 14 ice-cream sticks. Have another count out 10 of the 14 and place a rubber band around them. Have the two students start at 10 and count aloud to tell the total number. (10, 11, 12, 13, 14; 14) Have a student write the total number on the board. (14) Repeat this procedure for all numbers 11 through 20. Note that 20 sticks are bundled as two groups of 10.

- Shuffle the large number cards 0 through 20. Have a student draw a card and lead classmates in counting on to 20 from that number. Repeat. Then, have students lead classmates in counting backward to zero from the drawn number.

- Give students copies of Reproducible Master RM7. Have them write numbers you dictate between 10 and 20.

- Have students work in pairs with one set of number cards and object cards for 0 through 20. Together the pair should choose four of the number cards from 0 to 20 and their matching object cards. Then, have pairs play Concentration with these number and object cards. Tell students to mix the eight cards and scatter them facedown. Have students choose two cards at a time. A match wins the cards, whereas cards not matched are laid facedown again. Have students use cards for five or six numbers at a time when they are able to do so.

- Place 20 counters in the paper bag and have a student guess how many counters can be taken out with one hand. Have the student then check his or her guess by taking a handful of counters. Have the student write the guessed and actual numbers on the board and count on to find out how close the guess was. When students realize that the actual number is dependent on hand size, encourage them to make a more accurate guess by first judging the size of the hand.

Now Try This!

124   Lesson 7-1 • Numbers 0 Through 20

Connect the dots in order from 0 to 20.

© Pearson Education, Inc./Dale Seymour Publications/Pearson Learning Group. All rights reserved. Copying strictly prohibited.

For Mixed Abilities

Common Errors • Intervention

Some students may need more practice with groups of 10 and ones. Have students work in pairs with 20 ice-cream sticks and 2 rubber bands. One partner bundles 10 sticks and then shows 1 bundle and some single sticks. The other partner writes the number. Partners can reverse roles to repeat the activity.

Enrichment • Number Sense

1. Have students find three different ways to lay out 10 objects that make it easy to see that there are 10 in all.

2. Have students name numbers between 10 and 20 as 1 bundle of ten and the remaining number of ones, such as 21, 1 bundle of ten and 1.

---

3 Practice

Using page 123 Discuss the picture in the example at the top of the page. Ask students how many bubbles are in the dashed circle. (10) Have them trace the circle around the bubbles to show that the set is a group of 10 bubbles. Have students begin at 10 to count aloud with you to find out how many bubbles there are in all. (10, 11, 12, 13, 14; 14) Have students trace the 14.

• Tell students to count 10 of the bubbles in Exercise 1 and to circle them. Tell students to begin counting at 10 and count on to find out how many bubbles are shown in all. (10, 11, 12, . . . , 16; 16) Tell students to write 16 on the line to tell there are 16 bubbles in all. Work through the next couple of exercises with students before having them complete the page independently.

Using page 124 Have students complete the page independently.

Now Try This! Tell students that they are to complete the picture by joining the numbers in order from 0 to 20. Help them find the zero and then have them complete the page independently. Once students have connected all the dots, have them color the picture in the colors of their choice.

4 Assess

Show an object card and have a student write the number on the board. Then, choose another student to write the number before and after it.

T124

7-2 Less Than and Greater Than

pages 125–126

1 Getting Started

Objectives

- To review counting and writing the numbers for 10 or less
- To circle a group having fewer or more objects than another group
- To define *greater* and *less*

Vocabulary

greater, less

Materials

*20 objects to count; *yarn; *"Spotted Worm" display

Warm Up • Mental Math

Have students tell the o'clock

1. after 2:00 (3:00)
2. after 12:00 (1:00)
3. before 10:00 (9:00)
4. after 8:00 (9:00)
5. before 1:00 (12:00)
6. after 11:00 (12:00)

Warm Up • Number Sense

Have students count orally to 10 from any number, 0 through 9. Then, name a number from 1 to 10 for students to write on the board.

Count the objects in each set. Write the number.
Circle the greater number.

Lesson 7-2 • Less Than and Greater Than **125**

2 Teach

Develop Skills and Concepts Place two circles of yarn next to each other on a table. Then, place 5 objects in one circle of yarn and 4 objects in the other circle. Have students tell the number in each circle. (5; 4) Ask which group has more. (5) Ask which group has less. (4) Encourage students to match the objects one-to-one if necessary. Continue for more groups of 10 or fewer objects.

- Draw a group of 6 Xs on the board and directly underneath, a group of 7 Xs. Draw a circle around each group. Have a student write the number of Xs in each group beside the group. Ask which group has more. (7) Have a student circle the number 7 to show it is more than 6. Continue for comparing more numbers through 10 and circling the number that is greater. When students are comfortable with the concept of more than, begin using the term *greater* interchangeably and gradually use only the latter term.

- Repeat the previous activity to have students find and circle the number that is less, representing the group that has the fewest objects.

- Repeat the previous two activities, having 10 or fewer students standing versus 10 or fewer students sitting. Have a student write the numbers on the board and circle the greater number or the number that is less. Have students compare the groups orally. (Possible answer: There are 9 students standing. There are 7 students sitting. 9 is greater than 7. There are more students standing.)

- Remove Spotted Worm's segments from 0 to 10, but leave Segment 5 in place. Show Segment 6 and ask if 6 is greater than or less than 5. (greater than) Have a student put the segment in its place. (right of 5) Continue for other numbers through 9.

It's Algebra! The concepts of this lesson prepare students for algebra.

T125

Count the objects in each set. Write the number.
Circle the lesser number.

126 Lesson 7-2 • Less Than and Greater Than

For Mixed Abilities

Common Errors • Intervention

Some students may need more practice with the concepts of less than and greater than. Have students work in pairs with ice-cream sticks. Write **6** and **9** on the board. Have one partner pick up 6 sticks and the other pick up 9 sticks. Then, have partners lay down one stick at a time, at the same time, until one partner runs out of sticks. Ask, *Who had more sticks? So, which number is greater?* (9) Then, *Who had fewer sticks? So, which number is less?* (6) Repeat with other pairs of numbers.

Enrichment • Number Sense

Have students write all the numbers through 10 that are greater than a given number, 0 through 9. Repeat for all the numbers less than a given number, 0 through 10.

ESL/ELL STRATEGIES

Compare pairs of numbers between 1 and 5 using *greater than* and *less than*. Ask students to repeat. For example, say, *Five is greater than two* or *Two is less than four*. Elicit synonyms for each phrase. (Possible answers: bigger than; smaller than)

3 Practice

Using page 125 Have students look at the example at the top of the page. Then, have them count the starfish in the first group and tell the number. (6) Tell them to trace the 6 to show that there are 6 starfish in the first group. Now, have students count the starfish in the second group, tell the number (8), and trace the 8. Ask which group has the greater number of starfish. (second group or 8) Tell students to trace the circle around the 8 to show that 8 is a greater number than 6.

• Work through Exercises 1 and 2 with students to be sure they count the objects, write the numbers, and then circle the number that is greater. Have students complete the page independently.

Using page 126 Have students look at the example at the top of the page. Then, have students count the shells in the first group and tell the number. (5) Now, tell them to trace the 5 to show that there are 5 shells in the first

group. Have students count the shells in the second group, tell the number (6), and trace the 6. Ask which group has fewer shells. (first group or 5) Tell students to trace the circle around the 5 to show that 5 is less than 6.

• Work through Exercises 1 and 2 with students to be sure they count the objects, write the numbers, and then circle the number that is less. Have students complete the page independently.

4 Assess

Have students point to a segment on Spotted Worm. Have them say the chosen number and identify a number that is less than and a number that is greater than this number.

T126

7-3 Ordinal Numbers Through Fifth

pages 127–128

1 Getting Started

Objectives
- To use ordinal numbers first through fifth
- To define ordinal number words *first* through *fifth*

Vocabulary
first, second, third, fourth, fifth

Materials
*yarn; *20 counters; *ordinal word cards first through fifth; *color word cards for red, green, blue, yellow, brown, and purple; *colored paper in red, green, blue, yellow, brown, and purple

Warm Up • Mental Math
Ask students which number is greater.
1. 7 or 5 (7)
2. 8 or 10 (10)
3. 9 or 3 (9)

Ask students which number is less.
4. 7 or 5 (5)
5. 8 or 10 (8)
3. 9 or 3 (3)

Warm Up • Number Sense
Place 7 counters in a yarn ring and have a student place fewer counters in another ring. Repeat for a greater number of counters in a third ring.

2 Teach

Develop Skills and Concepts Arrange 5 chairs in a line in the front of the room facing either left or right. Place a sheet of green paper on the first chair and a sheet of red paper on the last chair. Have students begin at the chair with the sheet of green paper and count the chairs. (5) Have a student sit in the chair with the green sheet of paper as you write **first** on the board. Read the word aloud and say, *This is the word we use to tell about one object when there is more than one object and all the objects are in a row.* Have a student sit in the chair with the red sheet of paper behind the first student. Write **second** on the board and say, *This is the word we use to tell about the object that comes after the first object.* Continue to develop the words *third, fourth,* and *fifth.* Now, have students say each word written on the board while pointing to the appropriate chair.

Circle the object in the position named by the ordinal number.   Lesson 7-3 • Ordinal Numbers Through Fifth **127**

- Read the poem "Lining Up" on page T123a. Reread the poem as students choose the correct ordinal word card and stand in line. Reverse the order of the line. Have students exchange word cards accordingly, noting that the middle student keeps the same position and word card.

- Have five sitting students from the same row name their positions in complete sentences. (Possible answer: I am third in this row.) Have students exchange seats and tell if they are in the same positions. (no; unless a student is in the middle of the row) Have students name their new positions in complete sentences.

- Give each of five students from across the room one of the ordinal word cards for *first* through *fifth.* Then, have these students arrange themselves in order. Be sure that the students are all facing the same way in line. Now, hand out one color word card to each of these students.

T127

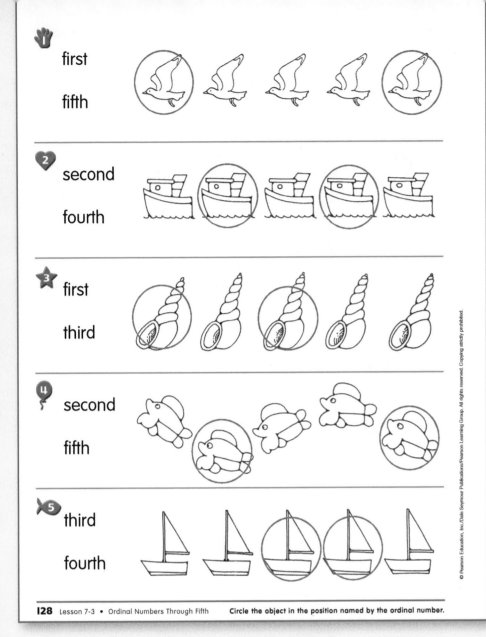

first

fifth

2 second

fourth

3 first

third

4 second

fifth

5 third

fourth

128 Lesson 7-3 • Ordinal Numbers Through Fifth   Circle the object in the position named by the ordinal number.

For Mixed Abilities

Common Errors • Intervention

Students may need more practice with ordinal numbers. Have them work in groups of five. Give each group the ordinal word cards for *first* to *fifth*. Have each member of a group take one card and then allow time for the groups to arrange themselves in one row in the correct order. Then say, *Will the first student in each group please stand up and show me your card?* Continue until all five students in each group have stood. Have the students in each group exchange cards, change direction, and repeat the activity.

ESL/ELL STRATEGIES

As you present the ordinal numbers, ask, *Which ordinal numbers have a -th sound at the end?* (fourth, fifth; also, sixth, seventh, eighth, ninth, and tenth, which are taught in Lesson 7-4) Then, write the three exceptions on the board: **first, second**, and **third.** Now, practice the pronunciation of each of the three exceptions.

Help students read the cards for color words. Have students in the class use complete sentences to describe the color and the student's position in line. (Possible answer: The first student in line has green.)

3 Practice

Using page 127 Have students look at the example at the top of the page and tell about the fish going to school. Have them recite the ordinal words aloud with you. Then, have students color the first fish green, the second fish blue, the third fish red, the fourth fish yellow, and the fifth fish brown. Now, ask the color of the fifth fish, the second, and so on. Tell students that these ordinal numbers tell the order in which the fish go to school. The first fish is number 1, the second fish is number 2, and so on.

• Tell students to look at the row of birds in Exercise 1 and the word before this group, *second*. Be sure students realize that they are to circle the object in each row that holds the indicated position. Also, point out to

students that the objects in each exercise are facing left, so they should start counting from the left each time. After completing all the exercises, have students color each circled object the color of the corresponding fish at the top of the page.

Using page 128 Tell students that for Exercises 1 to 5, they are going to circle two objects in each group because there are two ordinal numbers given for each set. Be alert to students who do not start counting from the left or those who start counting all over again from the first circled object.

4 Assess

Have students place one ordinal word card from first to fifth in order on five chairs. Now, scramble the cards and have different students place them in order, beginning at the opposite end of the row of chairs.

T128

7-4 Ordinal Numbers Through Tenth

pages 129–130

1 Getting Started

Objectives
- To use ordinal numbers sixth through tenth
- To define ordinal number words *sixth* through *tenth*

Vocabulary
sixth, seventh, eighth, ninth, tenth

Materials
10 different-colored crayons

Warm Up • Mental Math
Ask students which ordinal number follows the given ordinal number.

1. fourth (fifth)
2. first (second)
3. third (fourth)
4. second (third)

Warm Up • Number Sense
Have five students line up at the door, facing it. Point to the student closest to the door and say, *This student is first*. Point to each student in order and say aloud the student's ordinal position. Then, have the class say aloud the ordinal numbers first through fifth as the students bow or wave in turn.

Name _____

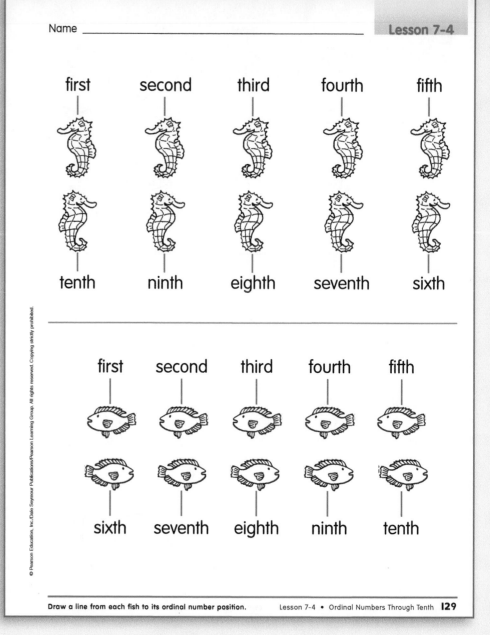

Draw a line from each fish to its ordinal number position.

Lesson 7-4 • Ordinal Numbers Through Tenth **129**

2 Teach

Develop Skills and Concepts Have ten students line up in front of the board, facing the same direction. Have the students count off from 1 to 10 as you write the numbers on the board above them. Tell the class that each person has a position in line. Point to the first student in the line and write **first** on the board above the 1. Continue in this way for the words *second* through *fifth*. Then say, *The position that comes after fifth is sixth. Who is sixth in line?* (Answers will vary.) Continue to develop the words *seventh* through *tenth*.

- Now, have different students from the class stand in line under the writing on the board. Ask the class for the position of random students in line. Next, erase the writing on the board and ask the first student to wave. Have the class decide if the correct student waved. Repeat for all the ordinal numbers.

- Have ten students stand in a line facing the board. Say, *Pretend these students are in line to write on the board.*

Have the students count off from first to tenth to show their order. Then, have the same ten students turn around and face another place in the room. Say, *Now, pretend that these students are in line for lunch.* Have the students count off from first to tenth. Ask the class, *Who was first in line to write on the board?* (Answers will vary.) Now ask, *Who is first in line for lunch?* (The answer will be a different student's name.) Then, conclude by asking, *Why do you think that different people are first in line?* (The direction of the line changed.) Make sure students understand that the direction of the line determines who or what is first.

- Have every student in the class line up facing the same direction. Have students count from first to tenth, starting in the front of the line. As soon as the counting starts over with first, that student should break the remaining students away and start a new line. Continue until all students have counted aloud. Now say, *Will the first student in each line raise a hand?* Have the class decide if the correct person in each line raised a hand. Repeat this procedure through tenth.

T129

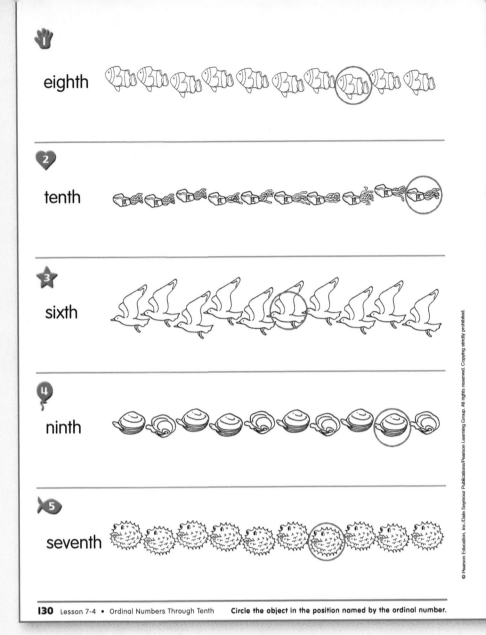

1 eighth

2 tenth

3 sixth

4 ninth

5 seventh

Circle the object in the position named by the ordinal number.

For Mixed Abilities

Common Errors • Intervention

Students may need more practice with the ordinal numbers. Have students work in groups of ten. Give the groups one set of ordinal number cards for *first* through *tenth*. Have each student in a group hold one card and then have these students organize themselves in two different lines facing two different directions.

Enrichment • Number Sense

1. Have student pairs describe and follow the steps to complete a simple task. For example, one student could tell the steps to sharpen a pencil while the other student follows the steps. Remind students to use ordinal numbers in their directions. Listen to make sure students are correctly using the words *first* through *tenth*.

2. Play Follow the Leader with the class. Demonstrate and say a series of ten actions using ordinal numbers. For example, say, *First, touch your toes. Second, hop on one foot. Third, snap your fingers. Fourth, clap your hands.* Continue until there are ten actions. Have students follow you and do each action.

3 Practice

Using page 129 Have students describe the seahorses at the top of the page. Now, have them point to each seahorse as they say the ordinal number aloud with you. Next, tell students what color to color each of the ten seahorses. Then, ask students to name the color of a seahorse by its position. For example, ask, *What color is the eighth seahorse?*

• Have students look at the fish at the bottom of the page. Ask students which direction all the fish are facing. (left) Then, have them draw a line from the first fish to the word *first*. Have neighbors swap papers to check each other's work. Once students swap back, have them complete the page independently.

Using page 130 Have students look at Exercise 1 and tell in which direction the fish are swimming. (left) Ask students to point to the second fish. Then, have them circle the eighth fish in line. Tell students that for Exercises 2 to 5, they should first look at the direction the animals are facing and then circle the animal in the position named by the ordinal number. Have students complete the page independently.

4 Assess

Ask students to describe how they know which object is the first in a line. (There is not another object in front of it. It is the first on the left.)

T130

1 Getting Started

Objectives
- To count 0 through 31 objects
- To write 11 through 31

Materials
*number cards 0 through 31; *yarn; *31 counters; *31 objects to count; Reproducible Master RM7; number cards 0 through 31; object cards 0 through 31

Warm Up • Warm Up
Have ten students line up and tell who is

1. first (Answers will vary.)
2. sixth
3. fifth
4. tenth
5. seventh
6. fourth

Warm Up • Number Sense
Distribute the number cards 11 through 15 and have students line up in order. Ask which number comes first (11), second (12), and so on through fifth (15). Now, distribute number cards for 16 through 20 so that students with the number cards 11 through 20 are in order. Ask which number is sixth (16), seventh (17), and so on.

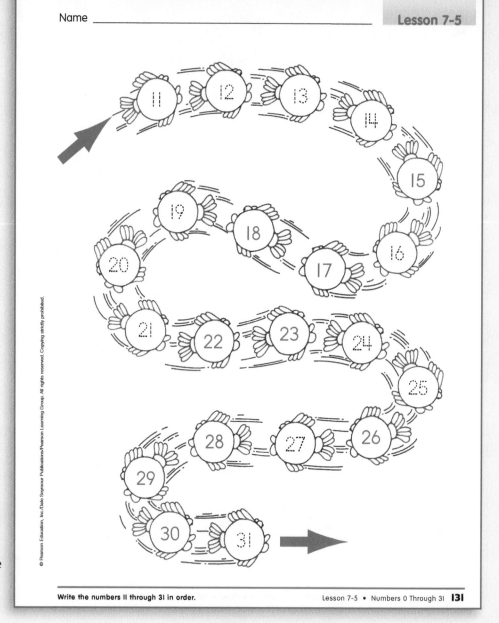

Write the numbers 11 through 31 in order.

2 Teach

Develop Skills and Concepts Lay out 2 circles of yarn with 10 counters in each. Ask students how many groups of 10 there are. (2) Write **20** on the board and tell students that 20 means two groups of 10. Then, lay out 1 more counter outside both rings and tell students that 20 and 1 more is 21 as you write **21** on the board to the right of 20. Continue to develop and write the numbers through 29.

- Place 1 more counter with the 29 and tell students that 29 and 1 more is 30 as you write **30** on the board. Ask them to tell if another circle of 10 can be formed. (yes) Place the ten single counters into a third yarn ring to help students see that three circles of 10 is 30. Lay out 1 more counter outside all three rings and tell students that 30 and 1 more is 31 as you write **31** on the board. Although students are not expected to write numbers beyond 31 this year, some may enjoy counting on to 100.

- Repeat the last activity by drawing sticks and circling groups of 10 on the board to have circled groups and numbers shown through 31.

- Distribute the following number cards: 0, 1, 1, 2, 2, 3, 4, 5, 6, 7, 8, 9. Place 10 objects inside a circle of yarn and have students with the 1 and 0 number cards form a 10 as another student writes 10 on the board. Place 1 more object in the circle and have students form the number with the number cards (11) and write **11** on the board. Continue until all numbers through 31 are formed and written.

- Have students count in unison from 0 to 31.

- Give students copies of Reproducible Master RM7. Have them practice writing numbers from 11 to 31.

- Distribute the number cards and sing verses of "We Have a Whole Lot of Numbers" from page 123a for students to show each correct number card.

- Have students match the number cards to object cards for 0 through 31.

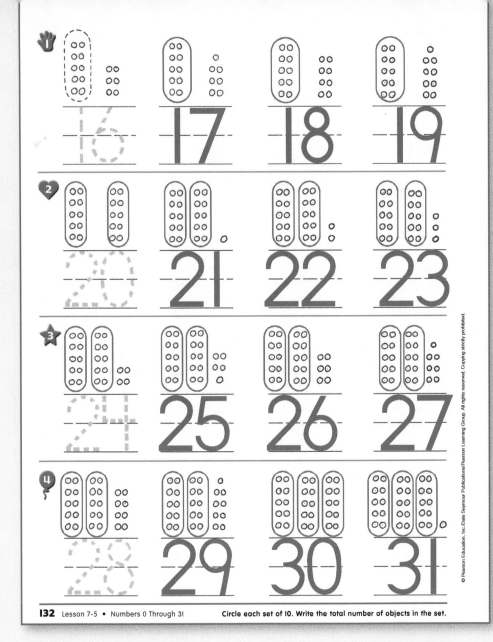

Circle each set of 10. Write the total number of objects in the set.

For Mixed Abilities

Common Errors • Intervention

Some students may need practice counting to 31. Deal out the number cards from 11 to 31. Have the student who has the number card for 11 place it on a desk, say the number, and write the number on the board. Continue in the same manner until all number cards 11 through 31 have been placed in order.

Enrichment • Number Sense

1. Distribute a jar of 31 countable objects to each student. Ask students to take a handful of the objects and estimate how many are in the handful. Have students record their estimates and then count to check. Repeat with different-sized handfuls.

2. Have students place counters on two fish on the path on page 131. Ask students to name the two numbers and tell which is greater. Repeat several times with different pairs of numbers. Make sure students understand that the numbers on the fish are in order, so the numbers on the fish at the beginning of the path are less than the numbers at the end.

3 Practice

Using page 131 Have students use their fingers to begin at the number 11 fish and follow the path to the fish whose number is 31. Tell students to trace the 11, then the 12, 13, and 14. Ask what number comes next. (15) Tell them to write 15 on the next fish and then continue to write the numbers in order through 31. Remind students that some of the numbers along the way are to be traced. This will help them check their work and ensure that they are following the direction lines. Supervise students as they complete the page independently.

Using page 132 Have students count the bubbles that are in the first circled group of Exercise 1 and tell the number. (10) Tell students to trace the circle around the 10 bubbles and then begin at 10 and count on to find out how many bubbles are in the whole group. (16) Have them then count aloud with you, *11, 12, . . . , 16.* Tell students to trace the 16 to show that there are 16 bubbles

in this group. Repeat the procedure to help students to find and circle the group of ten and then to find the total number of bubbles in the next three groups of Exercise 1. Have them complete the page independently.

4 Assess

Write **31** on the board. Have students draw a group of 31 stars. Then, have them circle the groups of 10. (three groups of 10 with one star remaining)

1 Getting Started

Objectives

- To say days of the week in order
- To use the correct words for the days of the week
- To associate the days of the week with the dates of a month
- To complete a monthly calendar
- To define *calendar*, *month*, *week*, and *day*

Vocabulary

calendar, month, week, day

Materials

*current monthly calendar showing full names of days of the week; *word cards for days of the week; *number cards 1 through 31; *small paper squares; *strip of paper with months in order top to bottom; word cards for days of the week; Reproducible Master RM19

Warm Up • Mental Math

Ask students to give each number.

1. 1 more than 10 (11)
2. 1 less than 30 (29)
3. 1 more than 20 (21)
4. 1 less than 20 (19)
5. 1 less than 10 (9)

Warm Up • Number Sense

Ask a student to name the number that comes after or before any number between 0 and 30.

Sunday	Monday	Tuesday	Wednesday	Thursday	Friday	Saturday
		1	2	3	4	5
6	7	8	9	10	11	12
13	14	15	16	17	18	19
20	21	22	23	24	25	26
27	28	29	30	31		

Mondays [7], [14], [21], [28]

Describe the calendar. Use the vocabulary *calendar*, *month*, *week*, and *day* and the names of the days of the week. Write the numbers for each Monday.

Lesson 7-6 • Calendar **133**

2 Teach

Develop Skills and Concepts Ask students, *What object do we use to tell us what time it is?* (clock) Tell students that a *calendar* helps us keep track of days. Point to the current month on the calendar, say the name of the month, and tell students that the name of the month is usually at the top of a calendar as you point to its location on the current monthly calendar.

- Name the first and last days of the week. Point to the day words in order as you say them and then have students recite them with you. Lay out the word cards for the days of the week in order and have students read them with you. Ask how many days are in a week (7), how many days of a week we go to school (5), which day of the week today is, which day of the week tomorrow will be, which day of the week yesterday was, and so on.

- Read the poem "Tommy and Bessy" on page T123a and have students tell the day for tomorrow in each verse. Repeat to have students hold up the word cards for the days.

- Read "Penelope Lundy" on page T123a for students to hold up the word cards for the days of the week. Discuss the growth stages from birth to school age.

- Discuss different months of the year as you flip through the calendar pages. Note with students that January is the first month and December is the last month. Focus on the current month and have students find today's date and tell which day of the week today is. Ask how many days are in this month and have students say the numbers in order from 1. Ask the day of the week for the first and last days of this month. Ask how many days in this month are Sundays, Thursdays, and so on.

- Ask students to tell the month of their birthday. Give students a square with their birthday month written on it. Attach the names of the months to the bulletin board and have students attach their squares next to

Sunday	Monday	Tuesday	Wednesday	Thursday	Friday	Saturday

Sundays ☐   Mondays ☐   Tuesdays ☐

Wednesdays ☐   Thursdays ☐   Fridays ☐

Saturdays ☐

134 Lesson 7-6 • Calendar

Write the numbers 1 through 31 for this month on the calendar. Count the number of times each day occurs in this month and write the number.

For Mixed Abilities

Common Errors • Intervention

Some students may need more practice with days of the week. Have students work in groups of seven with word cards for the names of the days of the week. Each member of the group should take a card. Then, the seven members, starting with the one who has Sunday, should arrange themselves in order from left to right.

Enrichment • Application

1. Have students find out how many months have 31 days, how many have 30 days, and how many have less than 30 days.

2. Have students make a calendar for their birthday month and put their picture on their special day.

3. Have students find out how many school days or weekend days are in this month.

4. Have students use a monthly calendar to find the date of the day before and the day after today's date.

---

the proper month. This bar graph may be used to tell which months have the most, least, or no birthdays.

3 Practice

Using page 133 Ask students to name the number for the first day of the month. (1) Tell students to circle the first day of the month on this calendar. Now, have them find the first Monday of the month and tell the number. (7) Tell students to write an X on the 7 for the first Monday and then write an X on each of the other Mondays in the month. Ask students to tell the numbers for all the Mondays. (7, 14, 21, 28)

• Continue similarly to have students identify and circle the following: first Sunday, second Thursday, third Friday, fifth Wednesday, and fourth Tuesday. Now, have them color each Thursday red, each Saturday yellow, and each Wednesday brown.

Using page 134 Use the current month's calendar as a model. Ask students to name the day of the week for the first day of the month. Assist them in finding that box on their calendar page. Tell students to write a 1 in that box. Tell students to continue to write the numbers in order as shown on the classroom calendar.

• Now, have students count the number of Sundays and write this number in the box beside the word. Continue to have them count the number of a particular day and write the number beside its respective day. Have students complete the page as independently as possible. Note that answers will vary depending on the month used.

4 Assess

Have students place the number cards in order on Reproducible Master RM19 to create the current month.

T134

pages 135–136

1 Getting Started

Objective
- To use groups of 10 to identify and write numbers through 31

Materials
*current monthly calendar;
*cards showing bundle or single-stick combinations;
31 ice-cream sticks; 3 rubber bands;
number cards 10 through 31

Warm Up • Mental Math
Ask students to tell which is less.

1. 1 dime or 19 pennies (1 dime)
2. 13 pennies or 2 nickels (2 nickels)
3. 9 pennies or 1 dime (9 pennies)
4. 2 nickels or 12 pennies (2 nickels)
5. 1 nickel or 1 dime (1 nickel)
6. 23 pennies or 2 dimes (2 dimes)

Warm Up • Calendar Activity
Name numbers on this month's calendar and have students tell the day of the week for each.

2 Teach

Develop Skills and Concepts Have students count out 15 ice-cream sticks. Tell them to now count 10 of the sticks and put a rubber band around the sticks. Write **15** on the board and tell students that 10 and 5 more is 15. Tell students that 15 is 1 ten and 5 ones as you point to the 1 and 5, respectively.

- Have students lay out 1 more stick. Write **16** on the board and tell students that 1 ten and 6 ones is 16. Continue to develop the numbers through 19. Then, have students lay out 1 more stick to make 20 sticks and tell how many sticks are not bundled. (10) Tell them that 10 ones can be bundled for another ten. Bundle the ten sticks and then ask how many tens there are now. (2) Ask how many ones or sticks are left over. (none or 0) Tell students that 20 is 2 tens and 0 ones as you write **20** on the board. Continue to develop the numbers through 31, writing each number on the board and stressing the number of tens and the number of ones.

- Have students lay out two bundles of 10 and tell how many sticks there are in all. (20) Have a student write the number on the board as you stress that the 2 means 2 tens and the zero means no ones left over. Continue having students show and write more numbers through 31. Have students read each number as tens and ones.

- Distribute cards showing various bundle or single-stick combinations and have students hold the cards to act out the number for classmates to name and write on the board.

- Name any number, 10 to 31, as its number of tens and ones and have students show the appropriate number card.

3 Practice

Using page 135 Tell students to find the first example in the top left corner of the page. Tell them to count the sticks, trace the circle around them, and trace the number. (10) Tell them that the second example in the top right corner shows the same number of sticks, but they have been bundled. Have them trace the second 10.

- For Exercises 1 to 10, tell students to start at 10 for each bundle and then count on to count the loose sticks. Ask what number they will start with if they have one bundle (10), two bundles (20), and three bundles (30).

Write the number that tells how many there are.

136 Lesson 7-7 • Place Value

Write the number that tells how many there are.

Common Errors • Intervention

Some students may need practice relating groups of 10 to numbers. Have them work in pairs with 31 ice-cream sticks and 3 rubber bands. Give each pair shuffled number cards from 10 to 31. Have pairs arrange the cards in order and make a model with the sticks for each number.

Enrichment • Number Sense

1. Have students work in pairs with number cards 10 through 31. Have students take turns showing a number card for the partner to tell the number of tens and ones.

2. Have students work in groups to cut numbers through 31 from advertisements. Have students place all numbers with 1 ten in order on a sheet of paper, numbers with 2 tens on another sheet, and numbers with 3 tens on a third sheet.

3. Give students 3 dimes and 10 pennies to show the numbers 10 through 31. Have them tell the number of dimes, or tens, and pennies, or ones, in the amount.

4. Have students draw coins, write their amounts, and explain why a dime is greater than a penny, a nickel is less than a dime, and so on.

Work through Exercises 1 to 3 with students, noting that bundles of 10 replace the need to circle 10. Have students complete the page as independently as possible.

Using page 136 Tell students that the exercises on this page are like those they have just completed on page 135. Remind them to begin at 10 if there is one bundle and count on to find the number of sticks in all. Ask students how many sticks there are in all in Exercise 1. (14) Have them trace the number. Then, ask students what number they will begin to count from when two bundles are shown. (20) Repeat for three bundles. (30) Have students complete the page as independently as possible.

4 Assess

Have each student choose a number between 20 and 31. Then, give each student 31 ice-cream sticks and 3 rubber bands to show the number. Remind students to bundle groups of 10 sticks as needed.

1 Getting Started

Objective
- To put 0 through 30 in order

Materials
*number cards 0 through 30;
*calendar; 31 counters;
object cards 0 through 31;
Reproducible Master RM19

Warm Up • Mental Math
Have students name the day
1. after Tuesday (Wednesday)
2. before Friday (Thursday)
3. last in the week (Saturday)
4. before Sunday (Saturday)
5. first in the week (Sunday)
6. before the last day of the week
   (Friday)

Warm Up • Number Sense
Have students group their counters
into tens and ones to tell each
number you write on the board.
Have a student then lead the class by
writing a number on the board to be
shown in groups of tens and ones.

2 Teach

Develop Skills and Concepts
Distribute the number cards 1
through 30 randomly to students. Place the zero card on
the far left of a table as you tell students that the number
you have begins the number card display. Tell students
that you would like the student who has number 1 to
place the card beside the zero card. Now, tell them to
decide when their number is needed so that the numbers
are placed in order across the table.

- Distribute number cards so that each student has one
  card. This may mean you are using only the cards for 0
  through 22 or 10 through 31. Have the student with the
  smallest number begin a number train by finding the
  student with the next number. They hold hands and go
  on to find the next number to be attached to the train.
  Continue until all students are part of the train. Have
  students then call out their numbers in order to check
  the sequence. Repeat the activity for the object cards.

- Give students copies of Reproducible Master RM19. Tell
  each student to find his or her birthday month on the
  classroom calendar and the day of the week for the first
  day of that month. Tell students to begin at 1 on that day

and write the numbers in order to show their birthday
month. Have students color their special day on their
calendar and circle the day before and the day after it.

It's Algebra! The concepts of this lesson prepare
students for algebra.

3 Practice

Using page 137 Explain to students that a number line is
another way of showing numbers in order. Before they
begin this page, show them a complete number line from 1
to 30 in a continuous horizontal line. Tell students that
they will be working with the same kind of number line
except their line is broken at the edge of the page. Discuss
with students what numbers are missing from the first
row. (9, 10, 12) Have them write these numbers. When
they understand what is being asked of them, allow them
to independently complete the number lines to 30.

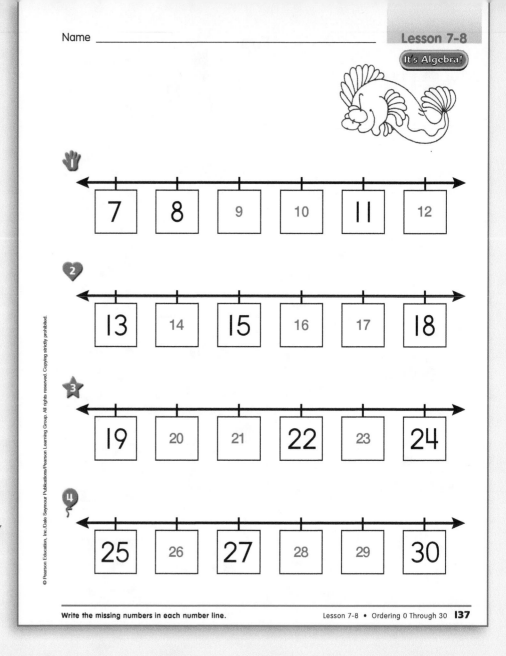

Write the missing numbers in each number line.   Lesson 7-8 • Ordering 0 Through 30 **137**

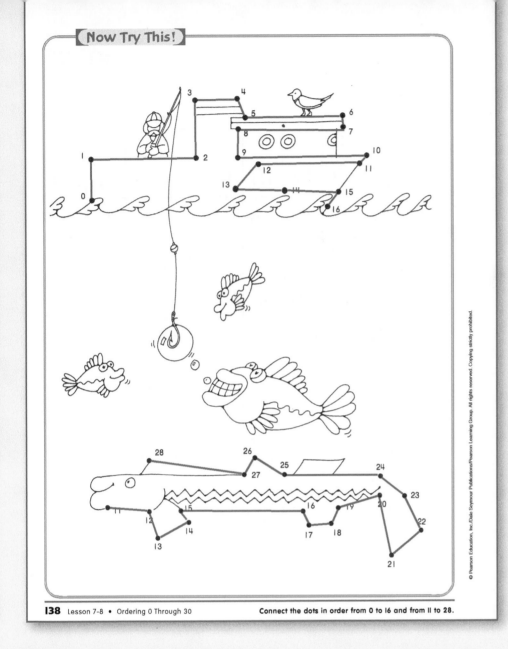

138 Lesson 7-8 • Ordering 0 Through 30   Connect the dots in order from 0 to 16 and from 11 to 28.

For Mixed Abilities

Common Errors • Intervention

Some students may need more practice counting from 0 to 30. Draw a zigzag design on the board with 30 dots. Have a volunteer number the dots while the class counts in unison. Have other class members take turns numbering other zigzag designs.

Enrichment • Number Sense

1. Give students a duplicated alphabet in order and have them number the alphabet letters in order from 1 to 26.

2. Have each student write the number that tells the date of his or her birthday. Tell students to then write all the numbers in order from 1 to their date to see how many days they must wait from the first day of that month. Students whose birthdays are early in the month could write the numbers through the end of the month.

Using page 138 Tell students that the fishing scene has two parts that need to be completed. Tell students that they are to join the numbers in order from 0 to 16 in the top part of the picture and from 11 to 28 in the bottom part of the picture. Have them complete the page independently and then color the picture in their choice of colors.

4 Assess

Write **26, _____, 28, _____, 30** on the board. Have students fill in the missing numbers so that the numbers 26 through 30 are in order. (27; 29)

T138

7-9 Problem Solving: Look for a Pattern

pages 139–140

1 Getting Started

Objectives

- To count from 1 to 100 using a hundred board
- To find patterns on a hundred board

Vocabulary

one hundred, hundred board

Materials

Reproducible Master RM19; copy of a hundred board; 1 counter; 10 cubes

Warm Up • Mental Math

Name the number that comes before

1. 20 (19)
2. 31 (30)
3. 19 (18)
4. 25 (24)
5. 7 (6)
6. 30 (29)
7. 15 (14)

Warm Up • Application

Have students count aloud the numbers 1 through 31. Then, give students a copy of Reproducible Master RM19 and have them write the numbers starting with 1 in order to show the dates for this month.

Name _____

1	2	3	4	5	6	7	8	9	10
11	12	13	14	15	16	17	18	19	20
21	22	23	24	25	26	27	28	29	30
31	32	33	34	35	36	37	38	39	40
41	42	43	44	45	46	47	48	49	50
51	52	53	54	55	56	57	58	59	60
61	62	63	64	65	66	67	68	69	70
71	72	73	74	75	76	77	78	79	80
81	82	83	84	85	86	87	88	89	90
91	92	93	94	95	96	97	98	99	100

Count from 1 to 100 out loud. Point to each number you say.   Lesson 7-9 • Problem Solving: Look for a Pattern **139**

---

2 Teach

Develop Skills and Concepts Distribute a copy of the hundred board to each student. Ask, *Why do you think this is called a hundred board?* (It shows the numbers to 100 in order.) Have students point to rows and columns to be sure students can differentiate between the two. Talk about how you read the numbers across the rows, from left to right. At the end of each row, go down one row and start at the far left. Discuss the number of rows and columns on the chart.

- Give each student 1 counter. Have students work in pairs. Each partner will take turns tossing the counter onto his or her chart and moving it slightly to cover only one number. Then, the other partner will have to name the number the counter is on. Students should also name the number that comes before, after, above, and below the number with the counter on it. Have partners continue to take turns.

- Have students color the hundred board to show a picture. Then, ask students to take turns describing their pictures to the class. Encourage them to name the numbers they colored and then see if there is a number pattern as well. (Possible answer: each diagonal was colored; this number pattern is +11 from left to right.)

It's Algebra! The concepts of this lesson prepare students for algebra.

3 Practice

Using page 139 Have students look at the hundred board. As a class, count aloud from 1 to 100 while students point to each number on the hundred board as they say it.

- Then, call out a random number and have students point to the number you name. Repeat multiple times.

Using page 140 Have students look at the hundred board. Ask if this hundred board is different in any way.

T139

1	2	3	4	5	6	7	8	9	⑩
11	12	13	14	15	16	17	18	19	⑳
21	22	23	24	25	26	27	28	29	㉚
31	32	33	34	35	36	37	38	39	㊵
41	42	43	44	45	46	47	48	49	㊿
51	52	53	54	55	56	57	58	59	60
61	62	63	64	65	66	67	68	69	70
71	72	73	74	75	76	77	78	79	80
81	82	83	84	85	86	87	88	89	90
91	92	93	94	95	96	97	98	99	100

Count by 2s. Color those numbers blue.
Count by 5s. Underline those numbers red.
Count by 10s. Circle those numbers.

140 Lesson 7-9 • Problem Solving: Look for a Pattern

Common Errors • Intervention

Some students may need to see the relationship between written numbers and a set with that many items. Have students count 100 items such as unit cubes. For each set of 10 they count, have them find the number on the hundred board. Students may also place the cubes in rows of 10 as they count. When they finish, they should have ten rows of 10 cubes in the same shape as the hundred board on page 139.

Enrichment • Logic

Have students look at the hundred board on page 139. Have them cover the entire board using two sheets of blank paper except for the column from 9 to 99. Ask, *What patterns do you see?* (Each number has a 9 as the ones digit and the tens digit increases by 1 from 1 to 9.) Then, have students cover the entire chart except for the row from 41 to 50. Ask, *What patterns do you see?* (Each number has 4 in the tens digit, except 50, and the ones digit increases from 1 to 9; the row ends with 50, the next 10.)

(no) Point out to students that all hundred boards will look the same. Tell students that they will count by 2s using the hundred board. Every number that is counted will be colored blue. Now, apply the four-step plan. For SEE, ask, *What are you asked to do?* (count by 2s using the hundred board; color these numbers blue) Then, *What information is given?* (hundred board; count by 2s) For PLAN, ask students how they will solve the problem. (color every other box blue, starting with 2) For DO, have students color every other box blue, starting with the number 2. (2, 4, . . . , 100) For CHECK, have students look at the colored hundred board. Ask, *What do you notice on the colored hundred board?* (Every other column is colored.) Ask students if the numbers in these columns are odd or even. (even) Remind students that when you start at 2 and count by 2s, you are counting all the even numbers.

• Now, work through coloring the 5s with students. You may wish to give students 5 unit cubes to help them count by 5s. Have students place one cube in each box, starting with the number 1. The fifth cube will land in the box with the number 5. Have students color this

box red. Then, have students place the cubes on the next set of numbers, starting in the box with the number 6. The last cube will land in the box for the number 10. Have students color this box red.

• Students may wish to use 10 cubes in order to count by 10s on the hundred board. Be sure students circle these numbers instead of coloring them.

• You may wish to ask students why the numbers that were selected when counting by 10s are circled and colored in red and blue. (The number 10 can be broken into groups of 2, 5, or 10.) Then, you may wish to use the 10 cubes to demonstrate the different groupings for 10: five groups of 2, two groups of 5, or one group of 10.

4 Assess

Have students open their books to page 139. Write **75** on the board. Have students write the numbers that come before, after, above, and below 75 on the hundred board. (74; 76; 65; 85)

T140

Chapter 7 Test and Challenge

pages 141–142

1 Getting Started

Test Objectives

- **Items 1–2:** To count and write numbers for 10 or less; to circle a group having fewer or more objects than another group (see pages 125–126)

- **Item 3:** To use ordinal numbers first through fifth (see pages 127–128)

- **Item 4:** To use ordinal numbers sixth through tenth (see pages 129–130)

- **Items 5–10:** To count 0 through 31 objects; to write 11 through 31; to use groups of 10 to identify and write numbers through 31 (see pages 131–132, 135–136)

Challenge Objective
- To create and read a bar graph

Materials
*1 paper bag; *2 paper circles and 1 paper triangle

2 Give the Test

Prepare for the Test Sing a verse of "We Have a Whole Lot of Numbers" from page T123a. Have a student write the number on the board and then read it in tens and ones. Continue for other numbers through 30.

- After students have mastered counting groups with objects through 31 and writing the numbers, have them complete the Chapter 7 Test. You may wish to review test-taking rules with students. Remind them that they are not to talk to each other and they are to finish the test on their own. Decide ahead of time whether or not to allow the use of manipulatives such as counters.

Using page 141 Students should be able to complete this page independently. For Exercises 1 and 2, students will count the bubbles in the group, write the number, and then circle the number that is less and box the greater number. For Exercises 3 and 4, students are to circle the fish and seagull in the position given by the ordinal number. For Exercises 5 through 10, students will count the bubbles or bundles and sticks and then write the number that tells how many there are in all.

T141

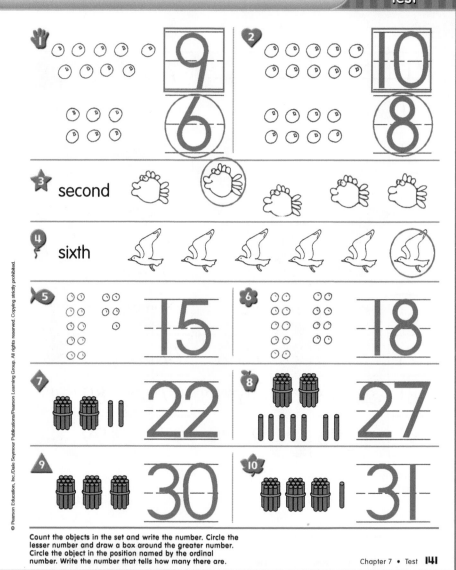

Name _____

Count the objects in the set and write the number. Circle the lesser number and draw a box around the greater number. Circle the object in the position named by the ordinal number. Write the number that tells how many there are.

Chapter 7 • Test **141**

3 Teach the Challenge

Develop Skills and Concepts for the Challenge Place 2 circles and 1 triangle into a paper bag. Tell students that you are going to draw a bar graph to give them information about what is in the paper bag. Draw 1 circle and 1 triangle toward the bottom of the board. Draw five boxes above each shape to make a bar graph. Ask students what shapes are inside the bag. (circles and triangles) Now, shade in 2 boxes above the circle and 1 above the triangle. Then, ask students how many of each shape they think is inside the bag. (2 circles and 1 triangle) Have students explain how they got their answers. Then, show students the shapes that are in the bag.

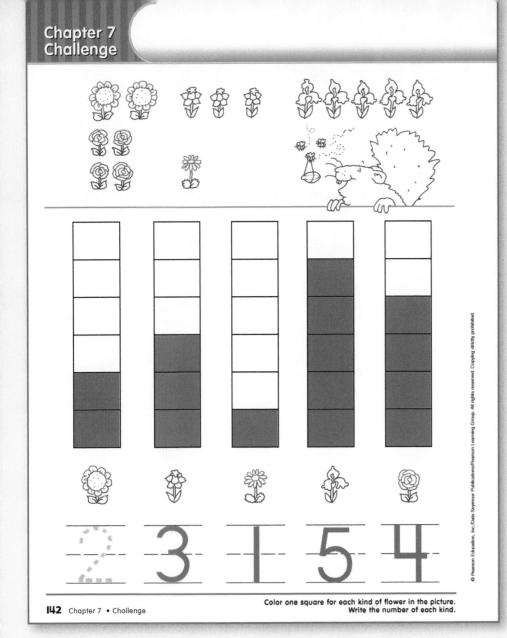

Color one square for each kind of flower in the picture.
Write the number of each kind.

142   Chapter 7 • Challenge

Common Errors • Intervention

Some students may need more practice deciding which number is greater or less. Have students work in pairs with number cards 15 through 20 and object cards 15 through 20. The partners should split each deck and receive 3 of the number and object cards. One player puts a number card down. The other player puts an object card down and has to tell if the number of items in the set is greater than or less than the other person's number. Students should not do equal groups. Both partners must agree for the turn to be over. Players then switch roles and continue to play until all the cards have been used.

Enrichment • Number Sense

1. Tell students that you bought an item costing 4¢ with a dime. Have them draw the coins you would have left. (6¢) Have them repeat the purchase with a nickel and draw the change. (1¢)

2. Have students draw dimes, nickels, and pennies to show different ways to make 21¢.

3. Have students find out which classmate's birthday is furthest from the first of the month.

4 Assess

Alternate Chapter Test You may wish to use the Alternate Chapter Test on page 211 of this manual for further review and assessment.

Challenge To assess students' understanding of bar graphs, lay out 4 circles and 4 triangles. On the board, draw a bar graph for students to fill in. Have students go to the board to color in different squares of the bar graph representing the 4 circles and the 4 triangles. Have other students in the class explain if they feel the completed graph is correct and why.

Using page 142 Discuss the different flowers in the picture at the top of the page. Have students find the sunflowers in the top left of the page and tell how many there are. (2) Tell them to find the sunflower at the bottom of the graph and color one square above it for each sunflower in the picture. Tell students that 2 squares are lightly shaded to help them get started. Now, tell them to write the number 2 on the line under the graph to tell that 2 sunflowers are shown on the graph.

• Continue for the next group of flowers in the picture. Then, have students color the number of squares on the graph and write the number to tell about them. Now, have students complete the page independently.

pages 143–160

Chapter Objectives

- To identify, count, and write the number of equal parts in a figure
- To recognize parts of a figure as not equal
- To define *whole, part, equal*, and *not equal*
- To identify one-half, one-fourth, and one-third of a figure
- To define *one-half, one-fourth*, and *one-third*
- To draw a picture to solve problems
- To split groups with no more than ten objects into two or three equal groups
- To identify a container that holds more or less than another
- To measure lengths using nonstandard units
- To define *length*
- To identify an object that is heavier or lighter than another
- To define *heavier* and *lighter*
- To compare scenes for hotter or colder

Materials

You will need the following:
*rectangular crackers perforated to break into 4 parts; *apples; *tape; *colored chalk; *ruler; *rectangles and squares sectioned into halves, fourths, and thirds; *1-cup measuring cup; *various-sized containers; *5 pairs of objects to weigh; *drinking cups in 2 sizes; *newsprint or butcher paper

Students will need the following:
Reproducible Masters RM20, RM21, RM22, RM23, and RM24; crayons; scissors; construction paper; paper triangles, circles, squares, and rectangles; rectangles, squares, and circles lined for thirds; 10 cubes; 10 counters; 2 sheets of plain paper; paper pre-lined into equal thirds; paper pre-lined into 3 unequal parts; containers in various sizes; water, sand, rice, or beans; objects to be measured; 2-inch plastic straw; paper clips or interlocking cubes; two paper strips in different lengths of 2 inches or more; one index card labeled "heavier" and another labeled "lighter"

Bulletin Board Suggestion

Enlarge a paper clip on a copying machine. Make a border by placing equal length copies of the paper clips end to end. Have students count the paper clips from corner to corner to find the length and width of the board. Be sure to have no more than 31 paper clips in all. Display halves, fourths, and thirds of a rectangle, square, circle, and triangle for students to assemble each figure and tell about each part.

Verses

Poems and songs may be used any time after the pages noted beside each.

Arithmetic Pie (pages 149–150)

Arithmetic pie is tasty and sweet.
As soon as it's hard, we'll each have a treat.
I'll take the whole pie and cut it into
Two equal parts, one for me, one for you.

Each part is one-half; what a big piece of pie!
Maybe one-half for you but not me! I might die!
My tummy would hurt! One-half is absurd!
Perhaps we'd be smarter to each eat one-third.

With three equal parts there'd be one for Mother.
We'd each eat one-third and she'd eat the other.
But what about Dad? He drools over pies!
I'll cut it in fourths; each piece the same size.

Here's one-fourth for me and one-fourth for you,
And equal size pieces for Mom and Dad, too.
Hey! Where are you going? Come back here and eat!
I'm serving my mud pie arithmetic treat.

Arithmetic pie, arithmetic pie,
You're made out of mud so no one will try
A half, third, or fourth, not even a lick;
What a waste of all of this arithmetic.

How Do I Measure? (pages 155–156)

How far is it from me to you?
  To find out I will have to do
Some measuring. But first decide
  What unit to use as a measuring guide.
Shall I use my feet? If so I'll find
  The distance by putting one foot behind
The other, and count the number of steps
  From me to you. I count footsteps.
Or I could count the number of times
  Any object is used such as one of these dimes,
A paper clip, book, a pencil or pen—
  I place the object from end to end
And count the times that it will fit;
  Then I've measured the distance in that unit.
But people's feet vary and pencils get used
  So the distance depends on whose feet I choose.
And that's why a paper clip can be used
  It's the same no matter whose paper clip is whose.

Big Mouth Bass

Deep in the ocean, there lived a very nice family of fish named Mr. and Mrs. C. Bass and their son, Big Mouth. It is true that Big Mouth's mouth was very big. He was especially embarrassed about it when he went to school.

Mrs. Longtooth Shark was Big Mouth's teacher. She was much nicer than her name sounded. She loved all her students very much. On Sunday, the first day of the week, she thought about them a lot.

On Monday, the second day of the week, and Tuesday, the third day of the week, Mrs. Longtooth Shark tried to teach 20 little fishes how to see a fish hook. She didn't want them to get caught.

On Wednesday, the fourth day of the week, and Thursday, the fifth day of the week, Mrs. Longtooth Shark took 20 little fishes into a dark cave. There, she tried to teach them how to swim in the dark without bumping their heads.

And on Friday, the sixth day of the week, Mrs. Longtooth Shark had a much larger class. Eleven more students from another class joined Mrs. Longtooth Shark's class. There were 31 students altogether. And everybody loved the lesson because Mrs. Longtooth Shark taught them how to blow bubbles!

Big Mouth felt so awful about his big mouth. At first he almost never said anything in class. His mouth was so big, no matter how many times he tried not to catch a hook in it, he did! Mrs. Longtooth Shark would help him get unhooked and tell him to be more careful.

When they swam in the dark cave, Big Mouth's jaws kept bumping the walls. Sometimes his head got stuck!

But on Friday, Big Mouth had lots of fun. Even though there were 31 students in the class, he always blew the biggest and greatest bubbles!

And all day Saturday, the seventh day of the week, Big Mouth swam around smiling the biggest smile in the ocean!

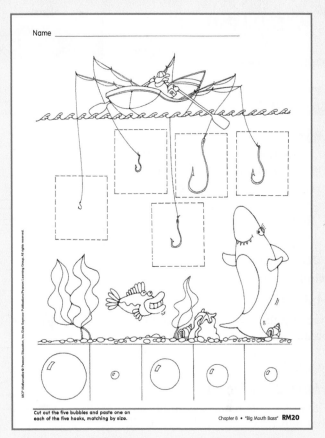

"Big Mouth Bass," p. RM20

Activities

Use Reproducible Master RM20, "Big Mouth Bass," to help students relate the story to the mathematical ideas presented.

- Give students construction paper in five different colors and a fish pattern. Have them trace and cut out 31 colored fishes in all. On each fish, have students write a different number from 1 to 31. Have students put a paper clip on each fish and place the fish in a large clear jar labeled "Fish." Make a fishing pole with a small magnet on the end of the line. Now, lay out a row of five containers, each labeled with one of the five color words. Have students fish, name the caught fish's number, and place it in the proper container by color. Have students tell the color in the first container and so on. Change the container arrangement and repeat. Now, have students name the number on the fish in each container and then place the fish in order along a table.

- Reread the story as one student points on the class calendar to each day of the week and another student holds up the word card for that day.

1 Getting Started

Objectives

- To identify, count, and write the number of equal parts in a figure
- To identify parts of a figure as not equal
- To define *whole, part, equal,* and *not equal*

Vocabulary

whole, part, equal, not equal

Materials

*rectangular crackers perforated for breaking into 4 parts; paper cutouts of triangles, circles, squares, and rectangles

Warm Up • Mental Math

Have students name the number that tells how many there are in all.

1. days in a week (7)
2. tens in 31 (3)
3. letters in this month's name (Answers will vary.)
4. students in the room
5. Saturdays in this month

Warm Up • Number Sense

Write a number from 10 to 31 on the board. Have a student write the next number and tell the number of tens and ones.

Name _____

Fractions and Measurement

Chapter **8**

Lesson 8-1

Write the number of equal parts.

Lesson 8-1 • Equal Parts **143**

---

2 Teach

Develop Skills and Concepts Show students a whole, rectangular cracker and ask its shape. Tell them that you are holding a whole cracker. Tell students that you want to share the cracker with 1 friend so that each of you has the same size part. Break the cracker into 2 equal parts and ask how many parts there are and if the parts are the same size. (2; yes) Tell children that we say the parts are equal because they are the same size. Have a student lay one part of the cracker on top of the other to see if they are equal, the same size. (yes) Ask if the whole cracker is in 2 equal parts. (yes)

- Show another whole, rectangular cracker and tell students that you want to share this whole cracker with 3 friends. Break the cracker into 4 equal parts and ask how many parts there are. (4) Have students lay the four parts on top of each other to check for equal parts.

- Match the edges of a circle to fold it into 2 equal parts. Open the circle and draw a line along the fold. Ask how many parts there are and if the parts are the same size. (2; yes) Cut along the line and ask if the 2 parts are equal. (yes) Lay one part on the other to show that the edges match. Now, show folding a circle into 2 parts that are not equal and ask students if the parts are equal. (no) Discuss how to tell that the parts are not equal. (Possible answer: One part is bigger when laid on top of the other.)

- Fold a circle into 4 equal parts and have students tell the number of parts and tell if the parts are equal. (4; yes) Have students fold circles into 4 equal parts. Then, show folding a circle into 4 unequal parts and have students do the same.

- Have students fold rectangles into 2 or 4 parts and tell how many parts there are and if the parts are equal or not equal.

Circle the figures with equal parts. Mark an X on the figures that do not have equal parts.

For Mixed Abilities

Common Errors • Intervention

Some students may have difficulty recognizing equal parts. Have them work with partners and various folded shapes, some showing equal parts and some not. Have them sort the shapes into two groups, one with equal parts and one with unequal parts.

Enrichment • Number Sense

1. Have students trace 2 circles, cut them out, and then fold the circles to show how they could share these circles or cookies equally with a friend. Students should then show how to share these cookies unequally with a bigger part for the friend.

2. Have students cut a premarked circle into 3 equal parts and fit them together to form a whole.

3. Have students draw coins to show how to share 20¢ equally with 1 friend. (Possible answer: 1 dime and 1 dime)

3 Practice

Using page 143 Ask into how many parts the picture at the top of the page is divided. (3) Ask, *Are these parts the same size?* (yes) Tell students that the parts are equal because they are the same size and shape. Neither part is bigger nor smaller than the other. Students should then trace the 3 to indicate the 3 equal pieces in the picture.

• Before students write the number of equal parts in Exercises 1 to 8, have them discuss into how many parts each picture is divided. Check to see that they understand that the parts in any given picture are equal to each other. Then, have students complete the page independently.

Using page 144 Have students look at the top of the page. Tell them that the butterfly is trying to decide if the two circles are divided into equal parts. Ask them to point to the first circle and tell how many parts there are. (2) Ask if the parts are equal. (no) Tell students to trace the dashed X on the circle to show that the parts of the

circle are not equal. Ask how many parts are in the second circle at the top of the page. (2) Ask if the 2 parts are equal. (yes) Tell students to trace the circle around the figure to show that the parts of the figure are equal.

• Work through Exercise 1 with the class to be sure students understand that they are to mark an X on a figure if its parts are not equal and circle a figure if its parts are equal. Have them complete the page independently.

4 Assess

Draw and cut out 1 square that has been divided into 4 equal parts and another square that has been divided into 4 unequal parts. Now, draw these squares on the board. Ask students into how many parts is each square divided. (4) Then ask, *Which square is divided into 4 equal parts?* (the first square) Have a volunteer fold the actual squares to show the divisions.

T144

pages 145–146

1 Getting Started

Objectives
- To identify one-half of a figure
- To define *one-half*

Vocabulary
one-half

Materials
*paper cutouts of triangles, circles, squares, and rectangles; *apples; *ruler; *colored chalk; Reproducible Master RM21; crayons; scissors

Warm Up • Mental Math
Have students name something in the room that fits the description.
1. used to tell time (clock, calendar)
2. used to tell what day it is (calendar)
3. above them (Answers will vary.)
4. to their right
5. the shape of a circle

Warm Up • Number Sense
Fold paper shapes into 2 or 4 equal or unequal parts and have students identify the shapes of the whole figures and tell if their parts are equal or not equal.

Circle any figure that is cut into equal halves. Mark an X on any figure that does not show two equal halves.

2 Teach

Develop Skills and Concepts Cut an apple into 2 equal parts. Ask if the parts are equal. (yes) Ask a student to pick up one-half of the apple. Have another student pick up one-half of the apple. Ask each student to tell what part of the apple he or she is holding. (one-half) Repeat for more students to identify one-half of the apple. Tell students that fractions are used to tell about equal parts of a whole. Ask them to tell how they know that each part of the apple is one-half. (It is 1 of 2 equal parts that make the whole apple.)

- Show students a large circle and ask the name of the shape. (circle) Fold the circle in half and draw a line along the fold. Ask how many parts there are. (2) Ask a student to show how to determine whether the parts are equal. (fold over or cut and place the parts on top of each other) Have a student write the number on the board that tells how many equal parts there are. (2) Ask students to name each part of the circle. (one-half)

- Show a square and ask a student to fold it into 2 equal parts. (2 rectangles or 2 triangles) Ask students to name the shape of each part, write the number on the board that tells how many equal parts there are (2), and then identify each part that is one-half.

- Draw around paper shapes to draw figures on the board. Use a ruler to divide some of the figures into halves and some into unequal parts. Have students mark an X on figures whose parts are not equal. Then, have students use colored chalk to color one-half of each shape whose parts are equal.

- Give students copies of Reproducible Master RM21. Have students color each half of the square blue, the rectangle red, the circle green, and the triangle yellow. Have students then cut out the shapes and match the colors to make the whole and show that each is one-half of the new figure.

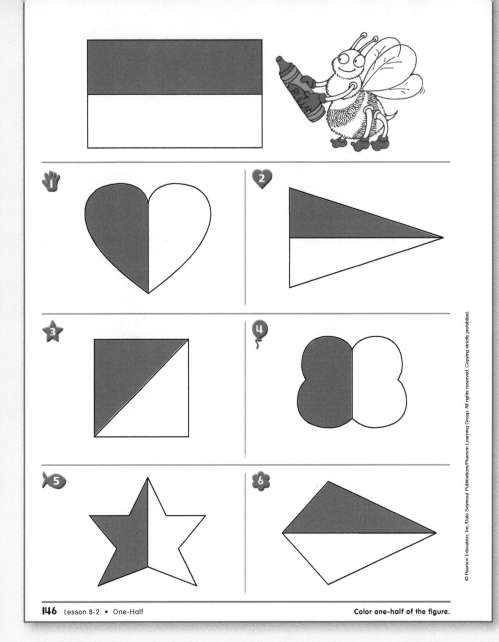

146 Lesson 8-2 • One-Half

Color one-half of the figure.

For Mixed Abilities

Common Errors • Intervention

Some students may have difficulty recognizing halves on page 146 because they think that the parts do not look alike and thus cannot be equal parts of the whole. Have students work with partners and a variety of shapes containing a line of symmetry, a line that separates the figure into mirror images. Have them fold the shapes to form 2 equal parts and describe each as half.

Enrichment • Number Sense

1. Have students draw a picture that explains the term *half-moon*.

2. Help students draw around their bodies on large paper on the floor. Have students cut out the outline and fold it in half lengthwise to show halves. Discuss the halves as each having 1 arm, 1 leg, one-half of the head, and so on.

3. Have students color clothing, faces, and so on, on the shapes they cut out in the previous activity.

3 Practice

Using page 145 Have students look at the top of the page. Tell students that the caterpillar is trying to circle the figure that is divided into 2 equal parts and to mark an X on the figure that is divided into 2 unequal parts. The caterpillar puts a circle around the picture on the left and an X through the one on the right. Be sure that students know what the circle and X represent as they trace them.

• For Exercises 1 to 8, before students draw circles and mark Xs, have them describe the pictures that are cut into halves and those that are cut into 2 unequal parts. Encourage them to verbalize the reasons for their decisions.

Using page 146 Have students look at the top of the page. Tell students that the bee is telling them to color one-half of each figure. Discuss with them why one-half of the rectangle at the top of the page is colored blue and one-half is not colored. Also, be sure that students notice

that every figure in Exercises 1 to 6 is divided into 2 equal parts. Now, have them complete the page independently by coloring one-half of each figure. Allow students to choose the colors they will use.

4 Assess

Give each student a rectangle and ask him or her to fold it into 2 equal parts. Then, ask students to color one-half of the rectangle.

T146

pages 147–148

Name _____

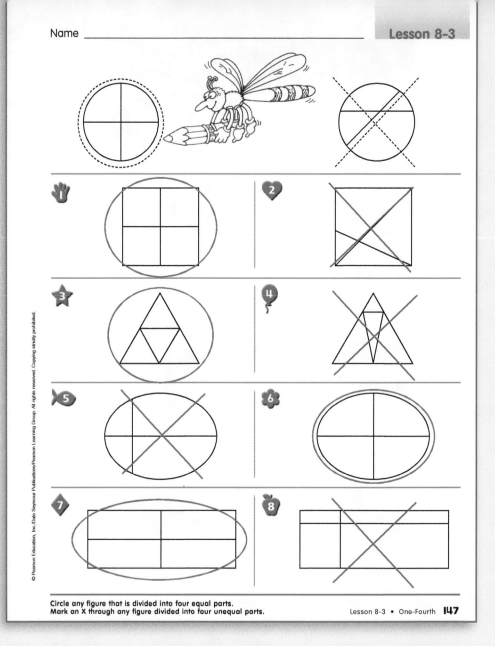

Circle any figure that is divided into four equal parts.
Mark an X through any figure divided into four unequal parts.

Lesson 8-3 • One-Fourth **147**

1 Getting Started

Objectives
- To identify one-fourth of a figure
- To define *one-fourth*

Vocabulary
one-fourth

Materials
*paper cutouts of triangles, circles, rectangles, and squares; *ruler; *colored chalk; crayons; scissors; Reproducible Master RM22

Warm Up • Mental Math
Ask students to tell the number.

1. before 31 (30)
2. second in their phone number (Answers will vary.)
3. fourth in their phone number
4. between 19 and 21 (20)
5. after 29 (30)
6. third when counting from zero (2)

Warm Up • Number Sense
Fold paper figures into 2 equal or 2 unequal parts. Have students tell the shape of the whole figure and tell if the figure is divided equally into parts that are each one-half.

2 Teach

Develop Skills and Concepts Have students fold a square to show 2 equal parts. Ask for the fraction that tells about each equal part. (one-half)

- Show students how to fold this square again to form 4 equal parts. Have them open the figure to see the parts and tell if they are equal. (yes) Ask how many equal parts there are. (4) Tell students that each of the 4 equal parts is called one-fourth. Now, fold a square into 4 parts that are not equal. Open the figure and point to one part as you tell students that the part is not one-fourth because the 4 parts are not equal.

- Fold more squares, some into fourths, 4 equal parts, and some into 4 unequal parts, and have students tell which figures are in fourths and which are not. When they identify the figures that are divided into fourths, point to the 4 parts of each figure for students to identify that each part is one-fourth.

- Have students fold rectangles, squares, and circles into fourths. Have students draw a line on the fold and cut

the parts to check for fourths. Triangles are more difficult to cut into fourths, but some children may enjoy experimenting with them.

- Draw around paper shapes on the board and use a ruler to divide some into fourths and some into 4 unequal parts. Have students mark an X on those figures with unequal parts. Then, have students use colored chalk to color one-fourth of each figure that has 4 equal parts.

- Give students copies of Reproducible Master RM22. Have students cut out the shapes and match them to make a whole circle or a whole square. Ask students how many parts there are for each figure and if they are equal. (4; yes) Have students then color one-fourth of the square orange and one-fourth of the circle brown.

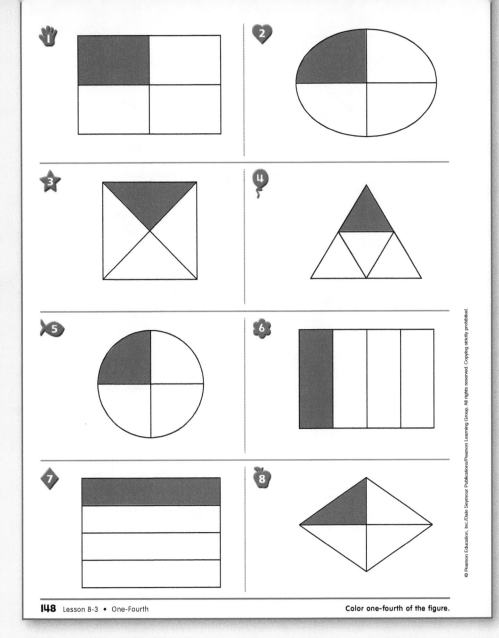

148 Lesson 8-3 • One-Fourth

Color one-fourth of the figure.

For Mixed Abilities

Common Errors • Intervention

Some students may have difficulty understanding fourths and halves. Have them work with a partner to fold a sheet of paper into halves; that is, each side matches the other. Next, have them color one-half red. Ask, *How many halves make one whole?* (2) Then ask, *How many parts did you color?* (1) Have students fold another sheet of paper in half and then fold it in half again to make fourths. Have them color one-fourth blue. Ask, *How many fourths make one whole?* (4) Then ask, *How many did you color?* (1)

Enrichment • Number Sense

1. Have students fold various-sized sheets of paper into 4 equal parts.

2. Tell students to fold a sheet of paper into fourths and write a different number from 0 to 11 on each fourth. Use the number grid to play a form of Bingo, with students laying a counter on each number they have as you draw numbers from shuffled number cards.

3 Practice

Using page 147 Tell students that they are to decide if each picture on this page is divided into 4 equal parts, fourths, or 4 unequal parts. Tell them that they should circle the pictures that show 4 equal parts and mark an X through the pictures that have 4 unequal parts. Now, have students look at the example at the top of the page. Ask if the insect should draw a circle around the picture on the left. (yes) Tell students to help the insect draw the circle. Tell them to help the insect mark the other picture with an X. Be sure students verbalize why they are using each mark. Have students complete the page independently.

Using page 148 Have students point to the rectangle in Exercise 1 and tell how many parts are in it. (4) Ask if the parts are equal. (yes) Tell students to color one-fourth of the rectangle. It may be helpful to have students color each of the 4 parts a different color. Work through another exercise, if necessary, before having students complete the page independently. Be sure students understand that each figure in Exercises 1 to 8 has 4 equal parts.

4 Assess

Give each student a square and ask him or her to fold it into 4 equal parts. Then, ask students to color one-fourth of the square.

T148

pages 149–150

1 Getting Started

Objectives
- To identify one-third of a figure
- To define *one-third*

Vocabulary
one-third

Materials
*paper cutouts of triangles, circles, squares, and rectangles; rectangle and square lined for thirds; crayons; Reproducible Master RM23

Warm Up • Mental Math
Have students name the fraction.
1. 1 of 2 equal parts (one-half)
2. 1 of 4 equal parts (one-fourth)

Have students name the number.
3. the o'clock after 12:00 (1:00)
4. 1 penny more than 2 dimes (21¢)

Warm Up • Number Sense
Have students fold a square, circle, rectangle, and triangle into halves and name each part. (one-half) Have students fold all but the triangle again to show fourths.

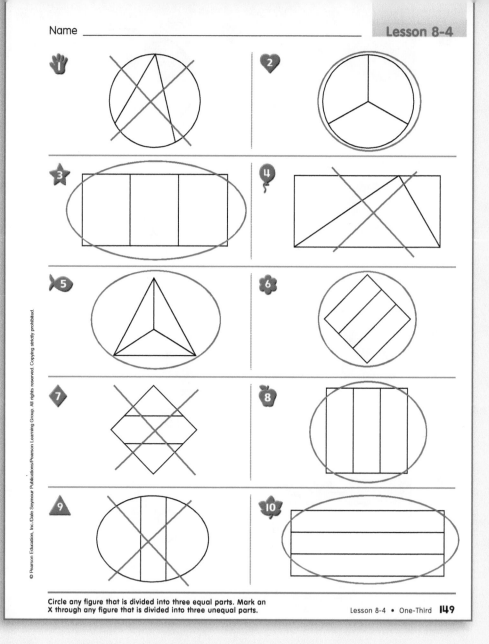

Circle any figure that is divided into three equal parts. Mark an X through any figure that is divided into three unequal parts.

Lesson 8-4 • One-Third **149**

2 Teach

Develop Skills and Concepts Fold a rectangle into 3 equal parts and ask students to tell the number of parts. (3) Ask if they are equal. (yes) Tell students that 1 of 3 equal parts is called one-third. Give each student a pre-lined rectangle and square to fold into thirds. Have them color one-third of each figure orange. Then, have students color one-third of each figure yellow and the remaining one-third black.

- Fold a figure to show 3 unequal parts. Have students tell if the parts are equal. (no) Ask if each part is one-third. (no) Help students tell why. (Parts must be equal for one of them to be one-third.) Now, show them a figure pre-lined for thirds and repeat the questions. Help students tell why each part is one-third. (The 3 parts of the whole are equal.)

- Display various shapes that are divided into 2, 3, or 4 equal parts and various shapes that are divided into 2, 3, or 4 unequal parts. Have students identify the figures with the 2, 3, or 4 unequal parts, and so on.

- Give students copies of Reproducible Master RM23. Have them color each third a different color and then cut out the parts. Ask them to hold up one-third of the circle or one-third of the rectangle.

3 Practice

Using page 149 Tell students that on this page they are to circle a figure if it is divided into 3 equal parts and mark an X on any figure that is not divided into 3 equal parts. Remind them that when a figure is divided into 3 equal parts, each part is called one-third. Work through Exercise 1 with students before assigning the page to be completed independently.

Using page 150 Ask students to tell how many equal parts are in each figure on this page. (3) Ask what each part of a figure is called when there are 3 equal parts. (one-third) Tell students that they are to color one-third of each figure on this page. Supervise as they complete the page independently. You may then want to have them use sentences to tell about the colors they used.

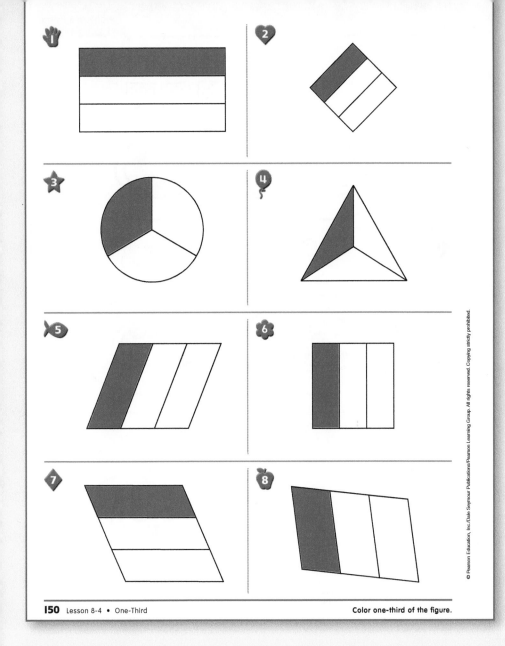

Color one-third of the figure.

For Mixed Abilities

Common Errors • Intervention

Some students may need more practice with thirds. Have them work with partners. Give each student a worksheet showing squares, rectangles, and circles. Tell students that these figures represent brownies, candy bars, and pizzas, and they are to draw lines to show how they would cut each into equal shares for 3 people. After drawing the lines, have them color one of the pieces, that is, one-third. Accept reasonable approximations of thirds.

Enrichment • Number Sense

1. Help students fold $8\frac{1}{2} \times 11$-inch sheets of paper into thirds to fit into business-sized envelopes.

2. Tell students to draw coins to show how much money they would have if they had one-third of 3 nickels, 3 pennies, or 3 dimes. (1 nickel, 5¢; 1 penny, 1¢; 1 dime, 10¢) Ask students which they would rather have, one-third of the nickels, pennies, or dimes. (dimes)

3. Distribute to students a circle spinner divided evenly into thirds with one-third shaded. Have them spin the spinner ten times and record how many times the spinner lands on the shaded part. Tell students that since only 1 part is shaded and 2 parts are unshaded, they are more likely to land on the unshaded part because there is more of it.

4. Have students draw 3 objects and color one-third of them. (1 object)

4 Assess

Give each student a rectangle pre-lined for thirds. Ask him or her to fold it into 3 equal parts. Then, ask students to color one-third of the rectangle.

ESL/ELL STRATEGIES

Help students avoid confusion by pointing out that two of the words used in the fractions, *fourth* and *third*, are the same as the related ordinal numbers. However, the fraction word for two equal parts of a whole, *half*, is different from the ordinal number, *second*.

T150

8-5 Problem Solving: Act It Out

pages 151–152

1 Getting Started

Objectives

- To draw a picture to solve problems
- To split groups of no more than ten objects into two or three equal groups

Materials

*rectangles and squares sectioned into halves, fourths, and thirds; 10 cubes; 10 counters; 2 sheets of plain paper

Warm Up • Mental Math

Ask students if the answer is yes or no.

1. 2 dimes is 30¢. (no)
2. 1¢ and 2 nickels is 11¢. (yes)
3. 1 of 3 equal parts is one-third. (yes)
4. This month is May. (Answer will vary.)
5. 30 comes next after 28. (no)
6. 7 days are in 1 week. (yes)

Warm Up • Number Sense

Show students rectangles and squares folded into halves, fourths, and thirds. Ask students to identify the rectangle that shows fourths, the square that shows halves, and so on.

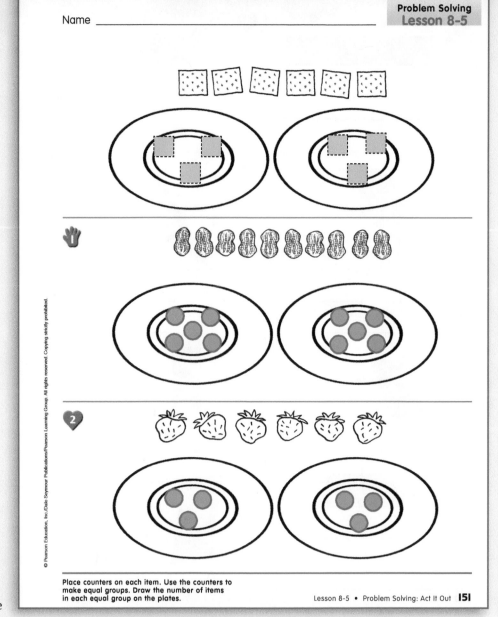

Place counters on each item. Use the counters to make equal groups. Draw the number of items in each equal group on the plates.

Lesson 8-5 • Problem Solving: Act It Out **151**

2 Teach

Develop Skills and Concepts Have students work in pairs. Give each pair of students 10 cubes. Have students make two different towers from the cubes. For example, one tower could have a height of 1 cube and the second could have a height of 9 cubes. Have students continue until both towers are the same height. (5 cubes in each tower) However, have students draw a picture of each of their tries and the final set of matching towers. Repeat this procedure with 8 cubes and 6 cubes. Then, repeat this task using an odd number of cubes and discuss why this will not work. (There will always be 1 cube too many.)

3 Practice

Using page 151 Give each student 10 counters to use to complete the exercises. Have students look at the example at the top of the page. Now, tell students that

they are to put the same number of counters from the given set on each plate. You may wish to give each student two sheets of paper to represent the plates.

- Now, apply the four-step plan. For SEE, ask, *What are you asked to do?* (make equal groups from the 6 objects) Then, *What information is given?* (6 crackers; 2 plates; equal groups) For PLAN, ask students how they will solve the problem. (Possible answer: I will start by putting 1 counter on each plate until I use up all 6 counters.) For DO, have students use their plan to make two equal groups from the 6 crackers. (3 in each group) For CHECK, have students count the first group of crackers and then count on the objects from the second group to be sure there are 6 crackers in all. (yes) Have students trace the 3 crackers in each group in the example.

- Work through Exercise 1 with the students if necessary. Then, have students complete Exercise 2 independently. Be sure the number of objects the students draw on each plate matches the number of counters they have arranged on their two sheets of paper.

T151

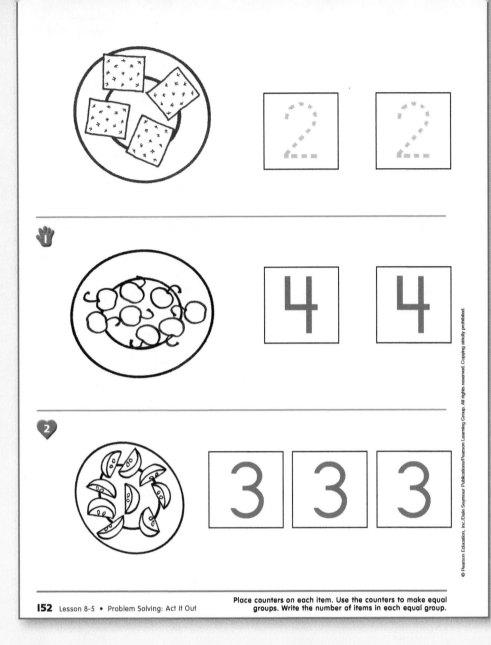

152 Lesson 8-5 • Problem Solving: Act It Out

Place counters on each item. Use the counters to make equal groups. Write the number of items in each equal group.

Common Errors • Intervention

Some students might have difficulty finding and using all the relevant information given in each problem. After students look at each exercise, have them write the number of objects they start with on top or next to the plate with the objects. Then, when students finish, have them count all the objects on each sheet of paper to be sure the number of objects totals the number they have written.

Enrichment • Logic

1. Have students experiment with the fractional parts from Reproducible Master RM24 to see how many fourths fit onto one-half. (2)

2. Have students draw 2 objects and color one-half (1), 3 objects and color one-third (1), and 4 objects and color one-fourth (1).

3. Tell students to draw coins to show how much money each friend would have if 4 nickels were split between 2 friends and if 4 nickels were split between 4 friends. (2; 1)

Using page 152 Have students look at the example at the top of the page. Have students use 4 counters and the two sheets of paper to make the two equal groups. Have students explain how they got their answers. (2 in each group; possible answer: I started with 4 counters and placed one on each plate until there were none left.) Then, have students trace the two dashed 2s to represent the 2 crackers in each of the matching groups. Have students use the counters and the two sheets of paper to complete the page independently.

4 Assess

Draw four squares on the board. Then, draw 3 Xs in each square. Tell students that you made four equal groups from one big group of Xs. Ask students to find the number of Xs you started with. (12; students can count the Xs in the first group, then count on with the Xs in the second group, and so on until all 12 Xs are counted.)

1 Getting Started

Objective
- To identify a container that holds more or less than another

Vocabulary
capacity

Materials
*1-cup measuring cup; containers in various sizes; sand, rice, or beans

Warm Up • Mental Math
Tell students to name the fraction.
1. 1 of 2 equal parts (one-half)
2. 1 of 3 equal parts (one-third)
3. 1 of 4 equal parts (one-fourth)

Warm Up • Number Sense
Have a student identify any object in the room and choose a friend to name an object that is bigger or smaller. Repeat for all students to participate.

2 Teach

Develop Skills and Concepts Show a large and a small container and ask students if the containers are the same size. (no) Ask which is larger and then which is smaller. Have students sort a number of containers into pairs so that each pair has a large and a small container.

- Show a large and a small container and have students identify the larger one and the smaller one. Ask students which container they think will hold more (larger) and which will hold less (smaller).

- Have students work in pairs with each pair having a large and a small container. Provide rice, sand, or beans, and tell students to fill the container they think is the smaller of the two containers with rice, sand, or beans. Then, have them pour that amount into the container they think is larger to see if their guesses were correct. Have students place their larger containers together in a group and their smaller containers in another group. Now, help them discuss how some of the larger containers are still larger than others in the group, and so on. Have students pour the fill to test their classifications. Repeat to compare sizes of the smaller containers. Help students see that size comparisons depend on what is being compared.

- Have students use a 1-cup measuring cup to tell the number of cups of rice, sand, or beans it takes to fill

various containers. They may then enjoy lining up several containers in order by size or finding equal-sized containers.

- Hold up two drinking cups of different sizes. Ask students which cup would be best if a person were very thirsty. (larger) Help students discuss how different situations present needs for different sizes of containers. Discuss why cereal and other foods come in different-sized containers, why there are different sizes of plates and glasses, and so on.

3 Practice

Using page 153 Have students look at the example at the top of the page. Tell students that the bee is trying to decide which container holds more. Ask them if the first or second container would hold more. (second) Tell them to trace the circle around the second container to show that it would hold more than the first container.

Circle the container that holds more.

Lesson 8-6 • Capacity **153**

Common Errors • Intervention

If some students are having difficulty deciding which container holds more, have them practice by pouring sand into two different-sized containers and tell which holds more or less and why.

Enrichment • Number Sense

1. Have students use copies of Reproducible Master RM24 to find out if one-half is larger than one-third, one-fourth is smaller than one-third, and so on.

2. Provide milk cartons in three sizes. Have students fill them in order to line up the cartons from the one that holds the least to the one that holds the most.

3. Have students bring in and tell about food containers for the same food in two different sizes.

154 Lesson 8-6 • Capacity

Circle the container that holds less.

Using page 154 Tell students that on this page they are to circle the container that holds less. Ask students which of the two glasses in the example would hold less. (first) Tell them to trace the circle around the first glass. Have students complete the page independently.

4 Assess

Place two containers, one larger than the other, on a table in the front of the room. Ask students, *If we fill the smaller container with rice and then pour this amount into the larger container, will it spill over?* (no) Have students explain their answers. (Possible answer: Because it is smaller, it holds less rice than the bigger one.)

pages 155–156

1 Getting Started

Objectives
- To measure lengths using nonstandard units
- To define *length*

Vocabulary
length

Materials
*various-sized containers; objects to be measured; 2-inch plastic straw; 2 paper strips in different lengths of 2 inches or more; paper clips

Warm Up • Mental Math
Ask students to name each shape.
1. a wheel (circle)
2. an exit sign (rectangle)
3. a dime (circle)
4. a board eraser (rectangle)
5. a floor tile (Possible answer: square)
6. a pennant (triangle)

Warm Up • Geometry
Show an assortment of containers. Pick one and ask a student to identify another container that would hold more. Repeat for all students to identify a container that holds more or less than the one shown.

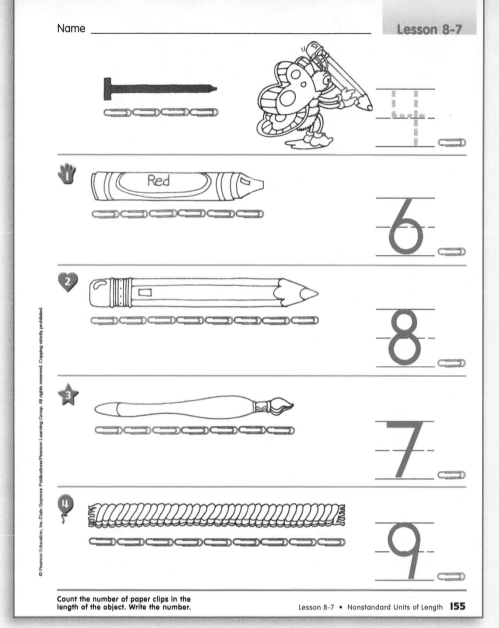

Name _____

Lesson 8-7

Count the number of paper clips in the length of the object. Write the number.

Lesson 8-7 • Nonstandard Units of Length **155**

2 Teach

Develop Skills and Concepts Tell students that you want to know how long one section of the board is as you point to the section and mark its two ends. Show an eraser and tell students that you will see how many erasers long this section of board is. Place the beginning of an eraser on the starting mark and make a mark on the board at the other end of the eraser. Then, place the beginning of the eraser at the second mark and make another mark on the board to show the length of 2 erasers. Continue to measure to the end of the section. Note that it will most likely not come out even. Ask students about how many erasers long the board is as you count them together. Tell them that the length of the board is about _____ erasers long. Tell students that all measurements are approximate, so they should use the word *about* in front of the measurements. Measure and mark off a piece of the board that is 2 erasers long and tell students that this piece of board is about 2 erasers long.

- Have students help measure the same section of board using a sheet of construction paper. Then, use pencils in two different lengths to measure the same board section. Have students orally tell the length of the board in each unit of measure.

- Give each student a 2-inch piece of straw and two strips of paper and tell students to measure their strips of paper to see about how many straws long they are. Tell students to write that number on one side of each strip of paper. Have students exchange strips with a friend to check each other's measurements. Repeat the activity measuring in paper clips.

- Have different students measure the width of the doorway by counting the number of foot lengths from one side to the other. Students should see that the number of units varies with the unit of measure used.

- Set up areas for pairs of students to experiment measuring objects using various units. Have students compare their measurements.

T155

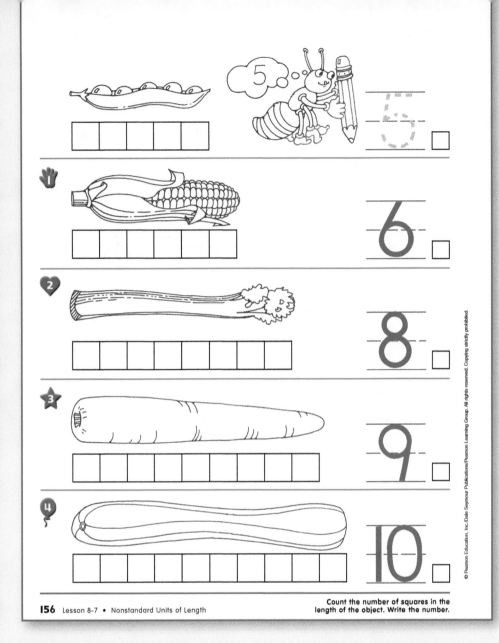

156 Lesson 8-7 • Nonstandard Units of Length

Count the number of squares in the length of the object. Write the number.

For Mixed Abilities

Common Errors • Intervention

Some students may need more help with the concept of length. Have students work in pairs to measure the length of a sheet of paper using their hand. Have them draw around their hand with fingers closed to measure the width or length of their sheets of paper.

Enrichment • Geometry

1. Have students find the length of the classroom by counting floor tiles or by using a sheet of construction paper as the unit of measure.

2. Have students use two different length pencils to measure the length of their desktop or table and compare the findings.

3. Tell students to find the length of a room at home using the length of their foot as the measuring unit.

---

3 Practice

Using page 155 Have students identify each of the objects on the page. (nail, crayon, pencil, paintbrush, rope) Ask what unit is being used to find the length of each of the objects. (paper clip) Ask about how many paper clips long the nail is. (4) Tell students to trace the 4 to tell that the nail is about 4 paper clips long. Have them say with you, *The nail is about 4 paper clips long.*

• For Exercises 1 to 4, students are to write the number that tells about how many paper clips long each object is. Have them complete the page independently.

Using page 156 Tell students that on this page they are to tell the length of each object in little square units. Ask about how long the peapod is. (about 5 squares) Have students say with you, *The peapod is about 5 squares long.* Tell them to trace the 5 to tell the length of the peapod is about 5 squares long. Help students with Exercise 1, if necessary, before having them complete the page independently.

4 Assess

Give each student a cube. Then, have students find objects in the classroom to measure using the cube. Once every student has finished, have students name their object and tell its approximate length in cubes.

T156

8-8 Weight

pages 157–158

1 Getting Started

Objectives
- To identify an object that is heavier or lighter than another
- To define *heavier* and *lighter*

Vocabulary
heavier, lighter

Materials
*5 pairs of objects to weigh; one index card labeled "heavier" and another labeled "lighter"

Warm Up • Mental Math
Have students name the object that holds more.
1. spoon or mug (mug)
2. juice box or eyedropper (juice box)
3. pail or scoop (pail)
4. jar or bathtub (bathtub)
5. bowl or swimming pool (swimming pool)

Warm Up • Geometry
Have students look at pairs of classroom objects that are different sizes and identify which is bigger and which is smaller, such as pencils. Repeat several times.

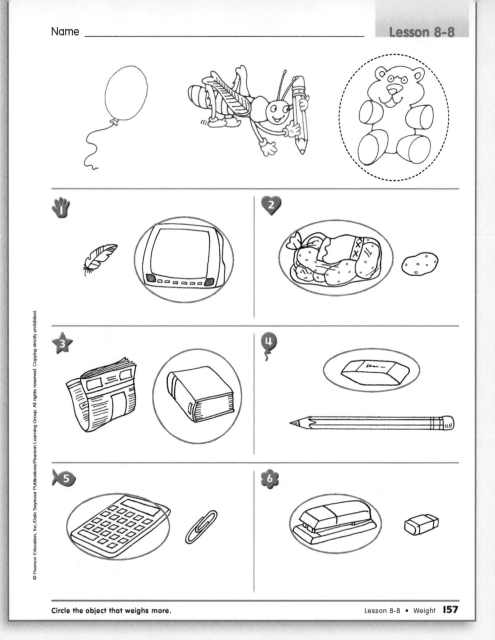

Name _____

Circle the object that weighs more.

Lesson 8-8 • Weight **157**

2 Teach

Develop Skills and Concepts Have students name some objects that are very large (Possible answers: elephant, truck) and some that are very small (Possible answers: paper clip, brick). List the objects on the board in two columns. Ask, *Are all large objects heavy, and are all small objects light?* (no) Ask volunteers to name examples of large objects that are light (Possible answers: cloud, beach ball) and heavy objects that are small (Possible answers: brick, paperweight).

- Have students act out lifting objects that are either heavy or light. For example, if it is a heavy object, encourage them to hunch over as if they are lifting the heavy object. They could pretend to struggle to move the object from the ground to above their head. The rest of the students have to guess if the object being lifted is heavy or light. Repeat until volunteers have acted out at least one heavy object and one light object.

3 Practice

Using page 157 Have students look at the objects in the example at the top of the page. Ask students if they can tell what the grasshopper is trying to decide. (which object is heavier, or weighs more) Ask them if the balloon or the teddy bear is heavier. (teddy bear) Tell them to trace the circle around the teddy bear to show that it is heavier than the balloon.

- Encourage students to think about each object pictured in the exercises and to imagine if it would be something that is heavy or light. Then, they should compare the objects and circle the one that is heavier. Have them complete the page independently.

T157

158 Lesson 8-8 • Weight

Circle the object that weighs less.

For Mixed Abilities

Common Errors • Intervention

Some students will be unable to look at two objects or pictures of two objects and estimate which is heavier and which is lighter. For those students, encourage repeated experiences holding pairs of objects and using a pan balance to compare the weights.

Enrichment • Geometry

1. Give students a pair of classroom objects that are different weights. Have them look at the objects and estimate which is heavier and which is lighter. Then, have students check their estimate by putting each object in one pan of a pan balance. Explain that the side of the balance that is lower holds the heavier object and the higher side holds the lighter object. Repeat with a different pair of objects.

2. Display a paper clip, a board eraser, and a stapler. Have students take turns holding the objects two at a time to determine the order from lightest to heaviest object. Have students check the order by placing the objects on a pan balance.

ESL/ELL STRATEGIES

Before introducing the lesson, use classroom objects to preteach the terms *heavy* and *light*. Hold up objects and describe them, such as *This dictionary is heavy. This eraser is light*. Ask students to repeat the sentences and then make up original sentences using other objects.

Using page 158 Tell students that for each exercise on this page, they are to look at each pair of objects and circle the one that is lighter. In Exercise 1, ask students which of the two objects weighs less. (the smaller container) Have students circle the smaller container and say, *This object weighs less and is lighter*. Then, have students complete the page independently.

4 Assess

Give each student two index cards on which you have written **heavier** and **lighter** and one pair of classroom objects that are different weights, such as a book and a pencil. Have students hold the pair of objects, one in each hand, to compare the weights. Then, have them label the objects with the index cards to identify which is heavier and which is lighter.

Chapter 8 Test and Challenge

pages 159–160

1 Getting Started

Test Objectives

- **Item 1:** To identify one-half, one-fourth, and one-third of a figure (see pages 145–150)

- **Items 2–3:** To identify a container that holds more or less than another (see pages 153–154)

- **Item 4:** To measure lengths using nonstandard units (see pages 155–156)

- **Items 5–6:** To identify an object that is heavier or lighter than another (see pages 157–158)

- **Item 7:** To draw a picture to solve problems; to split groups of no more than ten objects into two or three equal groups (see pages 151–152)

Challenge Objective

- To compare scenes for hotter or colder

Vocabulary

temperature

Materials

paper square; crayons

Name _____

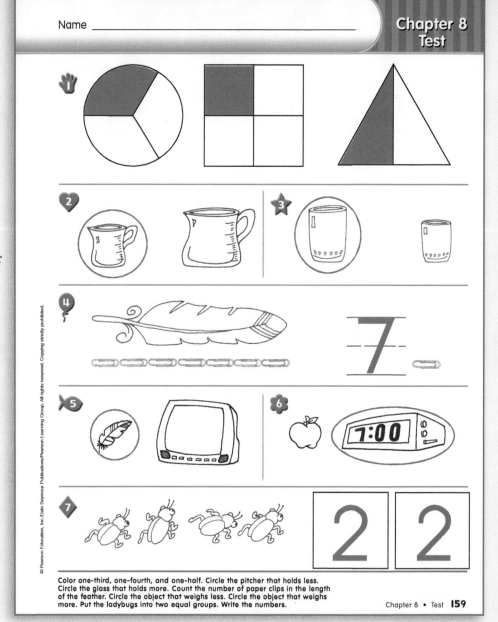

Color one-third, one-fourth, and one-half. Circle the pitcher that holds less. Circle the glass that holds more. Count the number of paper clips in the length of the feather. Circle the object that weighs less. Circle the object that weighs more. Put the ladybugs into two equal groups. Write the numbers.

Chapter 8 • Test **159**

2 Give the Test

Prepare for the Test Give each student one square. Ask students to fold the square in half to make two equal parts, halves, from the whole. Then, have students explain why these are halves. (They are 2 equal pieces that make 1 whole; they are equal because they fold over onto each other and make the same shape.)

- Ask students to find two objects in the classroom and decide which of the two is heavier. Have students report their choices to the class.

- After students have mastered fractions, deciding capacity and weight, measuring lengths using nonstandard units, and separating objects into equal groups, have them complete the Chapter 8 Test. You may wish to review test-taking rules with students. Remind them that they are not to talk to each other and they are to finish the test on their own. Decide ahead of time whether or not to allow the use of manipulatives such as counters.

Using page 159 Students should be able to complete this page independently. In Exercise 1, tell students to find the figure that shows halves (the last figure) and color one-half blue. Have students find the figure that shows thirds (the first figure) and color one-third red. Tell them to now find the figure that shows fourths (the second figure) and color one-fourth yellow. In Exercise 2, tell students to circle the container that holds less and in Exercise 3, circle the container that holds more. In Exercise 4, tell students to measure the length of the feather using the paper clips and write that number on the line. In Exercise 5, tell students to circle the object that weighs less and in Exercise 6, circle the object that weighs more. In Exercise 7, students are to put the ladybugs into two equal groups and write the number that tells how many is in each group.

3 Teach the Challenge

Develop Skills and Concepts for the Challenge Discuss temperature by asking what kinds of clothing students

T159

For Mixed Abilities

Common Errors • Intervention

Some students may need more practice measuring with nonstandard units. Have them work in pairs with one 15-inch-long strip of paper. Have them take turns using the lengths of their pencils to measure and mark the length of each strip. Be sure students understand that each pencil has a different length, causing different answers for the length of the same strip of paper.

Enrichment • Application

1. Have students walk the path from their bed at home to the nearest light switch and tell the distance in footsteps.

2. Provide 1-cup, 2-cup, and 4-cup measuring containers. Have students experiment with beans or rice to find out how many of the smaller containers are needed to fill each of the larger ones.

3. Show students a thermometer and tell them that it is a tool used to measure temperature. Explain that when something is hot, the temperature is a greater number and when something is cold, the temperature is a lesser number.

Describe each picture using the vocabulary *hot* and *cold*. Circle the scene that shows the hotter weather.

might wear on a hot day and then on a cold day. Have students cut pictures from magazines that show aspects of cold and hot weather. Have them use their pictures to talk about activities people like to do when the weather is hot. They may also find pictures of plants that are seen during hot weather. Repeat for discussions of cold weather activities and plants seen in cold weather. Help students note that some trees lose their leaves in cold weather and some do not.

Using page 160 Discuss the two pictures in Exercise 1. Ask students what clues are given in each picture to tell us which one is the hot weather scene. (Possible answers relate to clothing items such as sunglasses, sporting equipment, and trees.) Ask a student to use the clues in sentences to describe the hot weather scene in this exercise. For Exercise 2, have students circle the scene that shows the hotter weather.

• Continue to discuss the two pictures in each exercise and have students circle the scene that shows the hotter weather. For Exercise 3, you may wish to point out that you most often drink a cold drink in the summer and

most often drink a hot drink in the winter. You could also say that hot weather makes a glass with a cold drink sweat. For Exercise 4, point out to students that a wood-burning stove is used in cold weather to make that area warmer and a fan is used in warm weather to make the area cooler.

4 Assess

Alternate Chapter Test You may wish to use the Alternate Chapter Test on page 212 of this manual for further review and assessment.

Challenge To assess students' understanding of temperature, ask them to draw a scene that depicts hot weather. Then, have students work in pairs. Have partners describe their scene, being sure to describe the clothing, the activities, the colors, and so on. (Possible answer: beach scene; people wearing bathing suits and swimming)

9 Addition

Chapter Objectives

- To join two groups and write the number that tells how many in all
- To join two groups and write the number for 1 more
- To add 1 or 2 more for sums through 9
- To find sums through 9
- To use the plus and equal signs in number sentences
- To define *plus, equals, add,* and *addition*
- To add two groups of pennies for sums through 6¢
- To identify that 3 + 2 and 2 + 3 have the same sum
- To draw a picture to solve problems

Materials

You will need the following:
*magazines or catalogs; *3 plastic bags; *cardboard pennies; *number cards 1 through 6; *large number cards 1 through 6; *cards with the plus sign and equal sign; *objects in different lengths; *ruler; *object cards 1 through 6; *yarn; *6 objects to count; *dominoes; *large number cube with 2 dots on 2 faces, 3 dots on 2 faces, 4 dots on 2 faces; *large number cube with 1 dot on 3 faces and 2 dots on 3 faces; *items priced 1¢ through 6¢; *color word cards; *overhead projector or flannel board materials

Students will need the following:
apples; strips of paper in different lengths; paper clip; 9 counters; Reproducible Masters RM25 and RM26; number cards 1 through 9; object cards 1 through 9; yarn; crayons; 6 beads in two colors; 2 number cubes; 6 pennies; card with 6 squares drawn on it

Bulletin Board Suggestion

Use yarn or paper to divide the bulletin board into four sections. Now, attach three plastic bags to the bulletin board. Have students cut pictures of people from magazines or catalogs. Then, have them place their pictures in one of the bags. Place cardboard pennies in another bag and several numbers 1 through 5 and the plus and the equal signs in the third. Display a number sentence or vertical problem in each section of the bulletin board and have students use the pictures to show the problem. Vary the activity by displaying a scene for them to post the number sentence or vertical problem.

Verses

Poems and songs may be used any time after the pages noted beside each.

Adding 2 (pages 167–168)

Until today I never knew how to add for 2 plus 2.
It's not so hard 'cause now I know
I start at 2 and then I go
To 3 and then I go 1 more and 2 plus 2 will equal 4.
To start at 4 and add 2 more,
I simply start to count at 4.
Then 5 and 6, that's 4 plus 2
And counting's all I had to do.

Adding (pages 173–174)

Guess what adding is.
It's making two groups one.
Lay out two groups of sticks and I
Will show you how it's done.

Two plus two is four;
Five plus one is six.
The trouble is, when you get done,
You still will have just sticks!

Sticks, sticks, and then more sticks,
These sticks become a bore!
It's time I learn to add some coins
So I can go to a store.

Five pennies and another add
Together to make 6;
6 cents has much more value than
A silly pile of sticks!

Adding 1 (pages 177–178)

Adding is a lot of fun
'Specially when I'm adding 1,
'Cause all I do for 4 plus 1
Is count 1 more past 4 . . . I'm done!
Now how 'bout 5 plus 1? Let's see . . .
Count 5 . . . then 6; add easily.
Try 8 plus 1. Count 8, then 9 . . .
Yes, 8 plus 1, it equals 9.
Adding is a lot of fun
'Specially when I'm adding 1.

Sad Bedtime (pages 177–178)

Now that I have learned how to add,
My bedtime at 8 is quite sad!
It should be at 9,
One more hour'd be fine,
But now to convince Mom and Dad!

Long Enough For School

Squirm Worm and Wiggle Worm were very excited. They were crawling to the Worm School to see if they were long enough to go to school. They hoped they were, but they had to be measured to find out. It seemed they had waited and waited and grown a lot.

Mr. Fuzzy Worm, the principal, measured each new worm to see if it was long enough to go to school. To do this, Mr. Fuzzy Worm took a piece of string that was the same size as a grown-up worm. Then he cut the string into three pieces that were the same size. Each piece was one-third of the whole string, or one-third of a worm's grown-up size.

Then he laid one of the three pieces out straight on the ground. Each little worm had to stretch out beside the short piece of string. If the worm was as long as the piece of string, then the worm was long enough for school.

"Boy! I hope I'm finally long enough," said Wiggle Worm as he crawled up to the piece of string.

"Stop wiggling, Wiggle Worm!" said Mr. Fuzzy Worm. "I can't measure you until you stop wiggling!"

So Wiggle Worm stopped wiggling. He laid out as straight and as still as he could.

"Very good!" said Mr. Fuzzy Worm. "You're finally long enough! You have completed one-third of your growth. You may start school today!"

Next, Squirm Worm crawled over and stretched out beside the short piece of string.

"Stop squirming, Squirm Worm!" said Mr. Fuzzy Worm. "I can't measure you until you stop squirming!"

Squirm Worm stopped squirming. He stretched out as straight and as long as he could.

"Very good!" said Mr. Fuzzy Worm. "You're exactly as long as the string! You, too, have completed one-third of your growth. You may start school today!"

Squirm Worm squirmed and giggled. Wiggle Worm giggled and wiggled. At last, they were long enough to go to school!

Name _____

"Long Enough for School," p. RM25

Activities

Use Reproducible Master RM25, "Long Enough for School," to help students relate the story to the mathematical ideas presented.

- Provide one apple for each pair of students. Have pairs discuss the best way to share the apple equally. Cut each apple in half for them. Have them describe what each half looks like, noticing the pattern the seeds make and so on. Emphasize that both halves look the same. Then, let them enjoy their snack.

- Collect many objects in different lengths. Provide strips of paper in different lengths, longer than the longest object or shorter than the shortest object. Give each pair of students two strips of paper in different lengths. Tell students that they are to find and claim an object that is shorter than their longer strip but longer than their shorter strip. Have the pairs of students tell about their objects.

pages 161–162

1 Getting Started

Objective

• To join two groups and write the number that tells how many in all

Materials

*ruler; *number cards 1 through 6; *object cards 1 through 6; *yarn; *6 objects to count; paper clip

Warm Up • Mental Math

Ask students to name the following:

1. the first letter of the alphabet (A)
2. 1 of 3 equal parts (one-third)
3. 1 penny less than 30¢ (29¢)
4. the numbers between 16 and 21 (17, 18, 19, 20)
5. today's day of the week (Answers will vary.)
6. next month's name

Warm Up • Geometry

Use a ruler to draw a rectangle on the board. Have students measure the length and width of the rectangle using a nonstandard unit of measure, such as a paper clip or a pencil.

2 Teach

Develop Skills and Concepts Shuffle the number cards 1 through 5 and place them randomly on a table. Tell students that you want to place the numbers in order from 1 to 5. Have a student select the first number and place it at the beginning of the cards. Continue until the number cards are in order. Shuffle the cards again and use this procedure to put the numbers in reverse order, 5 to 1. Repeat using the object cards 1 through 5.

• Place a large yarn loop on the floor. Call a student to stand in the circle. Ask how many students are in the circle. (1) Call 2 students to join the first student in the circle. Ask how many more students joined the group. (2) Ask how many students are in the circle now. (3) Have students repeat with you, *One and two more is three.* Have students act out more examples using the numbers 1 through 4 for sums of 5 or less. Help students summarize the action with statements such as "Three and one more is four."

1 and 1 is ⬜2⬜.

2 and 1 is ⬜3⬜.

2 and 2 is ⬜4⬜.

3 and 2 is ⬜5⬜.

Write the number that tells how many animals there are in all.

Lesson 9-1 • Joining Sets **161**

• Lay out 2 objects in a group and ask students to tell how many. (2) Lay out another group of 1 object and ask how many. (1) Move the 1 object to join the 2 objects as you say, *The two objects are joined by one more object.* Ask students to count with you to see how many objects are in the group now, *1, 2, 3.* (3) Have a student write 3 on the board. Have students say with you, *Two and one more is three.* Repeat for joining 2 and 2, 2 and 3, 2 and 4, and so on, using numbers 1 through 4 for sums of 6 or less.

• Help students discover that 3 and 2 is 5 and that 2 and 3 is 5 by having 2 students stand in a circle of yarn and 3 more join them. Reverse the action and have students summarize, "Two and three is five and three and two is five." Repeat for 4 and 1, 1 and 4, and so on.

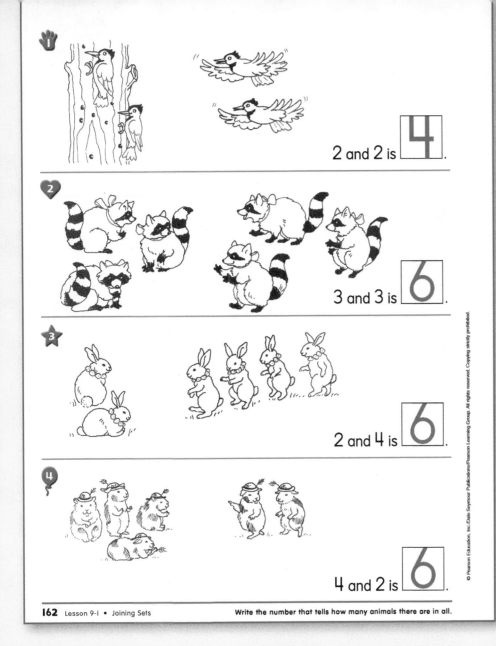

2 and 2 is 4.

3 and 3 is 6.

2 and 4 is 6.

4 and 2 is 6.

162 Lesson 9-1 • Joining Sets   Write the number that tells how many animals there are in all.

For Mixed Abilities

Common Errors • Intervention

Some students may need more help with the concept of joining. Draw a group of 2 balloons with strings on the board. Then, draw a separate group of 4 balloons. Tell students that the wind is blowing the 4 balloons toward the others as you draw 4 balloons next to the 2 balloons and erase the first group of 4 balloons. Have students count to tell how many balloons there are altogether. (6) Repeat with other combinations. Students can model each situation using counters.

Enrichment • Number Sense

1. Have students cut 4 pictures of objects from catalogs or magazines and paste them on paper to show that 3 and 1 is 4.

2. Have students draw a picture to show themselves holding 4 balloons, 2 in each hand.

---

3 Practice

Using page 161 Have students look at the example at the top of the page. Ask students, *How many birds are sitting on the limb?* (1) Ask, *How many birds are coming to join the 1 bird on the limb?* (1) Ask, *How many birds are there in all?* (2) Have students find the first number sentence and read aloud with you, *One and one is two.* Tell them to trace the 2 in the box to tell that there are 2 birds in all.

• Work similarly with students through Exercises 1 to 3, having them read each number sentence aloud with you before writing the number in the box.

Using page 162 Have students look at Exercise 1. Ask them how many birds are pecking at the tree. (2) Ask, *How many birds are coming to join the 2 birds at the tree?* (2) Ask, *How many birds are shown in all?* (4) Then, ask students for the number that is to be written in the box to tell how many birds there are in all. (4) Tell students to write the number 4 in the box. Have students read aloud with you, *Two and two is four.* Now, have students complete the page independently.

4 Assess

Have students look at Exercises 3 and 4 on page 162. Read the following problem to students and have them decide which exercise fits the problem:

There are 2 animals sitting. Then, 4 more animals join them. There are 6 animals in all. (Exercise 3)

T162

9-2 One More

pages 163–164

pages 163–164

1 Getting Started

Objectives
- To count 1 through 5 objects and write the number
- To join two groups and write the number for 1 more

Materials
*number cards 1 through 6; 6 counters; Reproducible Master RM26

Warm Up • Mental Math
Have students count aloud.
1. from 15 to 20 (15, 16, . . . , 20)
2. from 19 to 23 (19, 20, . . . , 23)
3. from 0 to 11 (0, 1, . . . , 11)
4. from 27 to 31 (27, 28, . . . , 31)
5. from 16 to 10 backward (16, 15, . . . , 10)
6. from 8 to 0 backward (8, 7, . . . , 0)

Warm Up • Skill Review
Have students tell the number that is 1 more than 6. (7) Repeat for a number that is 1 more than other numbers 0 through 30.

2 Teach

Develop Skills and Concepts Have 2 students stand. Ask how many students are standing. (2) Have another student stand and ask how many there are standing in all. (3) Encourage students to summarize the action with statements such as "Two and one more is three." Write **2 and 1 is** ☐ on the board and have a student write the sum in the box. (3)

- Have students lay out 3 counters and tell how many counters there are. (3) Write **3** on the board. Tell them to lay out 1 more counter to join the 3 counters as you write **3 and 1 is** ☐ on the board. Have students tell how many counters there are in all. (4) Have a student write the number on the board. (4) Have students read the sentence with you. Repeat with more examples for sums of 6 or less.

- Have students lay out counters to show the following problem:
Sean had 3 lollipops and then his mom gave him 1 more. How many lollipops does Sean have in all? (4)
Write **3 and 1 is** ☐ on the board and have a student write the sum in the box. (4) Have students read the number sentence with you.

Repeat the procedure for the following problems:
Mona bought 2 movie tickets and then she bought 1 more. How many tickets does Mona have now? (3)
Mom had 4 eggs and asked me to get 1 more. How many eggs did Mom have in all? (5)
Grandma ironed 5 shirts and then she ironed 1 more. How many shirts did Grandma iron in all? (6)

- Give students copies of Reproducible Master RM26. Ask how many petals are on the first flower. (2) Tell students that they are to trace the petals on the flower on the right to show that there are 2 and 1 more petal on the second flower. Ask how many petals are on the second flower. (3) Ask how many would be on the flower if there were 3 and 1 more petal. (4) Have students draw 4 petals on the second flower on the right. Have them complete the page by always adding 1 more petal to the flower on the left.

T163

Name _____

It's Algebra!

1 and 1 more is 2.

1 and 1 is 2 .

2 and 1 is 3 .

3 and 1 is 4 .

4 and 1 is 5 .

Write the number that tells how many animals are sitting, how many are joining them, and how many there are in all.

Lesson 9-2 • One More **163**

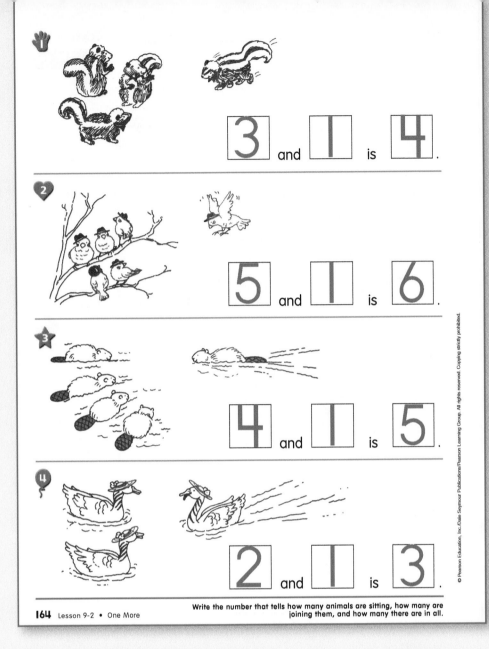

1 $\boxed{3}$ and $\boxed{1}$ is $\boxed{4}$.

2 $\boxed{5}$ and $\boxed{1}$ is $\boxed{6}$.

3 $\boxed{4}$ and $\boxed{1}$ is $\boxed{5}$.

4 $\boxed{2}$ and $\boxed{1}$ is $\boxed{3}$.

164 Lesson 9-2 • One More

Write the number that tells how many animals are sitting, how many are joining them, and how many there are in all.

For Mixed Abilities

Common Errors • Intervention

Some students may need more practice with the concept of 1 more. Display the number cards 1 through 6 in order on a table. Ask a student to come forward and point to the number that is 1 more than 4 and draw that number of circles on the board. (5) Repeat for the number that is 1 more than 3 (4), 1 more than 5 (6), 1 more than 2 (3), and 1 more than 1 (2).

Enrichment • Number Sense

1. Have students make up stories for 3 and 1 is 4, and so on. (Possible answer for 3 and 1 is 4: I had 3 stickers. John gave me 1 more. Now, I have 4 stickers in all.)

2. Have students draw a clock to show when Jamie went to bed if he is supposed to go to bed at 8:00, but his dad said he could stay up 1 more hour. (9:00)

It's Algebra! The concepts of this lesson prepare students for algebra.

3 Practice

Using page 163 Discuss the picture in the example at the top of the page, stressing that 1 swan is being joined by 1 more swan. Have students find the duck in the top right corner of the page. Tell them that the duck is watching the swans and saying, *"One and one more is two."* Have students trace the first 1 in the number sentence to tell that 1 swan is sitting still in the water and then the second 1 to show that 1 more swan is joining the first swan. Ask how many swans are shown in all. (2) Have students trace the 2 to tell that there are 2 swans in all. Have students read aloud with you, *One and one is two.*

• Help students complete and verbalize the action in Exercise 1 before assigning the page to be completed independently.

Using page 164 Have students use sentences, such as "Three skunks and one more skunk is four skunks," to tell about the action in each of the exercises on this page. Help students, if necessary, complete Exercise 1 before assigning the page to be completed independently.

4 Assess

Tell students that you bought 4 pencils. Then, you went back to the store and bought 1 more. Ask students how many pencils you have in all. (5) Have students write the addition sentence for this problem using the words *and* and *is*. (4 and 1 is 5.)

T164

pages 165–166

© Pearson Education, Inc./Dale Seymour Publications/Pearson Learning Group. All rights reserved. Copying strictly prohibited.

1 Getting Started

Objectives
- To add 1 more for sums through 6
- To use the plus and the equal signs in a number sentence
- To define *plus, equals, add,* and *addition*

Vocabulary
plus, equals, add, addition, addition sentence

Materials
*object cards 1 through 6; 6 counters

Warm Up • Mental Math
Tell students to name the number.
1. 3 and 1 more (4)
2. 1 more than 5 (6)
3. 1 less than 3 (2)
4. 5 and 1 more (6)
5. 1 less than 1 (0)
6. 1 more than 0 (1)

Warm Up • Number Sense
Choose names in the class that have 6 letters or less. Begin to print a name on the board and tell students that 1 letter and 1 more letter is 2 letters, 2 letters and 1 more letter is 3 letters, and so on, as you print each additional letter of the name. Have students count the letters in each name as you complete it.

2 Teach

Develop Skills and Concepts Have students lay out 5 counters and show 1 more counter joining the group. Write **5 and 1 is** ☐ on the board and have a student write the sum in the box. (6) Have students read the number sentence with you. Repeat for 3 and 1, 4 and 1, 2 and 1, and 1 and 1.

- Display the object cards for 1 to 6 on a table in random order. Have students identify the card with 5 objects, 3 objects, and so on. Continue the activity for other random arrangements.

- Show the card with 5 objects and have a student identify the object card that has 5 and 1 more object. (object card for 6) Write **5 and 1 is 6** on the board and have students read the number sentence orally with you. Repeat for more number sentences.

- Write the following on the board: **3 and 1 is 4.** Then, directly underneath, write **3 + 1 = 4.** Have students read the first number sentence orally with you. Now, point to the corresponding part of the number sentence as you say, *Three plus one equals four.* Reread the number sentence as you point to each number and sign, stressing the words *plus* and *equals.* Have students read the number sentence in unison with you. Say, *3 + 1 = 4 is an addition sentence because we add 1 more to a group of 3.* Use the words *plus* and *and,* and *equals* and *is* interchangeably.

- Have 2 students stand in a group as you write **2** on the board. Tell the class that you want to add 1 more student to the group as you write **+ 1 =** ☐ after the **2** on the board. Have 1 student join the group. Ask students to read the number sentence with you as you say, *Two plus one equals three,* stressing the words for the plus and the equal signs. Repeat the procedure for 5 + 1 = 6, 4 + 1 = 5, and 1 + 1 = 2.

Name _____

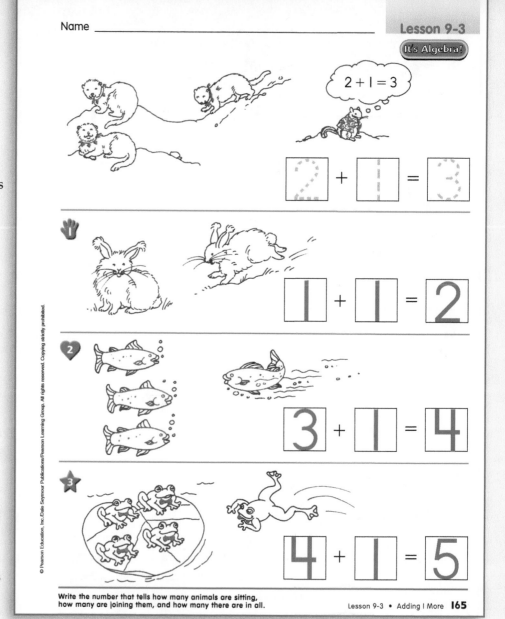

2 + 1 = 3

2 + 1 = 3

1 + 1 = 2

3 + 1 = 4

4 + 1 = 5

Write the number that tells how many animals are sitting, how many are joining them, and how many there are in all.

Lesson 9-3 • Adding 1 More **165**

T165

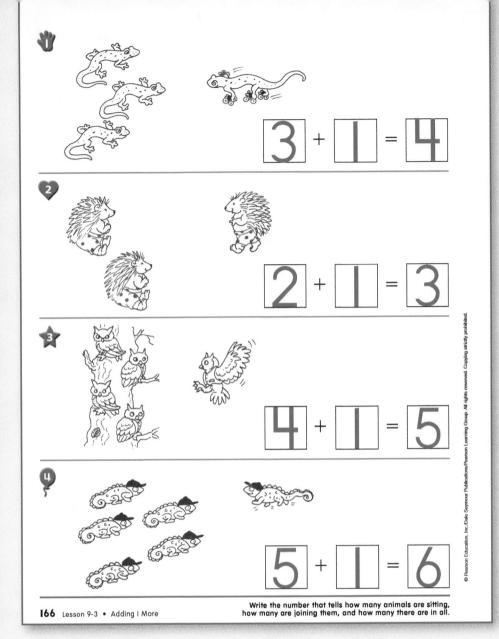

1. $3 + 1 = 4$

2. $2 + 1 = 3$

3. $4 + 1 = 5$

4. $5 + 1 = 6$

Write the number that tells how many animals are sitting, how many are joining them, and how many there are in all.

166 Lesson 9-3 • Adding 1 More

For Mixed Abilities

Common Errors • Intervention

Some students may need more practice understanding the plus sign. Have them act out addition sentences that you write on the board in which 1 is added to 1, 2, 3, 4, and 5.

Enrichment • Number Sense

1. Write three number sentences for adding 1 more to get sums through 6 on the board. Tell students to draw a picture showing one of the addition sentences.

2. Have students write addition sentences for dominoes whose sums are 6 or less when one of the addends is 1.

3. Have students tell a story for each addition sentence: $5 + 1 = 6$, $1 + 5 = 6$, $3 + 1 = 4$, and $1 + 3 = 4$.

ESL/ELL STRATEGIES

Use simple pictures and mathematical problems on the board to practice the terms *add*, *addition*, and *in all*. You can point to the items on the board and make statements such as *Let us add 1 and 2. We are doing addition. There are 3 birds in all.*

It's Algebra! The concepts of this lesson prepare students for algebra.

3 Practice

Using page 165 Have students use sentences such as "Two animals and one more animal is three animals" to tell about the action in the example. Help them note that the plus sign between the two numbers in the example means the numbers are to be added, or joined together. Have students trace the 2 and the 1 to tell that two animals are being joined by one more. Have students read aloud with you, *Two plus one*. Help them note that the equal sign is next. Have students read aloud with you, *Two plus one equals*. Ask how many animals there are in all. (3) Tell students to trace the number 3 in the box to tell that there are 3 animals in all. Have them read aloud with you, *Two plus one equals three*.

• Work similarly through Exercises 1 to 3, having students read the number sentences aloud with you as they complete each part.

Using page 166 Have students use sentences such as "Three lizards and one more lizard is four lizards" to tell about the action in each of the exercises on this page. If necessary, help students complete Exercise 1 before assigning the page to be completed independently.

4 Assess

Tell students the following story, have them find the sum, and then have them write the corresponding addition sentence:

There were four dogs playing in the park. One more dog joined them. Now, there are ____ dogs playing in the park. (5; $4 + 1 = 5$)

T166

9-4 Adding 2

pages 167–168

1 Getting Started

Objective
- To add 2 to a number for sums through 6

Materials
*object cards 1 through 6; *card with plus sign; *card with equal sign; 6 counters

Warm Up • Mental Math
Tell students to name the number that is 1 more.

1. 16 (17)
2. 0 (1)
3. 2 + 1 (4)
4. 30 (31)
5. 4 + 1 (6)

Warm Up • Number Sense
Display object cards 1 through 6 on a table. Use these cards as the addends for sums through 6, where one addend is always 1. Place cards with the plus and the equal signs in their proper places. Have a student identify and position the object card that tells how many objects there are in all. Then, have students write the corresponding addition sentence on the board.

2 and 2 more is 4.

$$2 + 2 = 4$$

$$1 + 2 = 3$$

$$3 + 2 = 5$$

$$4 + 2 = 6$$

Write the number that tells how many animals are sitting, how many are joining them, and how many there are in all.

2 Teach

Develop Skills and Concepts Give each student 6 counters. Tell students a story about Jane who has 2 goldfish and 2 hamsters. Tell students that we want to know how many pets Jane has in all. Have students show 2 counters as goldfish and 2 as hamsters to answer the question. Then, have a student write the addition sentence on the board for all students to read in unison. (2 + 2 = 4) Repeat this procedure for stories that present problems of 1 + 2 = 3, 4 + 2 = 6, 2 + 1 = 3, 5 + 1 = 6, and 3 + 2 = 5. Encourage students to make up the stories. Continue to use the words *and* and *plus*, and *equals* and *is* interchangeably.

- Have students act out the following number sentences: 2 + 4 = 6, 3 + 2 = 5, 4 + 2 = 6, and 2 + 3 = 5. As each addition sentence is acted out, have a student write the addition sentence on the board for classmates to read orally.

- Tell a student to write and solve 2 + 4 on the board as classmates solve the problem with their counters. (6)

Have students read the addition sentence aloud with you. Repeat for 2 + 3, 5 + 1, and 4 + 2.

- Read the poem "Adding 2" on page T161a. Have students lay out counters to work the problems in the poem.

It's Algebra! The concepts of this lesson prepare students for algebra.

3 Practice

Using page 167 Have students use sentences such as "Two butterflies and two more butterflies is four butterflies" to tell about the action in each exercise. Help students note that the plus sign between the two numbers in these exercises means the numbers are to be added. Have them write the two 2s for Exercise 1 to tell that two butterflies are being joined by two more.

T167

$$3 + 2 = 5$$

$$2 + 3 = 5$$

$$4 + 2 = 6$$

$$2 + 2 = 4$$

168 Lesson 9-4 • Adding 2

Write the number that tells how many animals are sitting, how many are joining them, and how many there are in all.

For Mixed Abilities

Common Errors • Intervention

Some students may need practice adding 2. Have them work in pairs using counters to solve 4 + 1. Then, have them add another counter and tell the total as you write **4 + 2 = ☐** on the board. Repeat for **3 + 1 = ☐** and **3 + 2 = ☐**. Encourage students to count on if necessary.

Enrichment • Number Sense

1. Have students use the number cards 1 through 4 to make four addition sentences whose sums equal 5. (1 + 4 = 5, 4 + 1 = 5, 2 + 3 = 5, 3 + 2 = 5)

2. Have students use number cards or object cards to explain why 2 + 2, 3 + 1, and 1 + 3 are all ways of showing 4. (Possible answer: In the end, all have four objects.)

Have students read aloud with you, *Two plus two equals*. Ask how many butterflies there are in all. **(4)** Tell them to write the number 4 in the box to tell that there are 4 butterflies in all. Have students read aloud with you, *Two plus two equals four*.

• Work similarly through Exercise 2 and then have students complete the page independently.

Using page 168 Tell students that they are to look at each picture, write the numbers that are to be added, and write the number that tells how many there are in all. Have them complete the page as independently as possible.

4 Assess

Have students draw a picture to solve the following addition sentence that you write on the board: **2 + 4 = ☐**. **(6)**

1 Getting Started

Objective
- To add vertically for sums through 6

Materials
6 counters

Warm Up • Mental Math
Have students name the number that comes between.

1. 4 and 6 (5)
2. 17 and 19 (18)
3. 29 and 31 (30)
4. 6 and 4 (5)
5. 3 and 5 (4)
6. 10 and 12 (11)

Warm Up • Number Sense
Have students write and solve the following addition sentences on the board: 3 + 2, 4 + 2, 5 + 1, 4 + 1, and 2 + 3. (5; 6; 6; 5; 5)

2 Teach

Develop Skills and Concepts Have students lay out 5 counters between two groups. Have students draw groups of sticks or circles on the board to show how they formed their two groups. As each example is shown, have volunteers write that addition sentence on the board. The following addition sentences should be included: 1 + 4 = 5, 4 + 1 = 5, 2 + 3 = 5, and 3 + 2 = 5.

- Repeat the above activity using 2, 3, 4, or 5 counters as the total number of counters. The following number sentences should be written on the board:

1 + 1 = 2	1 + 2 = 3	1 + 3 = 4	1 + 4 = 6
2 + 1 = 3	2 + 2 = 4	2 + 3 = 6	
	3 + 1 = 4	3 + 3 = 6	
		4 + 2 = 6	
		5 + 1 = 6	

Have students read each number sentence aloud.

- Draw 4 objects on the board. Write **4** beside the objects. Draw 2 more objects under the 4 objects and write **2** beside them. Verbalize your actions as you write the plus sign to the left of the 2 and draw a line under the 2. Point to each number and the plus sign as you read the problem, *Four plus two equals.* Ask students how many objects there are in all. (6) Have a student write 6 under the line and then have another student read the problem aloud as you point to each number and the plus sign. Write the addition sentence **4 + 2 = 6** on the board beside the vertical problem and tell students that these are two ways to write the same problem. Repeat the procedure for 5 + 1 = 6, 3 + 2 = 5, 2 + 2 = 4, and 2 + 4 = 6.

3 Practice

Using page 169 Have students look at the example, count the cakes in the first row, and tell how many there are. (2) Ask how many more cakes are in the second row. (2) Ask how many cakes there are in all. (4) Have students say with you, *Two plus two equals four.* Tell them to trace the 4 to show that 2 cakes plus 2 more cakes is 4 cakes in all.

- Help students work similarly through Exercises 1 to 3 before assigning the page to be completed independently.

T169

Name _____

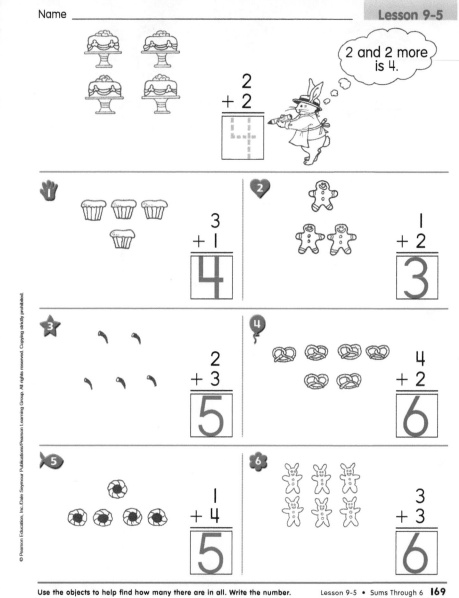

2 and 2 more is 4.

$$\begin{array}{r} 2 \\ + 2 \\ \hline \end{array}$$

1.
$$\begin{array}{r} 3 \\ + 1 \\ \hline 4 \end{array}$$

2.
$$\begin{array}{r} 1 \\ + 2 \\ \hline 3 \end{array}$$

3.
$$\begin{array}{r} 2 \\ + 3 \\ \hline 5 \end{array}$$

4.
$$\begin{array}{r} 4 \\ + 2 \\ \hline 6 \end{array}$$

5.
$$\begin{array}{r} 1 \\ + 4 \\ \hline 5 \end{array}$$

6.
$$\begin{array}{r} 3 \\ + 3 \\ \hline 6 \end{array}$$

Use the objects to help find how many there are in all. Write the number.

Lesson 9-5 • Sums Through 6 **169**

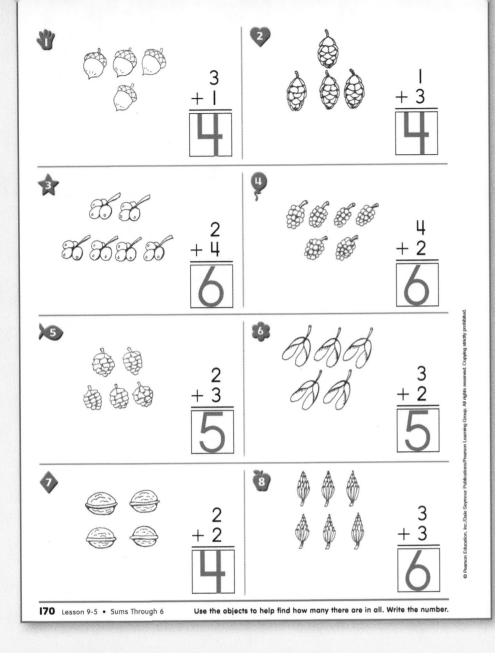

1. 3
    + 1
    4

2. 1
    + 3
    4

3. 2
    + 4
    6

4. 4
    + 2
    6

5. 2
    + 3
    5

6. 3
    + 2
    5

7. 2
    + 2
    4

8. 3
    + 3
    6

170 Lesson 9-5 • Sums Through 6   Use the objects to help find how many there are in all. Write the number.

For Mixed Abilities

Common Errors • Intervention
Some students may need more practice working addition problems in vertical form. Have them work with partners using object cards and number cards for 1 to 5 to form problems vertically for sums through 6. Have partners take turns finding the sums to the problems.

Enrichment • Number Sense
1. Have students draw the two groups of objects that show 3 + 2 = 5 so that, if turned upside down, the objects show that 2 + 3 = 5.

2. Have students change the position of number cards to show why 2 + 4 has the same sum as 4 + 2. Have them use the cards with another vertical addition problem pair to result in the same sum.

3. Have students complete addition sentences you write on the board by telling how many must be added to 5 to have 6 in all, and so on.

© Pearson Education, Inc./Dale Seymour Publications/Pearson Learning Group. All rights reserved. Copying strictly prohibited.

Using page 170 Talk through Exercise 1 with students, stressing that 3 and 1 more is 4 in all. Then, have students complete the page independently.

4 Assess

Draw 2 objects on the board. Then, draw 3 more objects directly underneath them. Now, have students write the corresponding vertical addition problem and the corresponding addition sentence. (2 + 3 = 5)

T170

9-6 More Sums Through 6

pages 171–172

1 Getting Started

Objective

• To practice addition facts for sums through 6

Materials

*dominoes; *yarn; *large number cubes described in the chapter opener on page T161a; 6 beads in two colors; number cards 1 through 6; 2 number cubes

Warm Up • Mental Math

Ask students how many there are in all.

1. $2 + 2$ (4)
2. $3 + 1$ (4)
3. $1 + 2$ (3)
4. $5 + 1$ (6)
5. $4 + 2$ (6)
6. $2 + 4$ (6)

Warm Up • Number Sense

Ask questions such as *If I have 3 black kittens and 3 white kittens, how many kittens do I have in all?* Have students act out the problem and then write and solve the addition sentence on the board. $(3 + 3 = 6)$ Continue for more word problems with sums through 6.

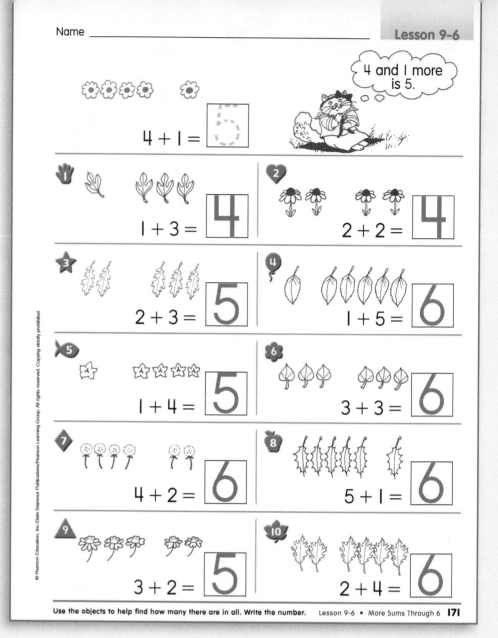

Name _____

Lesson 9-6

4 and 1 more is 5.

$4 + 1 = 5$

1. $1 + 3 = 4$
2. $2 + 2 = 4$
3. $2 + 3 = 5$
4. $1 + 5 = 6$
5. $1 + 4 = 5$
6. $3 + 3 = 6$
7. $4 + 2 = 6$
8. $5 + 1 = 6$
9. $3 + 2 = 5$
10. $2 + 4 = 6$

Use the objects to help find how many there are in all. Write the number.   Lesson 9-6 • More Sums Through 6   **171**

2 Teach

Develop Skills and Concepts Have students string 2 beads of one color and then 3 beads of another color on a piece of yarn. Ask how many beads are on the yarn. (5) Continue for other addends 1 through 5 for sums through 6.

• Show students a domino in the horizontal position with a total of 6 or fewer dots and addends of 1 through 5. Have students show the number cards that match the dots on each side of the domino. Then, have them show the number card that tells how many dots there are in all. Have a student write and solve the domino addition sentence on the board. Continue for other dominoes with sums through 6. Note that zero is not used as an addend at this time.

• Repeat the above activity for dominoes in a vertical position and have students write and solve the problems in vertical and horizontal notation.

• Have a student roll two of the large number cubes described in the chapter opener on page T161a and tell how many dots there are in all. Tell the student to write and solve the addition problem vertically on the board to show classmates the result of the roll.

• Repeat the previous activity with the two large number cubes, but have the student write an addition sentence on the board that tells about the roll. For example, if the student rolls a sum of 6 dots, then $\square + \square = 6$ should be written on the board. Classmates then guess the numbers rolled, and the student who guesses correctly wins the next roll. Allow students to use the number cards to help them prepare their guesses if necessary.

• Have students act out an addition problem such as silently placing 2 books in a stack and then adding 3 more. A classmate writes and solves the problem on the board to tell about the action, such as $2 + 3 = 5$.

172 Lesson 9-6 • More Sums Through 6

Draw a line from the domino to the number that tells how many dots there are in all.

For Mixed Abilities

Common Errors • Intervention

Some students may need practice with sums through 6. Have them work with partners and a worksheet of addition problems, a piece of string, and 6 beads in one color and 6 in another. Have them use the beads to string each of the addends in a different color and then tell how many beads there are in all in each problem.

Enrichment • Number Sense

1. Have students make their own number cube using heavy paper squares, wood cubes, and so on.

2. Have students make their own number cards or domino cards. Have them include addends 1 through 5 for sums through 6.

---

3 Practice

Using page 171 Have students look at the example and count the flowers in the first group. (4) Ask how many more flowers are to be added to the 4 flowers. (1) Ask how many flowers there are in all. (5) Have students trace the 5 that tells that there are 5 flowers in all. Tell students to begin at 1 and count the flowers to check their answer. (1, 2, 3, 4, 5) Have students look at what the cat is thinking and say with you, *Four and one more is five.* Have students then read the number sentence aloud with you, as *Four plus one equals five.*

- Work similarly through Exercises 1 to 3 before having students complete the page independently.

Using page 172 Tell students that they are to add the numbers on each domino to find out how many dots there are in all. They are then to draw a line from the domino to the number that tells how many dots there are in all. Ask a student to read the addition problem on the first domino on the left and tell how many dots there are in all. (1 plus 1; 2) Tell them to trace the line to the number 2. Have students count the number of dominoes and the boxed numbers. Be sure they understand that any boxed number, or answer, can be connected to more than one domino. Help students complete more as needed before assigning the page to be completed independently.

4 Assess

Draw a domino with 3 dots on the left side and 2 dots on the right side. Now, have students write the corresponding addition sentence to tell how many dots there are in all. (3 + 2 = 5) Be sure students have the addends in the correct order.

T172

1 Getting Started

Objective
• To add two groups of pennies for sums through 6¢

Materials
*large number cube described in the chapter opener on page T161a; *large number cards 1 through 6; 6 pennies; 6 counters

Warm Up • Mental Math
Ask students which is greater.
1. 8 or 7 (8)
2. 16 or 19 (19)
3. 27 or 17 (27)
4. 3 + 2 or 6 (6)
5. 31 or 21 (31)
6. 4 + 1 or 2 + 4 (2 + 4)

Warm Up • Number Sense
Have a student roll the large number cube and write a mystery addition sentence on the board to have classmates guess the roll, □ + □ = (number rolled).

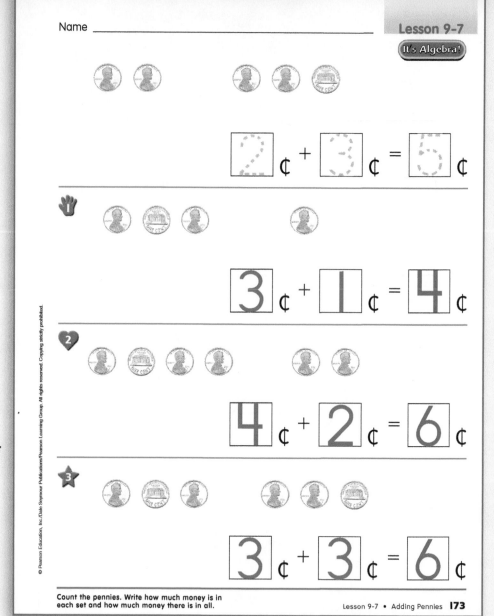

Name _____

Lesson 9-7

It's Algebra!

Count the pennies. Write how much money is in each set and how much money there is in all.

Lesson 9-7 • Adding Pennies **173**

2 Teach

Develop Skills and Concepts Have students lay out 3 pennies and then add 2 more to the group. Ask how many pennies there are in all. (5) Write **3¢ + 2¢ = 5¢** on the board and remind students that the cent sign is used to let anyone who sees the problem know that the problem is about money. Have them read the problem aloud with you, *Three cents plus two cents equals five cents.* Repeat laying out pennies to show 4¢ + 2¢ = 6¢, 1¢ + 5¢ = 6¢, 2¢ + 4¢ = 6¢, and 2¢ + 2¢ = 4¢. Have students write each addition sentence on the board when they are comfortable with doing so.

• Show the number cards for 2 and 4. Tell students to lay out two groups of pennies to match the two number cards on the table. Ask how much money this is in all. (6¢) Have a student write and solve an addition sentence on the board to tell about the action. (2¢ + 4¢ = 6¢) Have students read the addition sentence aloud, "Two

cents plus four cents equals six cents." Repeat to use the number cards 3 and 2, 1 and 5, and 3 and 3.

• Have a student lay out 6 pennies in two groups and choose a classmate to write and solve an addition problem on the board to tell about the action. The classmate should then form two groups of pennies for a sum of 6¢ or less and choose another classmate to depict the action on the board. Encourage students to write addition problems in either notation.

• Read the poem "Adding" on page T161a, stressing the first two lines about adding being the process of making two groups one. Have students lay out their counters or sticks and then lay out pennies beside the counters.

It's Algebra! The concepts of this lesson prepare students for algebra.

Count the pennies. Write how much money is in each set and how much money there is in all.

For Mixed Abilities

Common Errors • Intervention

Some students may have difficulty adding amounts of money. Have them work in pairs with a set of problems such as 3¢ + 2¢ = ☐ ¢. Have them use pennies to represent the addends and to count to find the sum. Then, have them write the total number of cents.

Enrichment • Number Sense

1. Have students draw coins to show how many more than 3 pennies they would need to buy an item costing 6¢, 5¢, or 4¢. (3¢; 2¢; 1¢)

2. Have students draw pennies in two groups to show five different ways to show 6 pennies. (1 penny and 5 pennies; 2 pennies and 4 pennies; 2 groups of 3 pennies; 4 pennies and 2 pennies; 5 pennies and 1 penny)

3 Practice

Using page 173 Have students look at the example at the top of the page. Then, tell students to count the pennies in the first set and tell the number. (2) Tell them to trace the 2 to tell that 2 cents is shown. Have students count the number of pennies in the second set (3), tell the number (3), and trace the 3 to show that 3 cents are to be added to the 2 cents. Tell students to trace the 5 to show that there are 5 pennies in all and then begin at 1 to count the pennies to check their addition. (1, 2, 3, 4, 5) Have them read the number sentence aloud with you, *Two cents plus three cents is five cents.*

• Work through Exercise 1 with students before assigning the page to be completed independently.

Using page 174 Have students look at Exercise 1. Tell students that they are to count the pennies in the first set and trace the number. (5) Tell them to count the number of pennies to be added and trace the number. (1) Ask how many pennies there are in all. (6) Have students write 6 to show that there are 6¢ in all.

• Tell students that for each exercise, they are to count the pennies to be added, write the numbers, and then write the number of pennies there are in all. Have students complete the page independently.

4 Assess

Draw 4 pennies in one group on the board. Then, draw 2 pennies directly underneath. Now, ask students to write the corresponding addition sentence. (4¢ + 2¢ = 6¢)

9-8 Sums Through 6¢

pages 175–176

© Pearson Education, Inc./Dale Seymour Publications/Pearson Learning Group. All rights reserved. Copying strictly prohibited.

1 Getting Started

Objective
• To find sums through 6¢

Materials
*items priced 1¢ through 6¢; 6 pennies; card with 6 squares drawn on it

Warm Up • Mental Math
Ask students if each total is less than or the same as 6.

1. 2 + 4 (same)
2. 5 + 1 (same)
3. 1 + 3 (less)
4. 4 + 1 (less)
5. 1 + 5 (same)
6. 3 + 2 (less)

Warm Up • Number Sense
Write **2 and 4 is 6** on the board and have a student write this number sentence a shorter way. (2 + 4 = 6) Then, have another student write this problem vertically. Repeat for more problems with sums through 6.

2 Teach

Develop Skills and Concepts Show students items priced 2¢ and 3¢. Have students lay out pennies to show the cost of the first item (2 pennies) and then form another group with 3 pennies to show the cost of the second item. Have them tell how much money the two items would cost altogether. (5¢) Write

_____¢ + _____¢ = _____¢ on the board and have a student complete the addition sentence. (2¢ + 3¢ = 5¢) Repeat for other problems with two items that have a total cost of 6¢ or less. Alternate writing the problems vertically and horizontally.

• Have students lay out a total of 5 pennies in two groups. Ask a student to select two priced items that could be bought with their pennies grouped as they are. Have them then write and solve an addition problem on the board to show the action. Now, ask if anyone has pennies in different groups and have that student write and solve a new problem on the board. Continue until all the various groupings have been acted out and shown on the board. Then, begin again with a total of 6¢, 4¢, and 3¢.

• Give students a card with 6 squares drawn on it. Tell students to write the amount of 1¢ on any one of the squares. Repeat for the amounts of 2¢ through 6¢ so that each square has a different value written on it. Now, show two priced items that will have a sum through 6¢ and have students stack the total number of pennies needed to buy the items on the square that tells the total cost. Repeat for more problems of two items for sums through 6¢.

It's Algebra! The concepts of this lesson prepare students for algebra.

3 Practice

Using page 175 Have students lay out 6 pennies for use in solving the problems on this page. Ask the cost of each toy in Exercise 1. (2¢) Tell students to lay out 2 pennies to show the cost of the first toy. Tell them to trace the 2 to

Write the cost of each toy and how much money both toys cost together.   Lesson 9-8 • Sums Through 6¢ **175**

Write the cost of each toy and how much money both toys cost together.

For Mixed Abilities

Common Errors • Intervention

Some students may need practice finding sums of money through 6¢. Write **6¢** on the board. Have different students go to the board, one at a time, and draw pennies in two groups to show addends whose sum is 6¢. Repeat with totals of 5¢, 4¢, and 3¢.

Enrichment • Number Sense

1. Have students work in groups to cut and price pictures of items from catalogs. Have them make a mini-store of items priced from 1¢ to 5¢. Have students shop with 6 pennies to buy two items from another group's store.

2. Have a student go to another group's store with 5¢ to see how many different combinations of two items can be purchased.

ESL/ELL STRATEGIES

Use sentences to introduce the noun and verb forms of *cost*. Point to pictures and say, *The cost of this toy is four cents. How much does that toy cost? It costs four cents.* Ask students to repeat each sentence and then make up their own sentences.

---

show that this toy costs 2¢. Have students lay out 2 more pennies to show the cost of the second toy, 2¢. Then, have students trace the second 2 to show that the second toy costs 2¢. Tell students to count the pennies to find out how many there are in all. (4¢) Have them say with you, *Two cents plus two cents is four cents.* Tell students that to find the total cost of two toys, they must add the two prices. Work through several more problems before assigning the page to be completed independently.

Using page 176 Tell students to lay out pennies to show the cost of the first toy (2¢) and write the number (2). Repeat for the cost of the second toy. (3¢; 3) Tell students to count the pennies to find out how many pennies are needed in all to buy both toys. (5) Have them write the amount (5) and then read the problem in unison with you, *Two cents plus three cents is five cents.* Have students complete the page independently.

4 Assess

Show one item priced at 4¢ and another priced at 2¢. Now, have students write a vertical addition problem showing the total cost of both items. (4¢ + 2¢ = 6¢)

T176

9-9 Sums Through 9

pages 177–178

1 Getting Started

Objectives
- To add 1 or 2 more for sums 7 through 9
- To find sums through 9
- To recognize that 6 + 2 and 2 + 6 have the same sum, 8

Materials
9 counters; object cards 1 through 9; number cards 1 through 9

Warm Up • Mental Math
Tell students to name the number.
1. 1 more than 6 (7)
2. 1 less than 9 (8)
3. 2¢ plus 4¢ (6¢)
4. today's date (Answer will vary.)
5. 3 dimes and 1 penny (31¢)
6. 3¢ plus 3¢ (6¢)

Warm Up • Number Sense
Have students lay out a total of 6 counters in two groups. Ask students to write addition problems on the board to show the different ways to have a sum of 6. (1 + 5, 2 + 4, 3 + 3, 4 + 2, 5 + 1)

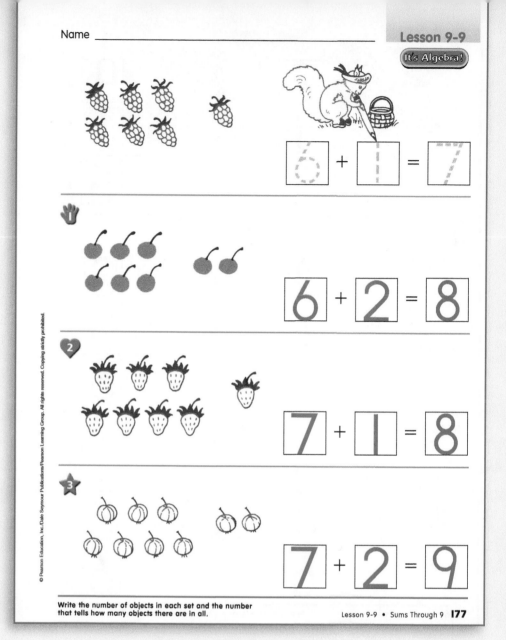

Write the number of objects in each set and the number that tells how many objects there are in all.

Lesson 9-9 • Sums Through 9 **177**

2 Teach

Develop Skills and Concepts Have students lay out a total of 7 counters in two groups so that one group has 2 counters. Ask how many counters are in each group. (2 and 5) Have students count on from the 5 to see how many counters there are in all. (7) Repeat for 7 counters in groups of 6 and 1 and then a total of 8 or 9 counters in groups such that 1 or 2 is one of the addends.

- Tell students to lay out a total of 8 counters in one group and then move 2 of the counters to form a group of 2. Ask how many counters are in each group. (6 and 2) Have a student write and solve an addition problem on the board to tell about the action. Repeat for 6 and 1, 2 and 6, 7 and 1, 2 and 7, and 8 and 1.

- Write 6 + 2 = ☐ on the board. Have students match object cards to the addends and then choose the object card that tells the sum. (8) Next, have students rearrange the addend cards to show that 2 + 6 = 8. Ask a student to write the addition problem on the board. Continue for 7 + 1 and 1 + 7, 8 + 1 and 1 + 8, and 7 + 2 and 2 + 7.

- Repeat the previous activity using the number cards.

- Draw 7 Xs in one group and 2 Xs in a second group on the board. Ask a student to write and solve an addition problem on the board to tell how many Xs there are in all. (7 + 2 = 9; 9) Repeat for groups with 8 and 1 and then 2 and 7.

- Read the poem "Adding 1" on page T161a, and then have students say it with you as they use counters to show the actions.

- Read the poem "Sad Bedtime" on page T161a, and have students show the number cards to tell the time for 1 hour more than 9 o'clock. (10 o'clock)

It's Algebra! The concepts of this lesson prepare students for algebra.

T177

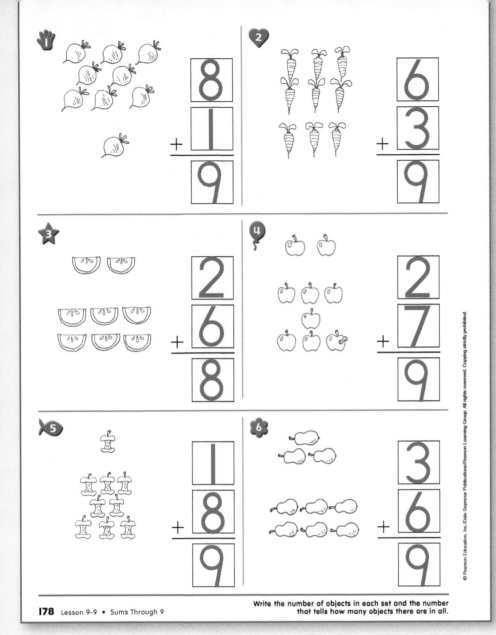

178 Lesson 9-9 • Sums Through 9

Write the number of objects in each set and the number that tells how many objects there are in all.

Common Errors • Intervention

Some students may need more help finding sums of 7, 8, and 9. Have 6 students stand in one group and 2 in another. Have students count on from 6 to find the sum. (8) Repeat for 6 and 1, 7 and 1, 7 and 2, and 8 and 1.

Enrichment • Number Sense

1. Have students experiment with different ways of putting 8 counters into two groups. (1 and 7, 2 and 6, 3 and 5, 4 and 4, 5 and 3, 6 and 2, 7 and 1)

2. Have students cut and paste pictures from magazines or catalogs to create a picture that shows 9 in all.

3. Have students draw clock faces to show their bedtime and the times for 1 hour more and 2 hours more.

---

3 Practice

Using page 177 Tell students that for each exercise on this page, they are to find the number of berries there are in all between the two groups. Now, have students look at the example at the top of the page. Ask students to tell how many berries there are in the first group. (6) Tell students to trace the 6 in the number sentence. Then, ask them how many more berries are to be added. (1) Tell students to trace the 1 to show that 1 berry is to be added to the 6. Tell them to count all the berries to find how many berries there are in all. (7) Have students trace the 7. Now, have students say with you, *Six and one more is seven* and then *Six plus one equals seven.* Have students complete the page independently.

Using page 178 Write **2 + 6 =** ☐ vertically on the board. Have students lay out 2 counters and 6 more and count to find the number in all. (8) Have a student write 8 on the board. Remind students that knowing the answer to the problem of 2 + 6 helps them to answer the problem of

6 + 2. Write **6 + 2 =** ☐ vertically on the board and have a student write the answer. (8) Note that this exercise will be helpful to students to work the new problems of 6 + 3 and 3 + 6 on this page.

• Work through Exercise 1 to have students count the vegetables in each group and then write the 8 and 1 as addends. Tell students to begin at 8 and count to see how many vegetables there are in all. (9) Have them write the number. (9) Work through Exercise 2 with students before assigning the page to be completed independently.

4 Assess

Have students create two groups using 5 counters. Now, have students write two vertical addition problems using these addends, switching the placement of the addends. (Possible answer: 3 + 2 = 5; 2 + 3 = 5)

T178

pages 179–180

1 Getting Started

Objectives

- To draw a picture to solve problems

- To find sums through 9

Materials

9 counters

Warm Up • Mental Math

Have students name the number.

1. $4 + 1$ (5)
2. $4 + 2$ (6)
3. $5 + 1$ (6)
4. $5 + 2$ (7)
5. $6 + 1$ (7)
6. $6 + 2$ (8)

Warm Up • Number Sense

Have students lay out 9 counters in two groups so that one group has 1 counter. Have students count how many counters are in the other group. (8) Write the addition problem on the board to show the groups, $8 + 1 = 9$. Repeat with 9 counters so that one group has 2 counters. ($7 + 2 = 9$)

2 Teach

Develop Skills and Concepts Write

$7 + 1 = \square$ on the board and have students read it aloud with you, *Seven plus one equals.* Under the addition problem, have a student draw a group of 7 circles and a separate group of 1 circle. Ask, *How can the drawing help you find the sum?* (I can count all of the circles in both groups to find the sum.) Then, ask a volunteer to count all of the circles and write 8 in the box. Have students read the completed addition sentence aloud with you, *Seven plus one equals eight.* Repeat with a different addition problem.

3 Practice

Using page 179 Have students look at the example at the top of the page.

- Now, apply the four-step plan. For SEE, ask, *What are you asked to do?* (draw a picture to show the addition) Then, *What information is given?* ($6 + 3$) For PLAN, ask students how they will solve the problem. (draw

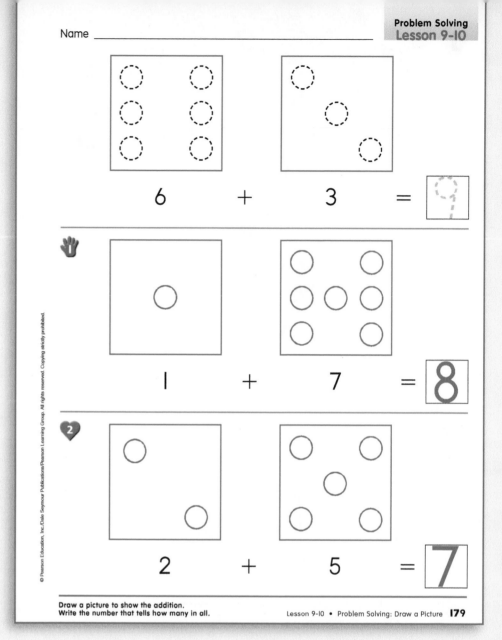

Name _____

$$6 \quad + \quad 3 \quad = \quad \boxed{9}$$

$$1 \quad + \quad 7 \quad = \quad \boxed{8}$$

$$2 \quad + \quad 5 \quad = \quad \boxed{7}$$

Draw a picture to show the addition.
Write the number that tells how many in all.

Lesson 9-10 • Problem Solving: Draw a Picture **179**

6 circles above the 6 and 3 circles above the 3; count all the circles) For DO, have students draw the 6 circles and the 3 circles in the appropriate boxes and then count the circles in the first box and count on using the circles in the second box to find that there are 9 circles in all. Have students trace the 9 in the box to show that there are 9 circles altogether. For CHECK, have students take a group of 9 counters and move 3 to the side. Ask, *How many counters are left in this group?* (6) Then, ask students if there are 9 counters in all. (yes) Now, have students read the addition sentence aloud, "Six plus three equals nine."

- Work through Exercise 1 with students if necessary. Then, have students complete Exercise 2 independently.

Using page 180 Have students look at Exercise 1 and explain what they have to do to complete the exercise. (Draw a group of 3 circles above each 3. Then, count all the circles to find the number of circles in all; 6) Now, have students complete the page independently using the same procedure.

T179

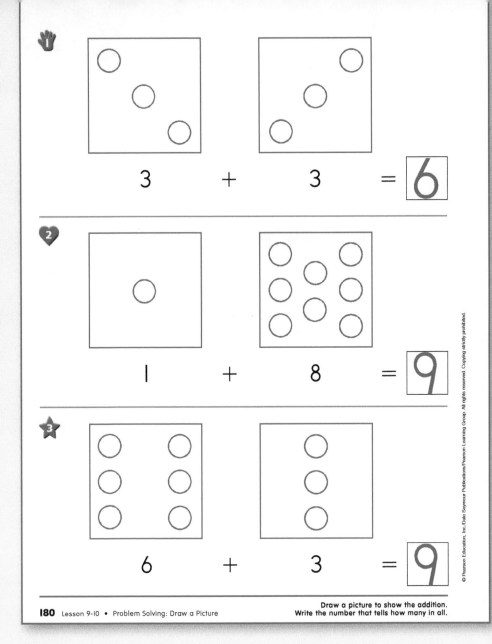

3   +   3   = 6

1   +   8   = 9

6   +   3   = 9

180   Lesson 9-10 • Problem Solving: Draw a Picture

Draw a picture to show the addition.
Write the number that tells how many in all.

4 Assess

Have students draw circles to find the sum of 5 + 4. (One group of five circles and another group of four circles equals nine circles.)

For Mixed Abilities

Common Errors • Intervention

Some students may find it helpful to complete an intermediate step of acting out the problem before drawing it. For example, to solve 2 + 4 = _____, have students use counters to show groups of 2 and 4. Then, have students draw the counters and count them to find the sum of 6.

Enrichment • Number Sense

1. Have students draw pictures to solve addition problems with three addends. Tell students this story:

   Lucky Duck has 3 pennies in his pocket. He takes 4 pennies from his bank. Then, he finds another penny on the floor. How many pennies does Lucky Duck have in all? (8 pennies)

2. Have students draw a group of 3 pennies, a group of 4 pennies, and 1 penny. Then, have students count all of the pennies and write 3 + 4 + 1 = 8.

3. Have students explore using estimation to determine if their answer is reasonable. Write **5 + 3** on the board and ask students to estimate the sum. Guide students to see that the sum would be a little more than 5 because 5 is the greater addend. Repeat using 5 as a benchmark so that students judge whether the sum would be greater than or less than 5. Be sure to encourage realistic estimates.

T180

Chapter 9 Test and Challenge

pages 181–182

1 Getting Started

Test Objectives

- **Items 1–4:** To join two groups for sums through 9; to write an addition problem using both vertical and horizontal forms (see pages 161–172, 177–178)
- **Item 5:** To add pennies for sums through 6¢ (see pages 173–176)

Challenge Objectives

- To find sums through 6
- To recognize the correct color names

Materials

*color word cards; 9 counters; 6 pennies; yellow, orange, red, and brown crayons

2 Give the Test

Prepare for the Test Have students work in pairs with 9 counters. Have one student in each pair make a group from the 9 counters, being sure not to use all 9. Have a second student write the corresponding addition problem. For example, for groups of 4 and 5 counters, students would write $4 + 5 = 9$ in either vertical or horizontal form. Once students have repeated this at least five times, have the pairs switch papers and check each other's work.

- After students have mastered finding sums through 9 and sums through 6¢, have them complete the Chapter 9 Test. You may wish to review test-taking rules with students. Remind them that they are not to talk to each other and they are to finish the test on their own. Decide ahead of time whether or not to allow the use of manipulatives such as counters.

Using page 181 Students should be able to complete this page independently. Tell students to lay out their counters for use in solving the exercises on this page. Have students look over the exercises and tell what is to be done in each. (find the number of items in each group and then the number there is in all) Then, have them complete the page independently.

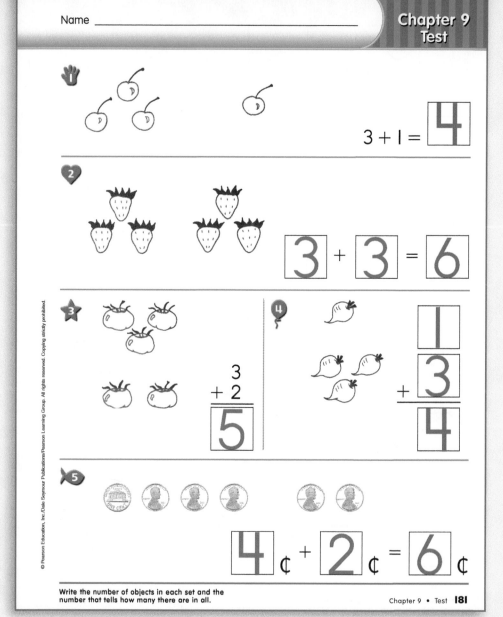

$3 + 1 = \boxed{4}$

$\boxed{3} + \boxed{3} = \boxed{6}$

$\begin{array}{r} 3 \\ + 2 \\ \hline \boxed{5} \end{array}$

$\begin{array}{r} \boxed{1} \\ \boxed{3} \\ + \boxed{3} \\ \hline \boxed{4} \end{array}$

$\boxed{4}¢ + \boxed{2}¢ = \boxed{6}¢$

Write the number of objects in each set and the number that tells how many there are in all.

Chapter 9 • Test **181**

3 Teach the Challenge

Develop Skills and Concepts for the Challenge Review the color words by having students show the crayon that is the color of the color word card you show. Then, write the following problems across the board: $2 + 3 = \square$, $4 + 2 = \square$, $3 + 1 = \square$, and $1 + 2 = \square$. Have students use their counters to work each problem. Have a student write each sum on the board.

- Now, write the following across the board: **3 yellow, 4 orange, 5 red,** and **6 brown.** Have students read each number-and-color pair. Tell them that you will point to one of the problems on the board and they are to hold up a yellow crayon if the sum of the problem is 3, an orange crayon if the sum is 4, and so on. Write another group of problems on the board for more practice.

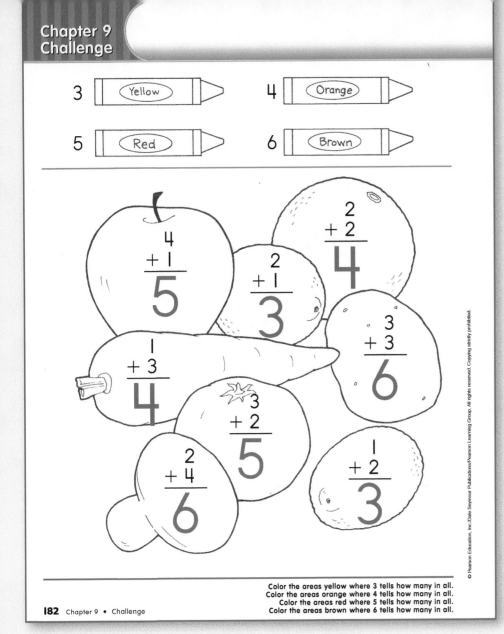

3 [Yellow]   4 [Orange]

5 [Red]   6 [Brown]

$$\begin{array}{r} 4 \\ + 1 \\ \hline 5 \end{array}$$

$$\begin{array}{r} 2 \\ + 2 \\ \hline 4 \end{array}$$

$$\begin{array}{r} 2 \\ + 1 \\ \hline 3 \end{array}$$

$$\begin{array}{r} 3 \\ + 3 \\ \hline 6 \end{array}$$

$$\begin{array}{r} 1 \\ + 3 \\ \hline 4 \end{array}$$

$$\begin{array}{r} 3 \\ + 2 \\ \hline 5 \end{array}$$

$$\begin{array}{r} 2 \\ + 4 \\ \hline 6 \end{array}$$

$$\begin{array}{r} 1 \\ + 2 \\ \hline 3 \end{array}$$

Color the areas yellow where 3 tells how many in all.
Color the areas orange where 4 tells how many in all.
Color the areas red where 5 tells how many in all.
Color the areas brown where 6 tells how many in all.

Common Errors • Intervention

Some students may need more practice finding sums through 9. Have students make fact cards for sums through 9, where 1 or 2 is one of the addends. Then, have students work in pairs and quiz each other with the fact cards.

Enrichment • Number Sense

1. Have students match vertical and horizontal fact cards for sums through 10, where 1 or 2 is one of the addends.

2. Have students experiment to find all the ways to group 9 counters into two groups. (1 and 8, 2 and 7, 3 and 6, 4 and 5, 5 and 4, 6 and 3, 7 and 2, 8 and 1) Have students write and solve all the matching addition problems.

3. Tell students to draw a picture to solve this problem:

    Mom used 2 skillets to cook 6 eggs. How many eggs were in each skillet? (3 eggs)

4. Have students tell a story that involves 4 + 3 = 7.

Using page 182 Have students point to the number 3 at the top of the page. Ask students to name the color word on the crayon beside the number 3. (yellow) Tell them to color the crayon yellow. Repeat for students to read the color words for the other three crayons and color the crayons accordingly.

• Tell students that they are to find the sum of the problem in each area and then color it yellow if the sum is 3, orange if the sum is 4, red if the sum is 5, and brown if the sum is 6. Allow students to use their counters as needed. Have them complete the page independently.

4 Assess

Alternate Chapter Test You may wish to use the Alternate Chapter Test on page 213 of this manual for further review and assessment.

Challenge Draw a set of 3 squares and another set of 5 squares on the board. Then, draw a set of 4 squares next to a set of 2 squares directly underneath. Tell a student to color each square with red chalk if it is in the group with a total of 7 squares. Then, tell another student to color each square with white chalk if it is in the group that has a total of 8 squares. Now, have the class decide if the correct colors were used.

pages 183–204

Chapter Objectives

- To understand take away as subtraction
- To define *subtracting, taking away, leaves,* and *opposite*
- To write the numbers in a subtraction sentence when 2 objects are taken away from 6 or less
- To use the minus and the equal signs in a number sentence
- To define *minus* and *subtract*
- To recognize related addition and subtraction facts
- To write differences in vertical problems with minuends through 6
- To cross out objects to be subtracted
- To count to find differences in vertical problems with minuends through 6
- To practice facts with minuends through 6
- To solve subtraction problems with minuends through 6¢
- To choose adding or subtracting to find sums through 9
- To find sums through 6 and solve related subtraction problems without pictures showing the action
- To solve mixed addition and subtraction problems
- To subtract 1 or 2 from 9 or less

Materials

You will need the following:
*large sheet of butcher paper; *magazines; *tape; *20 pennies; *items priced 5¢ or less; *hanger and 6 clip-on clothespins; *objects to count; *floor number line through 6; *fact cards for sums and minuends through 6; *fact cards for minuends 7, 8, and 9 where 1 or 2 is taken away; *yarn; *overhead projector or flannel board materials; *9 objects to count; *number cards 1 through 6

Students will need the following:
scissors; 6 pennies; Reproducible Masters RM27 and RM28; string and 6 beads; crayons; blocks; 3-by-5-inch index card; number cards 0 through 9; 9 counters; 6 blocks

Bulletin Board Suggestion

Draw a farm scene with some areas for animals to live on a large sheet of butcher paper. Have students cut pictures of pigs, cows, horses, sheep, and chickens from magazines. Tape four or less of each animal in or beside these areas. Make a bar graph using pictures and label "Pigs," "Cows," "Horses," "Sheep," and "Chickens" across the bottom to name the animals.

Provide four boxes above each of the animal names and have students color in the correct number of boxes to show how many of each animal are in the farm scene. Use the display to ask questions such as *Are there more pigs than sheep? How many cows and sheep are there altogether? How many animals are there in all? Are there fewer chickens than pigs? How many fewer chickens than pigs are there? Are there more horses or cows?* (Answers will vary.)

Verses

Poems and songs may be used any time after the pages noted beside each.

Tall and Small (anon.) (pages 183–184)

Here is a giant who is tall, tall, tall;
Here is an elf who is small, small, small;
The elf who is small will try, try, try
To reach to the giant who is high, high, high.

Six Little Monkeys (pages 185–186)

Six little monkeys jumping on the bed,
One fell off and bumped his head.
Momma called the doctor and the doctor said,
"How many monkeys are left on the bed?"

Repeat for five, four, three, two, and one.

Six Little Bunny Rabbits (pages 185–186)

Six little bunny rabbits sitting by a hive
One hopped away and then there were five.

Chorus: Bunny rabbits, bunny rabbits, brown and gray
        Bunny rabbits, bunny rabbits, hop away.

Five little bunny rabbits sitting by the door
One hopped away and then there were four.

Chorus.

Four little bunny rabbits looking at me
One hopped away and then there were three.

Chorus.

Three little bunny rabbits hiding in a shoe
One hopped away and then there were two.

Chorus.

Two little bunny rabbits, one plus one
One hopped away and then there was one.

Chorus.

One little bunny rabbit, his eating is done,
One hopped away and then there were none.

Silly Squirrel's Cake

One day Silly Squirrel went to the store to buy a cake. "How much is that pretty chocolate cake?" Silly Squirrel asked Mr. Rabbit.

"That cake is nine cents," he replied.

"Oh," said Silly Squirrel, "I only have three cents."

"Don't be silly, Silly Squirrel. To think you can get a nine-cent cake with only three cents is really silly."

So Silly Squirrel left. She thought and thought and thought about the cake. She wanted it very much.

Just then Kitty Cat came walking by.

"Hello, Kitty Cat," said Silly Squirrel. "How would you like to help me eat a pretty chocolate cake?"

"I would love it," said Kitty Cat.

"Do you have any money?" asked Silly Squirrel.

"Just three cents," said Kitty Cat.

"Then come with me," said Silly Squirrel. She knew that three cents and three cents added up to six cents. So Silly Squirrel and Kitty Cat went to look for someone else. Soon they saw Lucky Duck.

"Hello, Lucky Duck," said Silly Squirrel. "How would you like to help Kitty Cat and me eat a pretty chocolate cake?"

"I'd love it," said Lucky Duck.

"Do you have any money?" asked Silly Squirrel.

"Just three cents," said Lucky Duck.

"Good!" said Silly Squirrel. "Follow us!"

So Silly Squirrel led Kitty Cat and Lucky Duck straight to Mr. Rabbit's store.

"Mr. Rabbit," Silly Squirrel said, "I have come for the pretty chocolate cake!"

"Don't be silly!" Mr. Rabbit cried. "I told you that the chocolate cake is nine cents!"

"I know," said Silly Squirrel, "and Lucky Duck has three cents. Kitty Cat has three cents. That makes six cents. And I have three cents! That makes nine cents! We'll all buy the cake together . . . and we'll all three eat it!"

"Hmmm," said Mr. Rabbit as he handed Silly Squirrel the cake. "Maybe you aren't so silly after all!"

"Silly Squirrel's Cake," p. RM27

Activities

Use Reproducible Master RM27, "Silly Squirrel's Cake," to help students relate the story to the mathematical ideas presented.

- Give 3 pennies each to three students to act out the story as you reread it. Have students write and solve the addition problems on the board. (3¢ + 3¢ = 6¢; 6¢ + 3¢ = 9¢) Encourage students to count pennies to arrive at the sums. Allow time for more groups of three students to play the parts and write and solve their problems on the board. Vary the activity by changing the number of pennies for each character to 2¢.

- Give an equal number of students 1, 2, 3, or 4 pennies. Hold up an item priced at 5¢. Tell students to find a partner who has exactly the right amount of money to put with theirs so that the two of them could buy the item. Have students tell why they paired themselves. Have them write and solve an addition problem on the board to show why they could afford the item. Repeat the activity for items priced at 1¢, 2¢, 3¢, and 4¢. Students should see that some of them do not have to have a partner or pool their pennies to buy items and still have money left over.

pages 183–184

Objectives

- To understand take away as subtraction
- To define *subtracting, taking away, leaves*, and *opposite*

Vocabulary

subtracting, taking away, leaves, opposite

Materials

*yarn; *flannel board or overhead projector; *4 objects to count; number cards 0 through 9; 6 counters

Warm Up • Mental Math

Have students name the number.
1. 3 + 3 (6)
2. 8 + 1 (9)
3. 6 + 1 (7)
4. 2 + 6 (8)
5. 7 + 2 (9)
6. 5 + 1 (6)

Warm Up • Number Sense

Have students show the number card that is 1 less than a number from 1 to 9.

2 **Teach**

Develop Skills and Concepts Ask students to tell the opposite of cold. (hot) Ask for the opposite of yes. (no) Tell students that there are many opposites. Have students stand up. Ask them to do the opposite of stand up. (sit down) Have students perform more actions to show opposites such as face the front of the room (face the back), touch your left elbow (touch right elbow), and so on.

- Read the poem "Tall and Small" on page T183a. Have students act out the poem. Ask why an elf is the opposite of a giant. (An elf is small and a giant is large.)

- Arrange 5 chairs in a row. Have 5 students sit in the chairs. Ask students to tell how many students there are in all. (5) Have 2 students move away. Ask how many there are now (3) and how many went away (2). Tell students that the opposite of putting groups together is taking them apart, so the opposite of adding is subtracting.

- Write **5 minus 2 is** ☐ on the board. Read the subtraction sentence and then have students read aloud with you as you point to each number or word. Have students act out the problem. Have a student write the number in the problem that tells how many are left. (3) Repeat for 6 minus 1 (5), 4 minus 2 (2), and 5 minus 1 (4).

- Have 4 students stand in a yarn circle. Write **4** on the board. Ask 1 student to move away. After the 4, write **minus 1 is** ☐ on the board. Ask a student to write the number in the box to complete the subtraction sentence. (3) Have students read the sentence with you. Repeat for 5 minus 2 (3) and 6 minus 2 (4).

- Tell students that Jan had 5 kittens and gave 3 to friends. Tell students that we want to know how many kittens Jan has left. Have students use counters as you write **5 minus 3 is** ☐ on the board. Have a student write the answer. (2) Tell other stories for subtracting 1 or 2 from 6, 5, 4, or 3.

Lesson 10-1

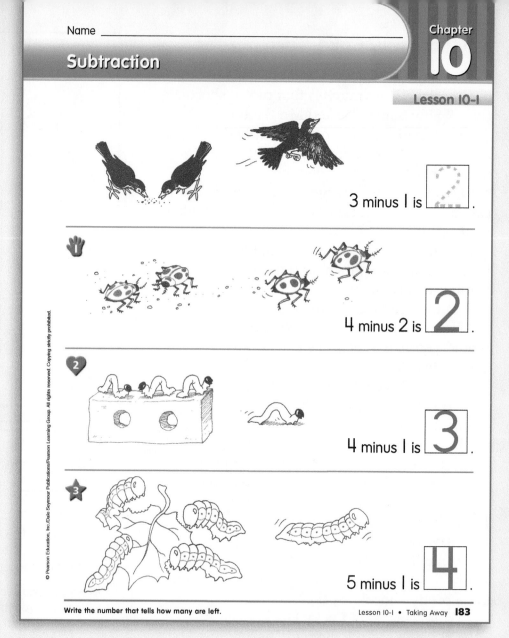

3 minus 1 is ☐ 2 .

4 minus 2 is 2 .

4 minus 1 is 3 .

5 minus 1 is 4 .

Write the number that tells how many are left.

Lesson 10-1 • Taking Away **183**

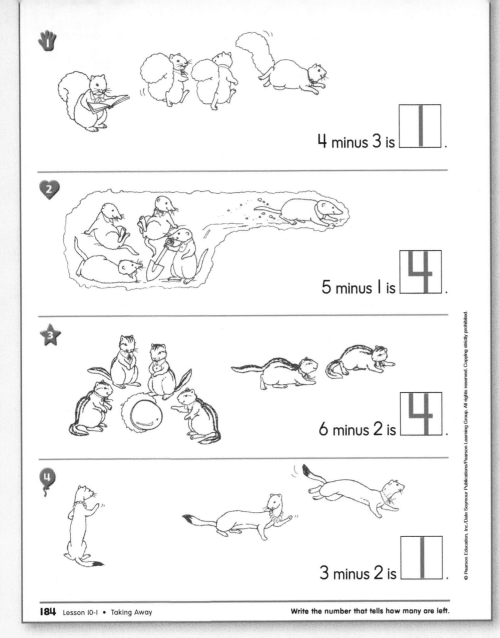

4 minus 3 is $\boxed{1}$.

5 minus 1 is $\boxed{4}$.

6 minus 2 is $\boxed{4}$.

3 minus 2 is $\boxed{1}$.

184 Lesson 10-1 • Taking Away

Write the number that tells how many are left.

For Mixed Abilities

Common Errors • Intervention

Some students may need help with the concept of subtraction. Line up 6 chairs and tell students to pretend that they are cars. Have students pretend to drive 1 or 2 cars away and tell how many are left. Write each subtraction sentence on the board and have students read it aloud with you.

Enrichment • Number Sense

1. Have students write an addition problem to show the opposite of $4 - 2 = 2$. ($2 + 2 = 4$)

2. Ask students to draw a picture that tells this story:

   There were 5 balls in a box, but 2 rolled away. (The picture should show 3 balls in the box and 2 rolling away from the box.)

ESL/ELL STRATEGIES

Discuss with students all the terms related to subtraction as you encounter them. Terms include *subtraction, minus,* and *take away.* Model activities on the board using each term; for example, "Four minus one is three."

3 Practice

Using page 183 Use the flannel board or overhead projector to show 4 objects. Ask students how many objects they see. (4) Move 1 object away and ask how many were removed. (1) Ask students how many objects are left. (3) Have them say with you, *Four minus one is three.* Repeat as necessary for 5 minus 1 (4), 3 minus 2 (1), and 4 minus 2 (2).

- Work through the example with students, asking how many birds in all are shown. (3) Ask how many birds are flying away. (1) Ask how many birds are left to eat the birdseed. (2) Have students trace the 2 to show that 2 birds are left. Have students read the number sentence aloud with you. Tell them to use their counters if necessary to complete the remaining exercises independently.

Using page 184 Remind students to first count the number of animals in all and look to see how many animals are going away. Tell them to then count the number of animals left and write this number in the box. Have students complete the page independently.

4 Assess

Place 3 objects on an overhead projector. Take 1 object away. Ask students how many objects are left. (2) Have students write the number sentence for this action. ($3 - 1 = 2$) Now, have the class read this subtraction sentence aloud. (Three minus one is two.)

T184

10-2 Subtracting 1

pages 185–186

1 Getting Started

Objective
- To write the numbers in a subtraction sentence when 1 object is taken away from 6 or less

Materials
*6 objects to count; *flannel board or overhead projector; 6 counters

Warm Up • Mental Math
Tell students to name the number.
1. 1 more than 20 (21)
2. 1 less than 19 (18)
3. 1 less than 5 (4)
4. 2 more than 6 (8)
5. 1 less than 30 (29)
6. 2 more than 5 (7)

Warm Up • Number Sense
Have students use counters to act out 3 take away 2 leaves 1, 6 take away 2 leaves 4, 5 take away 2 leaves 3, and 4 take away 2 leaves 2.

2 Teach

Develop Skills and Concepts Play Musical Chairs with 6 chairs in a circle. Have 6 students sit in the chairs. Start the music and have students stand and walk around the chairs as you remove 1 chair. Stop the music and have students sit. Ask students to tell what happened. (1 student has no chair.) Write **6 minus 1 is 5** on the board and have students read it aloud with you, *Six minus one is five.* Continue until there is only 1 student left. Write each subtraction sentence on the board and have students read it aloud with you.

- Have a student place 5 objects on a table. Take 1 object away and write **5 minus 1 is 4** on the board. Have students read the sentence with you. Repeat for 4 minus 1 (3), 3 minus 1 (2), and 2 minus 1 (1).

- Ask students to solve the following problem using counters:

  There were 4 apples on the tree. One apple became ripe and fell off the tree. How many apples were left on the tree? (3)

  Tell other story problems of 1 being subtracted from 6, 5, 3, or 2. Write each subtraction sentence on the board as it is solved and have students read it with you.

Write the number that tells how many there are in all, how many are leaving, and how many are left.

Lesson 10-2 • Subtracting 1 **185**

- Read the poem "Six Little Monkeys" on page T183a and have students answer the question at the end of each verse. Have students then recite the poem as they act it out.

- Read the poem "Six Little Bunny Rabbits" on page T183a and have students act out each verse.

  It's Algebra! The concepts of this lesson prepare students for algebra.

3 Practice

Using page 185 Use a flannel board or overhead projector to work through the example with students. Have a student tell how many cats are shown and place that number of objects on the board or projector surface. (2) Tell students to trace the 2 to show that there are 2 cats in all. Have a student tell how many cats are going away and subtract that number. (1) Tell students to trace the

T185

1 | 4 minus 1 is 3 .

2 | 3 minus 1 is 2 .

3 | 6 minus 1 is 5 .

4 | 5 minus 1 is 4 .

Write the number that tells how many there are in all, how many are leaving, and how many are left.

For Mixed Abilities

Common Errors • Intervention

Some students may need more practice with subtracting 1. Display the number cards 1 through 6 in order on a table. Ask a student to point to the number that is 1 less than 4. (3) Repeat for the number that is 1 less than 6 (5), 1 less than 5 (4), 1 less than 3 (2), and 1 less than 2 (1).

Enrichment • Number Sense

1. Make a spinner for numbers 1 through 6. Have students spin and tell the number that would be left if 1 is taken away from the number spun.

2. Have students think of times when 1 is taken away from a group. (Possible answers: musical chairs, losing a mitten or sock, eating 1 of a handful of raisins)

ESL/ELL STRATEGIES

Various uses of the verb *leave* may be confusing to some students. Explain as follows: *How many are leaving?* means "How many are in the process of going away right now?" *How many are left?* means "How many remain?" or "How many didn't go?"

1 to show that 1 cat is going away. Ask students how many cats are left. (1) Tell them to trace the 1 in the last box and then read the number sentence as "Two minus one is one." If necessary, work through Exercise 1 with the class before having students complete the page independently.

Using page 186 Have students look at Exercise 1 and point out the 4 raccoons with 1 of them walking away. Tell students that they are to write these numbers (4, 1) in the first two boxes and then in the last box write the number of raccoons left (3). Tell students that they may use their counters to help them complete the page independently.

4 Assess

Give students 6 counters. Tell them to take 1 counter away. Ask students how many counters are left. (5) Have students write the number sentence with you (6 − 1 = 5) and say it aloud, *Six minus one is five.*

Write the number that tells how many there are in all,
how many are leaving, and how many are left.

Lesson 10-3 • Subtracting 2 **187**

Lesson 10-3

It's Algebra!

1 Getting Started

Objectives
- To write the numbers in a subtraction sentence when 2 objects are taken away from 6 or less
- To use the minus and the equal signs in a number sentence
- To define *minus* and *subtract*
- To recognize related addition and subtraction facts

Vocabulary
minus, subtract

Materials
*hanger and 6 clip-on clothespins; *flannel board or overhead projector; 6 counters

Warm Up • Mental Math
Ask students how many there are in all.
1. 3 plus 2 (5)
2. 5 and 1 more (6)
3. 1 less than 4 (3)
4. 5 take away 1 (4)
5. 2 more than 2 (4)
6. 4 take away 2 (2)

Warm Up • Number Sense
Have students watch closely as you stack 5 books and remove 1. Have a student tell about the action as you write the problem on the board. (5 take away 1 leaves 4.) Repeat for 1 taken away from 6, 3, or 4.

2 Teach

Develop Skills and Concepts Show students a hanger as you place 6 clothespins on it. Ask how many clothespins they see. (6) Remove 1 pin and ask how many are left. (5) Write **6 minus 1 is 5** on the board and have students read it aloud. Tell students that the minus or subtraction sign can be used instead of the word *minus* as you write **6 − 1** under the subtraction sentence. Tell them that the equal sign can be used instead of the word *is* as you complete the number sentence on the board, **6 − 1 = 5**, and read, *Six minus one equals five.* Continue taking away one clothespin at a time and help students write each number sentence through 2 − 1 = 1. Note that 1 − 1 = 0 has not been introduced, though some may be ready to work

with zero objects left. Repeat this procedure to subtract 2 from 6, 5, 4, and 3.

- Have 5 students stand and then 2 sit down. Have students verbalize each part of the action. (There are 5 students. 2 students sit down. 3 students are still standing.) Have a student write the problem on the board. (5 − 2 = 3) Have students read aloud with you, *Five minus two equals three.*

- Write **5 − 2 = 3** on the board and help students tell a story about this subtraction sentence. Then, ask what addition sentence would tell about putting back the 2 they took away. (3 + 2 = 5 or 2 + 3 = 5) Repeat for 4 − 2, 5 − 1, 6 − 2, and 3 − 2.

It's Algebra! The concepts of this lesson prepare students for algebra.

T187

$$4 - 2 = 2$$

$$3 - 2 = 1$$

$$6 - 2 = 4$$

$$5 - 2 = 3$$

Write the number that tells how many there are in all,
how many are leaving, and how many are left.

188 Lesson 10-3 • Subtracting 2

For Mixed Abilities

Common Errors • Intervention

Some students may need more practice to understand subtraction sentences. Have students work with partners. One partner reads a subtraction sentence for subtracting 1 or 2 from 6 or less. The other partner uses counters to act out the sentence and show the answer. Partners can change roles to repeat the activity.

Enrichment • Number Sense

1. Have students work in pairs to write a subtraction sentence. Then, have them act it out for classmates to write and solve the subtraction sentence on the board.

2. Have students draw a picture showing 6 minus 2 equals 4.

3. Have students name a situation in which subtraction can be used. (Possible answer: There are 4 students in our group, but 2 are absent. Now, we have 2.)

3 Practice

Using page 187 Demonstrate the example on the flannel board or the overhead projector as students tell you how many objects to show (3) and how many to take away (2). Have students trace the 3 and the 2. Ask how many squirrels are left. (1) Tell them to trace the 1 to show this. Have them read the number sentence aloud with you, *Three minus two equals one.* Work through Exercise 1 if necessary before assigning the page to be completed independently.

Using page 188 Remind students that they are to look at each picture and write the number of animals in all, the number of animals going away, and the number of animals that are left. Tell students to complete the page independently.

4 Assess

Place 6 counters on the overhead projector. Take 2 counters away. Ask students how many counters are left. (4) Now, have students write the corresponding subtraction sentence. (6 − 2 = 4)

T188

10-4 Subtracting 3, 4, and 5

pages 189–190

1 Getting Started

Objectives
- To write differences in vertical problems with minuends through 6
- To recognize related addition and subtraction facts

Materials
6 blocks; string and 6 beads; Reproducible Master RM28; 6 counters

Warm Up • Mental Math
Ask students which is greater.

1. 2 + 2 or 5 (5)
2. 3 + 1 or 3 (3 + 1)
3. 0 or 6 (6)
4. 31 or 30 (31)
5. 5 + 1 or 4 (5 + 1)
6. 5 − 1 or their age (age)

Warm Up • Number Sense
Say a stanza of the poem "Six Little Monkeys" on page T183a, having 6 or fewer monkeys jumping on the bed and 1 or 2 falling off. Have students tell the number left and then write the subtraction sentence on the board.

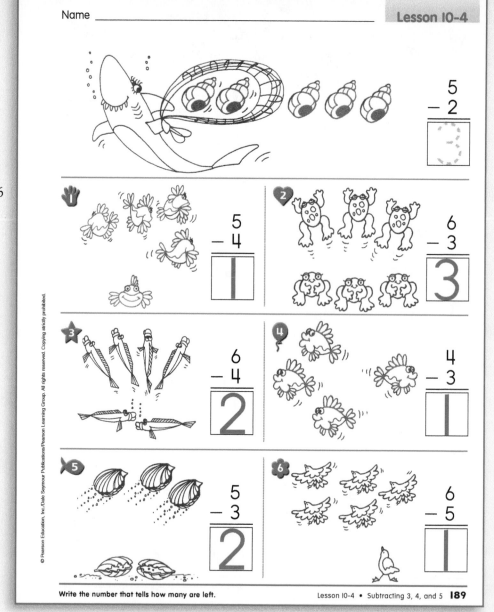

Write the number that tells how many are left.

Lesson 10-4 • Subtracting 3, 4, and 5 **189**

2 Teach

Develop Skills and Concepts Have students build towers of 6 or fewer blocks and take away some of the blocks. Have students write their subtraction sentences on the board so that all students' tower actions are shown. Have students read each subtraction sentence in unison.

- Have students string 5 beads. Tell them to take 4 beads away. Have a student write and solve the problem on the board to tell how many beads are left. (5 − 4 = 1) Have students read the number sentence aloud. Now, have students put the 4 beads back as you write **1 + 4 = 5** on the board. Have them read the addition sentence with you as *One plus four equals five*. Repeat for 6 − 4, 5 − 3, 6 − 5, and 6 − 3.

- Write **5 − 4 = ☐** on the board both horizontally as shown and vertically. Then, point to the number sentence and tell students, *Five minus four can be written like this*. Point to the vertical problem and tell students, *Subtraction problems can also be written like this*. Have students act out each problem with counters and then

ask volunteers to solve each problem on the board. (1) Write more horizontal and vertical subtraction problem pairs on the board for them to act out and then solve.

- Give students copies of Reproducible Master RM28. Tell students to find the first 3 circles. Tell them to place counters beside those circles to show 2 less than 3. (1) Continue similarly down the page to have students place 1 or 2 fewer counters than the number of shapes shown. Remove the counters. Now, have students draw 2 less than 3 circles in the space beside the circles. (1) Continue similarly to have them draw 1 or 2 fewer shapes than the number shown.

3 Practice

Using page 189 Encourage students to act out the example. Have a student place 5 objects on the table and take 2 away. Ask how many are left when 2 are taken away from 5. (3)

T189

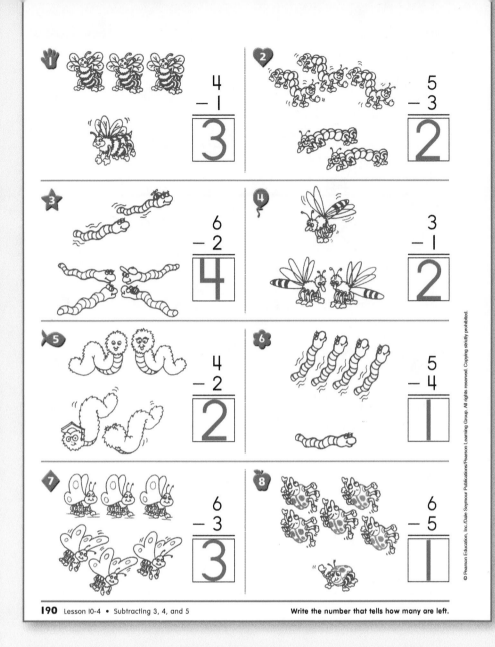

Common Errors • Intervention
Some students may need more practice working problems in vertical form. Have students read each subtraction problem on page 188 aloud with you. Have a volunteer then use the vertical form to write and solve the problem at the board as other students make sure it is correct.

Enrichment • Number Sense
Have students build a tower by adding 1 or 2 blocks at a time, and then unbuild the tower by removing any number of blocks at a time. Help students verbalize that addition builds and subtraction is the opposite, takes away. You may want to use the words *increase* and *decrease* in this activity.

190 Lesson 10-4 • Subtracting 3, 4, and 5

Write the number that tells how many are left.

Have students trace the 3 to show that 3 are left. Work through Exercise 1 with students if necessary and then have them complete the page independently.

Using page 190 Have students look at and read through each exercise in unison before assigning the page to be completed independently.

4 Assess

Have each student place 6 counters on his or her desk. Now, have students take away 4 of the counters and ask them to tell how many counters are left. **(2)** Tell students to write this as a vertical subtraction problem.

Name _____

1 Getting Started

Objectives
- To cross out objects to be subtracted
- To count to find differences in vertical problems with minuends through 6

Materials
6 counters; 3-by-5-inch index card

Warm Up • Mental Math
Have students name the number.
1. 6 minus 3 (3)
2. 3 plus 2 (5)
3. 1 of 3 equal parts (one-third)
4. 5 minus 2 (3)
5. their age and 1 more (Answers will vary.)
6. 1 less than their age

Warm Up • Number Sense
Write **5 − 1** vertically on the board. Have a student draw 5 objects on the board and then erase 1 to show the action and write the answer. (4) Repeat for more subtraction problems.

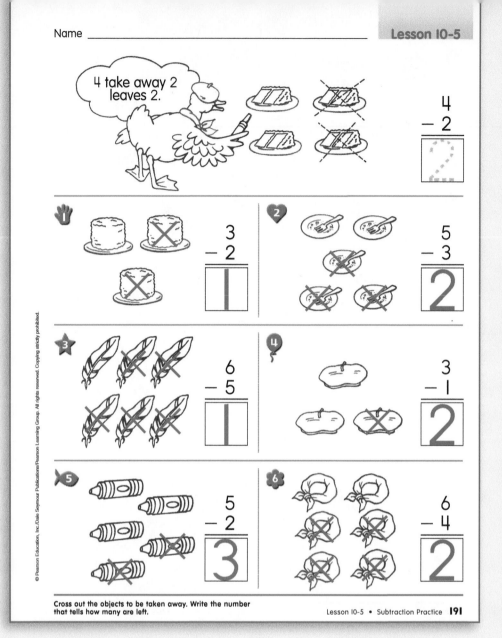

Cross out the objects to be taken away. Write the number that tells how many are left.

Lesson 10-5 • Subtraction Practice **191**

2 Teach

Develop Skills and Concepts Give 6 counters and 1 index card to each student. Tell students to lay out 4 counters and then use the card to cover up 2 of the 4 counters. Ask how many are left. (2) Repeat for more subtraction problems.

- Students should have 6 counters and an index card. Write **4 − 3** on the board and ask students how many counters they need to lay out. (4) Ask how many counters will be covered up with the index card. (3) Ask how many counters are left showing. (1) Continue the procedure for more subtraction problems.

- Write **5 − 2 = □** on the board. Draw 5 circles and ask how many circles there are. (5) Ask how many need to be taken away. (2) Have a volunteer cross out 2 of the circles. Ask how many circles are left. (3) Have a student write the answer in the box. (3) Repeat for 6 − 4 = □, 5 − 4 = □, 5 − 3 = □, and 6 − 5 = □, having students work each problem with their counters, also.

- Have students assist in making a list on the board of the subtraction facts with minuends through 6. Begin by telling students that we want to see what

subtraction problems they can do as you write **6 − 1 = □** on the board. Have students work the problem with counters. Have a student complete the problem on the board. (5) Tell students that they subtracted 1 from 6, so next they need to subtract 2 from 6 as you write **6 − 2 = □** under the first problem. Have them work the problem with counters. Have a student complete the problem on the board. (4) Continue until the following facts are displayed:

6 − 1 = 5	5 − 1 = 4	4 − 1 = 3	3 − 1 = 2	2 − 1 = 1
6 − 2 = 4	5 − 2 = 3	4 − 2 = 2	3 − 2 = 1	
6 − 3 = 3	5 − 3 = 2	4 − 3 = 1		
6 − 4 = 2	5 − 4 = 1			
6 − 5 = 1				

T191

1.
$$\begin{array}{r} 2 \\ -\ 1 \\ \hline 1 \end{array}$$

2.
$$\begin{array}{r} 4 \\ -\ 1 \\ \hline 3 \end{array}$$

3.
$$\begin{array}{r} 6 \\ -\ 2 \\ \hline 4 \end{array}$$

4.
$$\begin{array}{r} 5 \\ -\ 3 \\ \hline 2 \end{array}$$

5.
$$\begin{array}{r} 5 \\ -\ 2 \\ \hline 3 \end{array}$$

6.
$$\begin{array}{r} 6 \\ -\ 1 \\ \hline 5 \end{array}$$

7.
$$\begin{array}{r} 5 \\ -\ 4 \\ \hline 1 \end{array}$$

8.
$$\begin{array}{r} 4 \\ -\ 3 \\ \hline 1 \end{array}$$

Cross out the objects to be taken away. Write the number that tells how many are left.

192 Lesson 10-5 • Subtraction Practice

For Mixed Abilities

Common Errors • Intervention

Some students may need more practice with subtraction. Have them work in pairs with counters and a ring made of yarn on their work area. Have one partner place 6 counters inside the ring and cover 2 with an index card. Have the other partner write and solve the subtraction problem just acted out. (6 − 2 = 4) Have students reverse roles and continue for other subtraction problems.

Enrichment • Number Sense

1. Have students draw a picture of or cut and paste 4, 5, or 6 objects on a sheet of paper. Have students cover some of the objects for a partner to write and solve the subtraction problem.

2. Read the "Six Little Monkeys" poem on page T183a for the class to write and solve the problems.

---

3 Practice

Using page 191 Have students look at the example at the top of the page and tell how many pieces of cake are seen by the duck. (4) Have students point to the 4 in the problem. Ask how many pieces are crossed out. (2) Tell students to trace the dashed Xs through 2 of the pieces. Ask how many pieces are left. (2) Have students trace the 2 to show that 2 pieces are left. Have students read aloud with you, *Four minus two equals two.*

• Work through Exercises 1 and 2 with students since this is the first time they are asked to cross out the objects to be subtracted. When they seem comfortable with the exercises, have the page completed independently.

Using page 192 Remind students that they are to read the exercise to see how many objects are to be subtracted and then mark an X on that many objects before finding the answer. Have them complete the page independently.

4 Assess

Have students draw 6 circles on a sheet of paper. Next, have them write 6 − 1 = ☐ on the paper. Tell students to cross out the appropriate number of circles. (1) Now, tell students to find the answer and write this number in the box. (5)

10-6 Subtraction Facts

1 Getting Started

Objective
- To practice facts with minuends through 6

Materials
*6 objects to count; string and beads; 6 counters

Warm Up • Mental Math
Have students name the number.
1. between 19 and 21 (20)
2. 5 pennies and 1 more (6 pennies or 6¢)
3. 2 more than 4 (6)
4. 2 less than 5 (3)
5. 6 take away 4 (2)
6. next after 27 (28)

Warm Up • Number Sense
Draw 6 circles and write $6 - 4 = \square$ on the board. Ask a student to cross out the appropriate number of circles (4), solve the subtraction problem, and then write the answer in the box (2). Continue for other problems using 5 or 6 circles.

2 Teach

Develop Skills and Concepts Have a student stand beside a table that has 5 objects on it. Ask how many objects are on the table. (5) Tell the student to pick up 3 of the objects. Ask how many objects are now on the table. (2) Write $\square - \square = \square$ on the board and have a student complete the problem. (5 − 3 = 2) Repeat for 6 − 3, 5 − 4, 6 − 5, and 4 − 3.

- Write subtraction problems related to sums through 6 on the board. Have students string beads to show each action and then write the answer on the board.

- Have students use counters to show the action and then tell the number that completes each of the following stanzas:

  6 squeaky mice running out the door;
  2 fell behind, and then there were ___(4)___.

  5 five-year-olds swimming in the sea;
  2 stopped for lunch, and then there were ___(3)___.

  4 cackling witches scaring everyone;
  3 flew away, and then there was ___(1)___.

Encourage students to make up silly stanzas for
5 − 3 = 2 (Possible answer: 5 young ladies had much work to do; 3 ran off to play, and then there were 2.)
and 6 − 1 = 5 (Possible answer: 6 little children going for a ride; 1 went home, and then there were 5).

3 Practice

Using page 193 Have students look at the example at the top of the page. Ask students how many acorns in all are shown. (5) Ask how many acorns the mouse has drawn an X on. (2) Ask students why the mouse has marked an X on 2 of the acorns. (The problem says to subtract 2.) Tell students to trace the Xs on the 2 acorns and tell the number of acorns left. (3) Tell them to trace the 3. Have students read the number sentence aloud, "Five minus two equals three." Work similarly through Exercise 1 before having students complete the page independently.

T193

$5 - 2 = \boxed{3}$

1. $4 - 3 = \boxed{1}$

2. $3 - 1 = \boxed{2}$

3. $6 - 5 = \boxed{1}$

4. $5 - 3 = \boxed{2}$

5. $5 - 1 = \boxed{4}$

6. $4 - 2 = \boxed{2}$

7. $6 - 3 = \boxed{3}$

8. $5 - 4 = \boxed{1}$

9. $3 - 2 = \boxed{1}$

10. $6 - 4 = \boxed{2}$

Cross out the objects to be taken away. Write the number that tells how many are left.

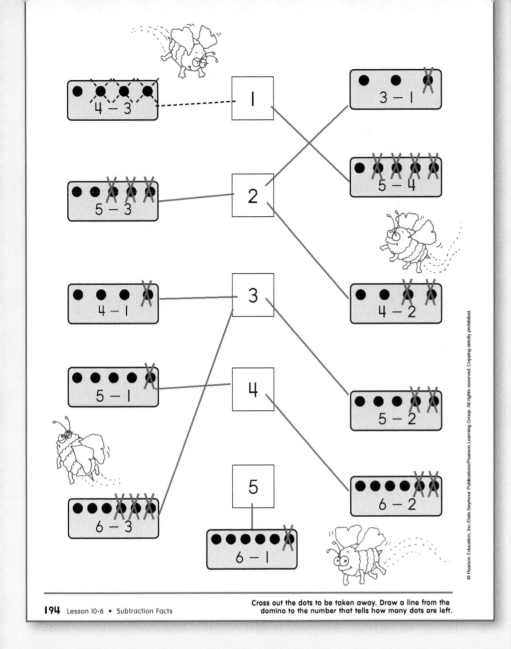

Cross out the dots to be taken away. Draw a line from the domino to the number that tells how many dots are left.

For Mixed Abilities

Common Errors • Intervention

Some students may need more practice subtracting from minuends through 6. Have 6 students act out the poem "Six Little Monkeys" on page T183a. Have classmates write and solve each problem presented in the poem. Change the numbers to include other subtrahends.

Enrichment • Number Sense

1. Have students draw around one hand on a sheet of paper, decorate only some of the fingers with rings, and then write a subtraction problem to tell how many fingers are not decorated.

2. Have students throw two number cubes and write a subtraction problem using the two numbers shown.

Using page 194 Tell students that they are to cross out the number of dots the problem tells them to subtract, find the number of dots that are left, and draw a line to that number. Work through the first problem written on the first domino in the top left and have students trace the dashed line to the number 1. Work through the next problem with them. Tell students that some of the number squares will be used more than once. Have students complete the page independently.

4 Assess

Draw 6 stars on the board. Now, cross out 3 of these stars. Have students write the corresponding subtraction fact horizontally and vertically. (6 − 3 = 3)

T194

10-7 Subtracting From 6¢ and Less

pages 195–196

1 Getting Started

Objective
• To solve subtraction problems with minuends through 6¢

Materials
*number cards 1 through 6; 6 pennies

Warm Up • Mental Math
Have students name the number.

1. 1 less than 10 (9)
2. on the right of 12 on a clock (1)
3. a dozen (12)
4. yesterday's date (Answers will vary.)
5. third in their phone number
6. their shoe size

Warm Up • Number Sense
Have students act out some silly rhymes from Lesson 10-6, pages T193–T194, for classmates to solve the subtraction problems stated in those rhymes.

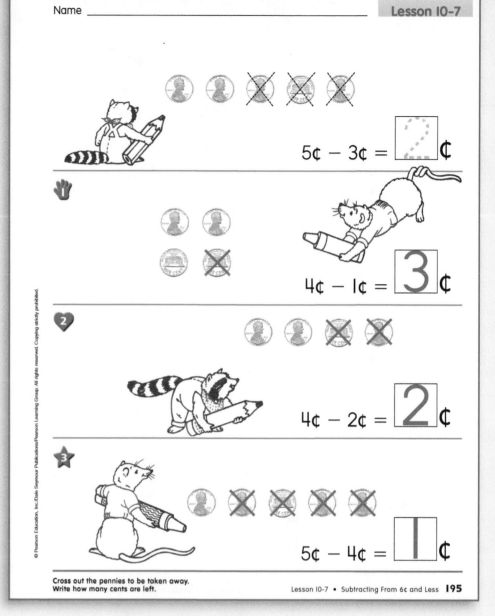

$5¢ - 3¢ = \boxed{2}¢$

$4¢ - 1¢ = \boxed{3}¢$

$4¢ - 2¢ = \boxed{2}¢$

$5¢ - 4¢ = \boxed{1}¢$

Cross out the pennies to be taken away.
Write how many cents are left.

2 Teach

Develop Skills and Concepts Place the number cards 1 through 6 in order along a table. Write $6 - 1 = \square$ on the board. Have a student match the number cards to the problem and then select the number card that answers the problem. (5) Continue for $5 - 4 = \square$, $3 - 2 = \square$, $6 - 5 = \square$, and $6 - 4 = \square$.

• Repeat the above activity using cent signs in each problem. Have students show each problem with pennies.

• Draw 6 circles on the board and write **1¢** in each circle. Ask how many pennies are shown. (6) Write $6¢ - 4¢ = \square¢$ on the board. Have a student cross out the number of pennies to be taken away (4) and write the number left in the box (2). Continue for other

subtraction problems that begin with 6 or fewer pennies. Have students work each problem with their pennies.

• Repeat the previous activity for solving vertical problems on the board.

• Make up story problems such as the following for students to solve with their pennies:

I have 5 cents and want to buy a toy that costs 3 cents. How much money will I have left? (2 cents)

Encourage students to make up story problems for classmates to solve. Write each problem on the board and have them read each completed problem in unison.

T195

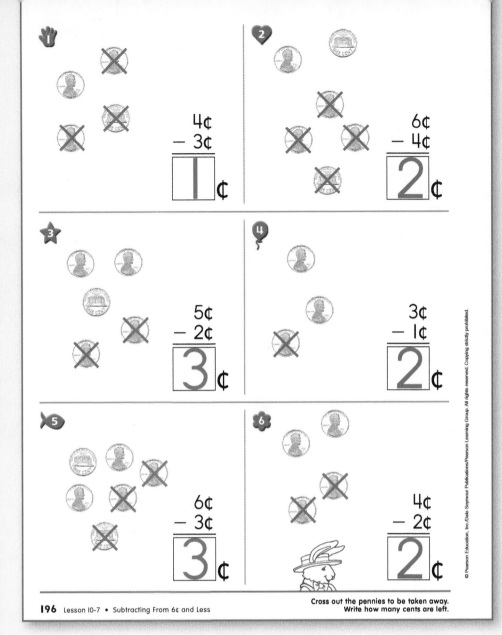

1.
$$\begin{array}{r} 4\text{¢} \\ -\ 3\text{¢} \\ \hline \boxed{1}\text{¢} \end{array}$$

2.
$$\begin{array}{r} 6\text{¢} \\ -\ 4\text{¢} \\ \hline \boxed{2}\text{¢} \end{array}$$

3.
$$\begin{array}{r} 5\text{¢} \\ -\ 2\text{¢} \\ \hline \boxed{3}\text{¢} \end{array}$$

4.
$$\begin{array}{r} 3\text{¢} \\ -\ 1\text{¢} \\ \hline \boxed{2}\text{¢} \end{array}$$

5.
$$\begin{array}{r} 6\text{¢} \\ -\ 3\text{¢} \\ \hline \boxed{3}\text{¢} \end{array}$$

6.
$$\begin{array}{r} 4\text{¢} \\ -\ 2\text{¢} \\ \hline \boxed{2}\text{¢} \end{array}$$

Cross out the pennies to be taken away.
Write how many cents are left.

196 Lesson 10-7 • Subtracting From 6¢ and Less

For Mixed Abilities

Common Errors • Intervention

Some students may have difficulty subtracting amounts of money. Have them work with partners and a set of problems such as $5\text{¢} - 2\text{¢} = \boxed{}\text{¢}$. Have them use pennies to represent the minuend, take away the amount indicated by the subtrahend, and then count the remaining pennies to find how many are left. One partner can check the work. If correct, both partners exchange roles for another subtraction problem.

Enrichment • Number Sense

1. Have each student cut a picture of a toy from a magazine or catalog and price the item 5¢ or less. Place all pictures in a box. Ask a student to draw a picture of an item to buy and to tell how many pennies are left from 6 pennies.

2. Give each student 1 nickel to use to buy items that are in the pictures used in the previous activity. Have students rotate playing the role of a cashier with at least 20 pennies to give change. Allow each student in the class to buy an item and check the change.

---

3 Practice

Using page 195 Ask students how many pennies are shown in the example at the top of the page. (5) Ask how many pennies must be taken away. (3) Tell students to trace the dashed Xs on these 3 pennies. Ask how many pennies are left. (2) Have students trace the 2. Have them read the money sentence aloud, "Five cents minus three cents equals two cents." Work through Exercise 1 with students before having them complete the page independently.

Using page 196 Work through Exercise 1 with students before having them complete the page independently.

4 Assess

Draw 5 circles on the board. Write **1¢** inside each circle. Tell students that these circles are pennies. Now, cross out 3 of these pennies. Ask students how many pennies are left. (2) Have students write the corresponding vertical subtraction problem. ($5\text{¢} - 3\text{¢} = 2\text{¢}$)

T196

10-8 Problem Solving: Choose an Operation

pages 197–198

1 Getting Started

Objective
- To choose adding or subtracting to find sums through 9

Materials
*floor number line through 6; 6 pennies; 6 counters

Warm Up • Mental Math
Have students tell the number(s).
1. 23 comes before _____. (24)
2. 16 comes after _____. (15)
3. 27 is between _____ and _____. (26, 28)
4. 4 o'clock comes after _____ o'clock. (3)
5. One-fourth is 1 of _____ equal parts. (4)
6. 6 is more than _____. (5, 4, 3, 2, 1, and 0)
7. 3 and 3 more is _____. (6)

Warm Up • Number Sense
Have students lay out 4 pennies and tell how many more are needed to have 6 pennies in all. (2) Repeat for laying out 1 through 5 pennies to have 6 in all. Continue to have 5 or 4 pennies in all.

Name _____

It's Algebra!

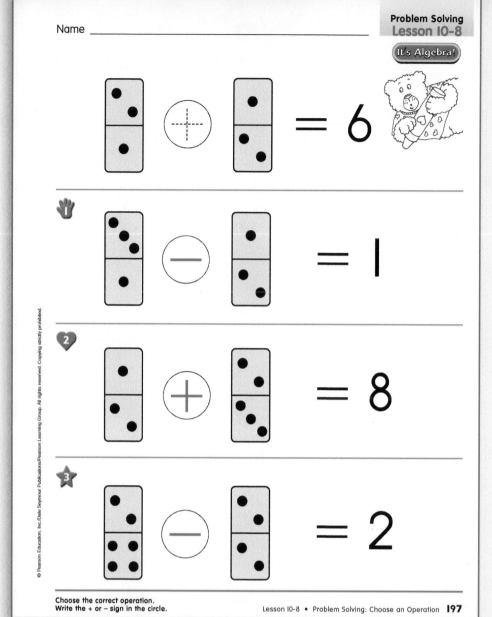

Choose the correct operation.
Write the + or – sign in the circle.

Lesson 10-8 • Problem Solving: Choose an Operation **197**

2 Teach

Develop Skills and Concepts Have students lay out 5 counters as you say, *Five busy beavers chewing on a tree; 2 swam away, and then there were ___(3)___.* Have students show the action with counters. Ask the number left when 2 are moved away from 5. (3) Ask if this is a subtraction problem or an addition problem. (subtraction) Ask if a minus or plus sign would be used. (minus sign) Have a student write and solve the problem on the board. (5 − 2 = 3) Repeat using verses like the following:

6 busy beavers gnawing on a door; 2 swam away, then there were ___(4)___. (6 − 2 = 4)

Leave all subtraction sentences on the board.

- Repeat the first rhyme above, having students move 2 counters away from 5. Have a student find the problem on the board that tells the action. (5 − 2 = 3) Now say, *Come back beavers! Come! Look alive! 2 beavers come back and now there are ___(5)___.* Have them show the action by moving the 2 counters back. Ask the number now. (5) Ask if this action showed addition or

subtraction. (addition) Ask if a plus or minus sign would be used. (plus sign) Have a student write and solve the problem on the board under 5 − 2 = 3. (3 + 2 = 5) Repeat using the following:

Come back beavers! Here are some sticks! The beavers come back and now there are ___(6)___.

Have a student write the addition sentence that shows this. (2 + 4 = 6)

It's Algebra! The concepts in this lesson prepare students for algebra.

3 Practice

Using page 197 Have students look at the example at the top of the page.

- Now, apply the four-step plan. For SEE, ask, *What are you asked to do?* (find the operation that was used to get an answer of 6) Then ask, *What information are you*

T197

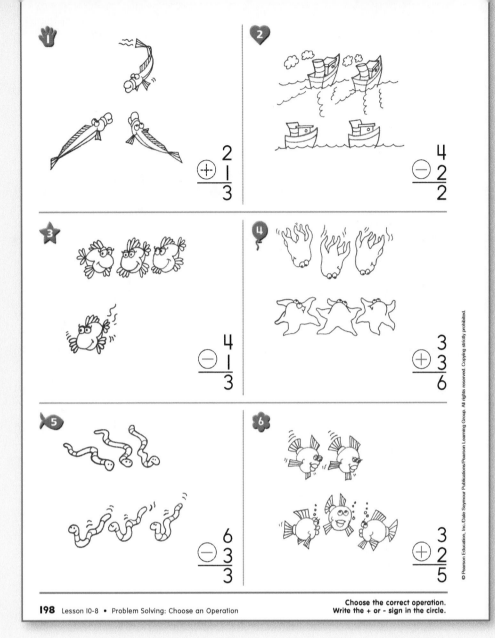

1

$$\begin{array}{r} \oplus\ \ 2 \\ 1 \\ \hline 3 \end{array}$$

2

$$\begin{array}{r} \ominus\ \ 4 \\ 2 \\ \hline 2 \end{array}$$

3

$$\begin{array}{r} \ominus\ \ 4 \\ 1 \\ \hline 3 \end{array}$$

4

$$\begin{array}{r} \oplus\ \ 3 \\ 3 \\ \hline 6 \end{array}$$

5

$$\begin{array}{r} \ominus\ \ 6 \\ 3 \\ \hline 3 \end{array}$$

6

$$\begin{array}{r} \oplus\ \ 3 \\ 2 \\ \hline 5 \end{array}$$

Choose the correct operation.
Write the + or − sign in the circle.

For Mixed Abilities

Common Errors • Intervention
Some students may need more help to see the difference between addition and subtraction. Write **4 + 2 = 6** on the board. Have a student walk the number line to show this problem. Then, write **6 − 2 = ☐** on the board and have the student walk backward to find the answer. **(4)** Repeat with other students for more sums through 6.

Enrichment • Number Sense
1. Have students draw placemats on a paper tablecloth to show 2 visitors came to have dinner with a family of 4.

2. Have students fold a pre-lined paper into 3 equal parts. Tell them to draw 5 flowers in a vase on the first third. Tell students that the middle third should show that 3 flowers wilted and were thrown away. Then, have students show that the 3 flowers were replaced in the last third.

---

given? (2 dominoes; each has 3 dots; an answer of 6) For PLAN, ask students how they will solve the problem. (I will place enough counters on my desk to represent each dot shown. Then, I will decide if I have to take away counters to get to 6 or join counters to get to 6.) For DO, have students follow their plan to notice that they will have to join, or add, all the counters to get 6 counters in all. For CHECK, have students regroup the 6 counters on their desks. Then, have students use a sheet of paper to cover 3 of the counters. Ask, *How many counters are left?* (3) Point out that the first 3 counters that were covered represent the dots on the first domino and the second set of 3 represents the dots on the second domino. Now, have students trace the addition sign in the circle.

• Work through Exercise 1 with students for subtraction. Then, have students complete the page independently. Be sure students write the sign in the circle for each exercise.

Using page 198 Tell students that on this page they are to write a plus or minus sign in each exercise to show the action. The students can find clues in the pictures and in the answers.

• Have students look at Exercise 1. Ask if the 1 fish is leaving or joining the other 2 fishes. (joining) Ask students if the picture shows adding or subtracting. (adding) Ask if 2 plus 1 equals 3. (yes) Tell them to trace the plus sign. Work similarly through the next several exercises before assigning the page to be completed independently.

4 Assess

Write **5 ◯ 2 = 3** and **5 ◯ 2 = 7** on the board. Then, have students use counters to decide which operation sign should be placed in each number sentence. (5 − 2 = 3; 5 + 2 = 7)

T198

10-9 Addition and Subtraction Practice

pages 199–200

1 Getting Started

Objectives
- To find sums through 6 and solve related subtraction problems without pictures
- To solve mixed addition and subtraction problems

Materials
*6 objects; *fact cards for sums and minuends through 6; 6 counters

Warm Up • Mental Math
Tell students to name the next number.
1. 3, 4, 5 (6)
2. 25, 24, 23 (22)
3. 17, 18, 19 (20)
4. 11, 10, 9 (8)
5. 28, 29, 30 (31)
6. 16, 15, 14, 13 (12)

Warm Up • Number Sense
Write **6** on the board and draw 3 Xs beside it. Have students tell how many more Xs need to be drawn to have 6 in all. (3) Have a student draw the 3 Xs. Write **3 + 3 = □** and **6 − 3 = □** on the board for students to complete. Repeat for more sums through 6.

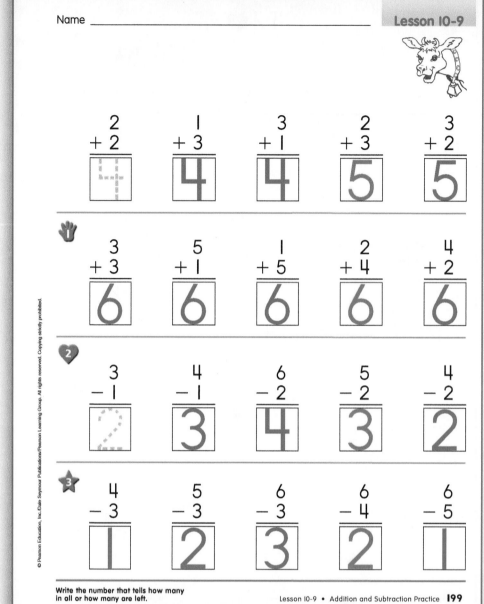

Write the number that tells how many in all or how many are left.

2 Teach

Develop Skills and Concepts Write the plus and the minus signs on the board and have students tell the meaning of each sign. (plus sign: add; minus sign: subtract) Now, write 6 ○ 4 = 2 and 4 ○ 2 = 6 on the board and have students tell which sign goes in each circle. (6 − 4 = 2; 4 + 2 = 6) Have a student write the correct sign in each circle. Repeat for more problems for students to choose the correct sign. Point out to students that addition problems should have an answer that is greater than the two numbers in the problem, unless an addend is zero. Also, point out that in a subtraction problem, the answer will be less than the two numbers in the problem, unless the subtrahend is zero.

- Write the following across the board:

6	5	3	6	5
− 3	− 4	− 2	− 4	− 2
(3)	(1)	(1)	(2)	(3)

2	2	3	1	3
+ 3	+ 4	+ 3	+ 5	+ 2
(5)	(6)	(6)	(6)	(5)

Help students notice that all the problems in the first row have the minus sign and those in the second row have the plus sign. Have them act out each problem using counters, write the answer, and then read the problem.

- Tell students that they will now see the problems mixed up with some minus signs and some plus signs. Write the problems in one row across the board, mixing the two groups. Have them read the problems in unison, being sure to read the correct sign in each. Have students work each problem with their counters and write the answers on the board. Leave the problems on the board.

T199

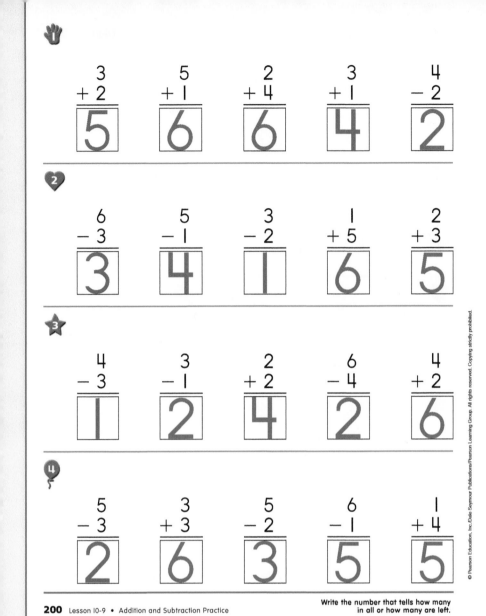

$$\begin{array}{r} 3 \\ + 2 \\ \hline 5 \end{array} \qquad \begin{array}{r} 5 \\ + 1 \\ \hline 6 \end{array} \qquad \begin{array}{r} 2 \\ + 4 \\ \hline 6 \end{array} \qquad \begin{array}{r} 3 \\ + 1 \\ \hline 4 \end{array} \qquad \begin{array}{r} 4 \\ - 2 \\ \hline 2 \end{array}$$

$$\begin{array}{r} 6 \\ - 3 \\ \hline 3 \end{array} \qquad \begin{array}{r} 5 \\ - 1 \\ \hline 4 \end{array} \qquad \begin{array}{r} 3 \\ - 2 \\ \hline 1 \end{array} \qquad \begin{array}{r} 1 \\ + 5 \\ \hline 6 \end{array} \qquad \begin{array}{r} 2 \\ + 3 \\ \hline 5 \end{array}$$

$$\begin{array}{r} 4 \\ - 3 \\ \hline 1 \end{array} \qquad \begin{array}{r} 3 \\ - 1 \\ \hline 2 \end{array} \qquad \begin{array}{r} 2 \\ + 2 \\ \hline 4 \end{array} \qquad \begin{array}{r} 6 \\ - 4 \\ \hline 2 \end{array} \qquad \begin{array}{r} 4 \\ + 2 \\ \hline 6 \end{array}$$

$$\begin{array}{r} 5 \\ - 3 \\ \hline 2 \end{array} \qquad \begin{array}{r} 3 \\ + 3 \\ \hline 6 \end{array} \qquad \begin{array}{r} 5 \\ - 2 \\ \hline 3 \end{array} \qquad \begin{array}{r} 6 \\ - 1 \\ \hline 5 \end{array} \qquad \begin{array}{r} 1 \\ + 4 \\ \hline 5 \end{array}$$

200 Lesson 10-9 • Addition and Subtraction Practice

Write the number that tells how many
in all or how many are left.

For Mixed Abilities

Common Errors • Intervention

Some students may need more practice with the facts. Shuffle the fact cards with minuends and sums through 6. Show a fact card for students to tell if there is a plus or minus sign. Have students use counters if necessary to solve the problem. Continue for more facts.

Enrichment • Number Sense

1. Have students draw pictures to show $6 - 2 = 4$ and $3 + 2 = 5$.

2. Have students tell a story of what might happen if we added instead of subtracted for the problem $4 - 2 = 2$. (Possible answer: I had 4 cookies and bought 2 more cookies. So, now I have 6 cookies.)

• Provide 6 objects on a table. Circle a problem on the board and erase its answer. Have a student tell if the problem says to add or subtract. Tell the student to use the objects on the table to act out the problem and write the answer on the board. Continue until all problems are acted out.

3 Practice

Using page 199 Note that this page asks students to work without the help of pictures. Ask students if they will add or subtract in the first problem in the example. (add) Have students read the problem in unison and then trace the sum of 4. Now, ask what is to be done in each of the problems in the example. (add) Have students read each problem in unison. Repeat for the problems in Exercise 1, having students tell what is to be done and then read each problem orally. Tell students to stop at the end of the row. You may want to have them color the line to help them remember to stop. Have students complete the problems in the example and Exercise 1 independently.

• Similarly, introduce the rows of subtraction problems in Exercises 2 and 3 and then have students complete both rows independently.

Using page 200 Ask students to go back to page 199 and tell how this page differs. (Problems of adding and subtracting are mixed up.) Ask students to place a finger on a subtraction problem in the first row. Repeat for students to identify a few more problems to be sure they are looking closely at each problem and its sign. Tell them to be very careful to look at the sign in each problem on this page before working the problem. Have them complete the page independently.

4 Assess

Have students write one addition problem and one subtraction problem using the numbers 4 and 2.
($4 + 2 = 6$; $4 - 2 = 2$)

T200

Subtracting From 9 and Less

pages 201–202

1 Getting Started

Objective
• To subtract 1 or 2 from 7, 8, or 9

Materials
*floor number line; *9 objects;
*fact cards for minuends through 9
where 1 or 2 is taken away; 9 counters

Warm Up • Mental Math
Have students name the number.

1. 4 + 2 (6)
2. 2 + 3 (5)
3. 3 − 1 (2)
4. 6 − 3 (3)
5. 3 + 2 (5)
6. 6 − 2 (4)

Warm Up • Number Sense
Draw a number line on the board for
numbers 0 through 9. Have a
student start at 6, add 1 more, and
show the sum on the number line.
(7) Continue for sums through 9
where 1 or 2 is one of the addends.

2 Teach

Develop Skills and Concepts Have
students show 6 − 4 on the number
line on the floor. (2) Repeat for other
subtraction facts for minuends
through 6.

• Give 9 counters to each student. Ask how many
counters each student has. (9) Tell students to move
1 counter away from the others as you write **9 − 1 = □**
on the board. Ask students how many counters are left.
(8) Have a student complete the problem on the board.
(8) Repeat for 9 − 2, 8 − 1, 8 − 2, 7 − 1, and 7 − 2.
Have students read each completed problem in unison.

• Write **9 − 2 = □** on the board and have a student act
out the problem using 9 objects on a table. Ask a
student to write the number left in the box. (7) Have
students read the problem in unison. Repeat for acting
out 8 − 2, 7 − 1, 6 − 2, and 7 − 2. Write some of the
problems on the board in vertical notation.

• Draw 9 circles on the board and write **9 − 1 = □**
under the circles. Ask students to read the problem in
unison and then tell how many circles need to be
crossed out. (1) Have a student cross out 1 circle and
tell how many circles are left. (8) Ask another student to
write the number in the box. (8) Continue having

students cross out objects to be subtracted and solve the
following problems in vertical or horizontal notation:
8 − 2 = □, 7 − 1 = □, 5 − 4 = □, 9 − 2 = □,
and 6 − 4 = □.

3 Practice

Using page 201 Have students look at the example at the
top of the page and count to see how many clocks there
are. (7) Have students place a finger on the 7 in the
subtraction sentence. Tell them to read the number
sentence to see how many clocks are to be subtracted. (2)
Tell students to trace the dashed Xs on the two clocks and
tell how many clocks are left. (5) Tell them to trace the 5
to show that 5 clocks are left. Have students read the
number sentence aloud, "Seven minus two equals five."

• Now, have students tell what they are to do in Exercises
1 to 4. (cross out the number of objects to be
subtracted and write the number left) Have students
complete the page independently.

T201

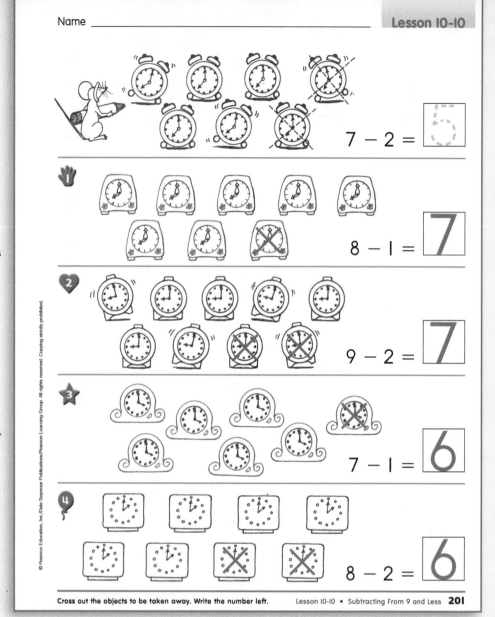

Name _____

Lesson 10-10

7 − 2 = 5

8 − 1 = 7

9 − 2 = 7

7 − 1 = 6

8 − 2 = 6

Cross out the objects to be taken away. Write the number left.   Lesson 10-10 • Subtracting From 9 and Less   **201**

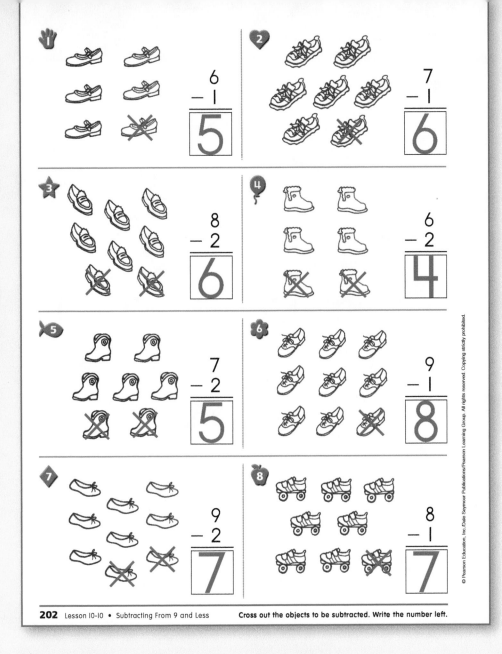

202 Lesson 10-10 • Subtracting From 9 and Less   **Cross out the objects to be subtracted. Write the number left.**

For Mixed Abilities

Common Errors • Intervention

Some students may need more practice with subtraction. Have students use counters to solve subtraction problems shown on the fact cards with minuends through 9 but subtrahends of 1 or 2. Then, have volunteers write and solve facts at the board as students take turns reading the minuend and subtrahend from the fact cards.

Enrichment • Number Sense

1. Have students draw a domino showing 5 dots and 2 dots. Tell students to tell about the domino by writing a subtraction problem using the domino. (7 − 2 = 5)

2. Have students make a subtraction fact card for each answer if 2 is taken away from 3, 4, and so on through 9.

Using page 202 Tell students that for Exercises 1 to 8, they are to cross out the number of objects to be subtracted and then write the number left. Have students complete the page independently.

4 Assess

Read the poem "Six Little Monkeys" on page T183a, changing the number to 9, 8, 7, or 5 monkeys with 1 or 2 monkeys falling off the bed in each verse. Have students write and solve the problem on the board for each verse.

Chapter 10 Test
and Challenge

pages 203–204

1 Getting Started

Test Objectives

- **Items 1–5:** To understand take-away subtraction; to find differences with minuends through 6 (see pages 183–194)
- **Item 6:** To solve subtraction problems with minuends through 6¢ (see pages 195–196)
- **Item 7:** To subtract 1 or 2 from 7, 8, or 9 (see pages 201–202)

Challenge Objective

- To recognize the plus and minus signs to solve mixed problems

Materials

7 counters; addition and subtraction cards with sums and minuends through 6; fact cards for minuends 7, 8, and 9 where 1 or 2 is taken away; brown, purple, red, and blue crayons

2 Give the Test

Prepare for the Test Have students lay out 7 counters on their desks. Then, have them move 2 counters away from the group. Ask students to tell how many counters are left. (5) Now, have students write the corresponding subtraction problem in both vertical and horizontal form. (7 − 2 = 5)

- After students have mastered subtracting from minuends through 6 or 6¢ and subtracting 1 or 2 from minuends of 7, 8, and 9, have them complete the Chapter 10 Test. You may wish to review test-taking rules with students. Remind them that they are not to talk to each other and they are to finish the test on their own. Decide ahead of time whether or not to allow the use of manipulatives such as counters.

Using page 203 Students are familiar with each exercise on this page and should be able to complete the page independently. In Exercises 1 to 5, students are subtracting from minuends through 6. They are to cross out the objects that are to be taken away first and then write the number that represents the amount of objects that are left. In Exercise 6, students are subtracting from a minuend of 5¢. Students are to cross out the number of

pennies that are to be taken away and then write the number of pennies that are left. In Exercise 7, students are subtracting 2 from the minuend 8. They are to cross out the correct number of acorns and then write the number that tells how many acorns are left.

3 Teach the Challenge

Develop Skills and Concepts for the Challenge Have students work in pairs with addition and subtraction fact cards with sums and minuends through 6. Students can also use addition and subtraction fact cards with sums and minuends through 9 only if one of the addends or the subtrahend is 1 or 2. Have partners split the cards and quiz each other. The student gets to keep the card if he or she answers the question correctly. Be sure students look at the operation sign on each card before trying to solve the fact.

$$5 - 3 = \boxed{2}$$

$$\begin{array}{r} 4 \\ -\ 1 \\ \hline \boxed{3} \end{array}$$

$$\begin{array}{r} 5 \\ -\ 4 \\ \hline \boxed{1} \end{array}$$

$$\begin{array}{r} 6 \\ -\ 3 \\ \hline \boxed{3} \end{array}$$

$$4 - 2 = \boxed{2}$$

$$\begin{array}{r} 5¢ \\ -\ 2¢ \\ \hline \boxed{3}¢ \end{array}$$

$$\begin{array}{r} 8 \\ -\ 2 \\ \hline \boxed{6} \end{array}$$

Cross out the objects to be taken away. Write the number left.

For Mixed Abilities

Common Errors • Intervention

Some students may not be able to write a subtraction problem in two forms. Have students work in pairs with 4 counters. Tell students to take 1 counter from the group. Ask, *How many counters are left?* (3) Have one partner write this as a subtraction sentence with words only, "Four minus one equals three." Then, have the other partner write the number sentence, $4 - 1 = 3$. Now, have the first partner write this vertically. Point out that the line in the vertical form represents the equal sign and the word *equals*.

Enrichment • Number Sense

1. Have students explain why 3 cannot be taken away from 2, 6 from 4, and so on. Allow students to use counters.

2. Have students draw objects to show all the numbers that can be subtracted from 9.

3. Have students draw 8 objects and write the number. Tell students to cross out 2 objects and write the number left, cross out 2 more objects and write the number left, and so on until no objects are left. Have them read the numbers in order from smallest to largest.

Crayon code:

2 Brown   3 Purple
4 Red   5 Blue

$$\begin{array}{r} 1 \\ +4 \\ \hline 5 \end{array} \qquad \begin{array}{r} 5 \\ -2 \\ \hline 3 \end{array}$$

$$\begin{array}{r} 5 \\ -1 \\ \hline 4 \end{array} \qquad \begin{array}{r} 2 \\ +3 \\ \hline 5 \end{array} \qquad \begin{array}{r} 6 \\ -1 \\ \hline 5 \end{array} \qquad \begin{array}{r} 2 \\ +1 \\ \hline 3 \end{array}$$

$$\begin{array}{r} 3 \\ +1 \\ \hline 4 \end{array} \qquad \begin{array}{r} 4 \\ -1 \\ \hline 3 \end{array} \qquad \begin{array}{r} 1 \\ +3 \\ \hline 4 \end{array}$$

$$\begin{array}{r} 6 \\ -3 \\ \hline 3 \end{array} \qquad \begin{array}{r} 4 \\ +1 \\ \hline 5 \end{array} \qquad \begin{array}{r} 1 \\ +2 \\ \hline 3 \end{array}$$

Write the answers. Color the picture using the number code.

Using page 204 Ask students to read the number before each crayon at the top of the page and then read the color word associated with that number. Tell students that they will do some coloring on this page after they work the problems. Tell them that they are to first find the answer to all the problems in the bubbles. Remind students that some problems have plus signs and some have minus signs. Ask them to find a problem that asks them to add 2 plus 3. Repeat for students to find several more problems to be sure they are looking closely at the sign in each problem. Tell students that they are to complete all the problems and then color the bubble brown if the answer is 2, purple if the answer is 3, red if the answer is 4, and blue if the answer is 5. Remind students that they will not color until they have worked all the problems. Supervise them as they complete the page independently.

4 Assess

Alternate Chapter Test You may wish to use the Alternate Chapter Test on page 214 of this manual for further review and assessment.

Challenge Have students write a subtraction problem using 5 and 3 that has 2 as the answer. Then, students should color the number according to page 204.
($5 - 3 = 2$; the 2 should be colored brown.)

 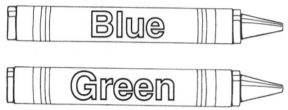

 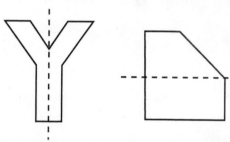

Circle the bug that is under the flower. Circle the bird that is in the middle. Color each crayon the correct color. Circle the ribbon that is the shortest. Color the triangles blue. Circle the square that is small and white. Circle the object with the same shape. Circle the shape with matching parts. Circle the shape that comes next in the pattern.

2   3   4   5

2   3   4   5

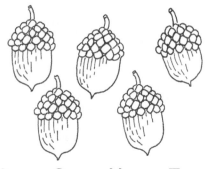

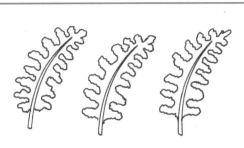

1   2   3   4

2   3   4   5

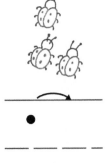

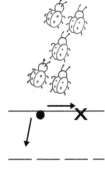

Draw a line from each ladybug to its flower to show one-to-one correspondence.
Then circle the group that has more. Circle the number that tells how many objects
are in the set. Write the numbers 1 through 5.

Name _____

0   1   2   3

3   4   5   6

6   7   8   9

6   7   8   9

6   7   8   9

6   7   8   9

_____

_ _ _ _ _

_____

_____

_ _ _ _ _

_____

_____

_____

_____

_____

_ _ _ _ _

_____

Circle the number that tells how many objects are in the set.
Count the objects in the set and write the number.

MCP Mathematics © Pearson Education, Inc./Dale Seymour Publications/Pearson Learning Group. All rights reserved.

1 9   10   11   12

2 9   10   11   12

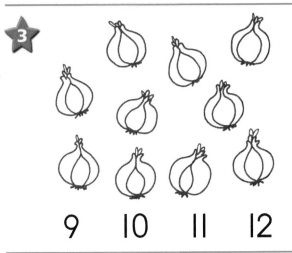

3 9   10   11   12

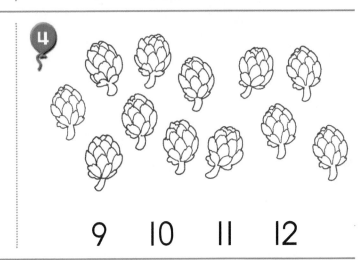

4 9   10   11   12

5

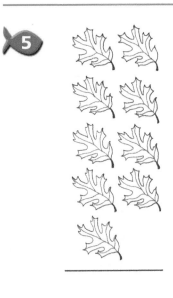

_____   _____   _____

_ _ _ _   _ _ _ _   _ _ _ _

_____   _____   _____

Circle the number that tells how many objects are in the set.
Count the objects in the set and write the number.

Circle the event that happens first. Circle the activity that takes less time. Circle the digital clock that matches the clock face. Count the dime and pennies. Then write how much money. Circle the coins needed to buy the item.

1

13 14 15 16

2

17 18 19 20

3

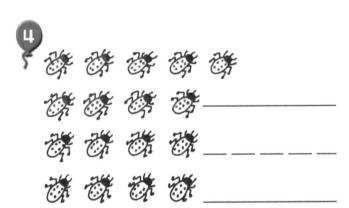

4

_____

_ _ _ _ _ _ _

_____

5

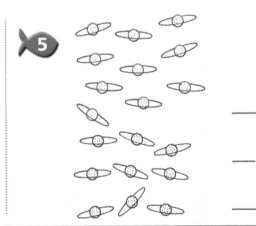

_____

_ _ _ _ _ _ _

_____

6

15¢ 17¢ 19¢

Circle the number that tells how many objects are in the set. Circle the coins
needed to buy the item. Count the objects in the set and write the number.
Circle the number that tells how much money there is in the set.

1

_____

_____

_ _ _ _ _ _ _ _

 2

_____

_____

_ _ _ _ _ _ _ _

3 fourth

4 ninth

5

_____

_ _ _ _ _ _ _ _

_____

6

_____

_ _ _ _ _ _ _ _

_____

7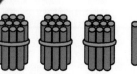

_____

_ _ _ _ _ _ _ _

8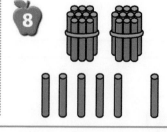

_____

_ _ _ _ _ _ _ _

9

_____

_ _ _ _ _ _ _ _

10

_____

_ _ _ _ _ _ _ _

Count the objects in the set and write the number. Circle the lesser number and draw a box around the greater number. Circle the object in the position named by the ordinal number. Write the number that tells how many there are.

1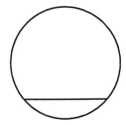

2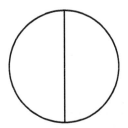

3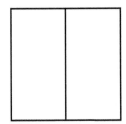

4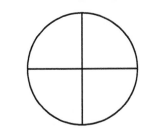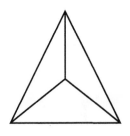

5

6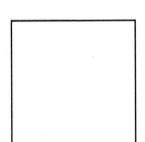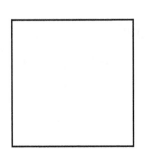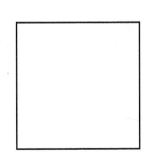

Circle the figure with equal parts. Mark an X on the figure that does not have equal parts.
Color one-half, one-fourth, and one-third. Circle the bucket that holds more. Circle the
object that weighs less. Count the number of paper clips in the length of the feather.
Put the ladybugs into three equal groups. Write the numbers.

$$4 + 1 = \boxed{}$$

$$\boxed{} + \boxed{} = \boxed{}$$

$$\boxed{} + \boxed{} = \boxed{}$$

Write the number that tells how many objects there are, how many are in each set, and how much there is in all.

$$4 - 3 = \boxed{}$$

$$\begin{array}{r} 6 \\ -\ 4 \\ \hline \boxed{} \end{array}$$

$$\begin{array}{r} 5 \\ -\ 1 \\ \hline \boxed{} \end{array}$$

  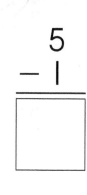

$$\begin{array}{r} 6 \\ -\ 5 \\ \hline \boxed{} \end{array}$$

$$3 - 2 = \boxed{}$$

 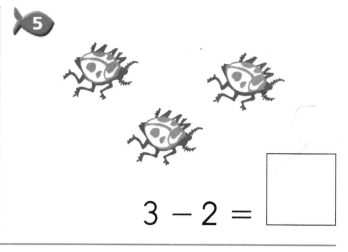

$$\begin{array}{r} 5¢ \\ -\ 3¢ \\ \hline \boxed{}¢ \end{array}$$

$$\begin{array}{r} 9 \\ -\ 2 \\ \hline \boxed{} \end{array}$$

Cross out the objects being taken away. Write the number left.

Chapter 1 Alternate Test

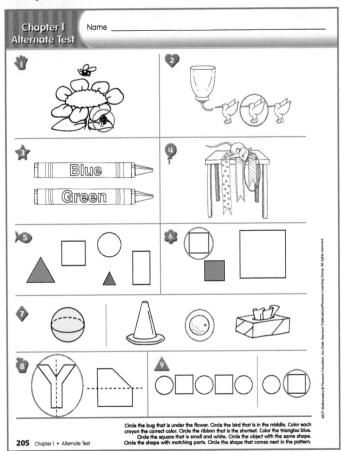

Circle the bug that is under the flower. Circle the bird that is in the middle. Color each crayon the correct color. Circle the ribbon that is the shortest. Color the triangles blue. Circle the square that is small and white. Circle the object with the same shape. Circle the shape with matching parts. Circle the shape that comes next in the pattern.

Chapter 2 Alternate Test

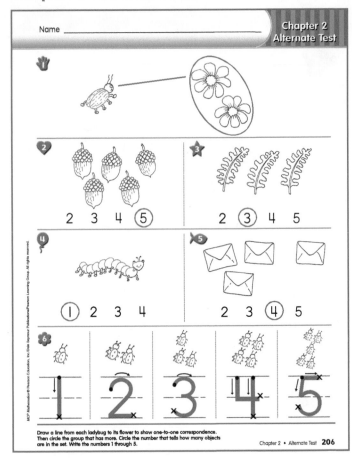

Draw a line from each ladybug to its flower to show one-to-one correspondence. Then circle the group that has more. Circle the number that tells how many objects are in the set. Write the numbers 1 through 5.

Test Objectives page 205

- **Item 1:** To describe position using *top, bottom, in, out, over, under, on top of, off, above, below, beside, outside,* and *inside* (see pages 1–2)
- **Item 2:** To describe position using *first, last, middle, between, front, behind, right,* and *left* (see pages 3–4)
- **Item 3:** To identify red, yellow, blue, orange, green, purple, brown, and black; to recognize each color word (see pages 5–6)
- **Item 4:** To compare using *shorter, shortest, taller, tallest, longer,* and *longest* (see pages 9–10)
- **Item 5:** To identify a circle, rectangle, square, and triangle; to identify and compare the attributes of plane shapes (see pages 11–12)
- **Item 6:** To sort and classify a set of shapes by more than one attribute (see pages 13–14)
- **Item 7:** To identify a sphere, cone, cube, and cylinder; to identify and compare the attributes of solid figures; to match objects to outlines of their shapes (see pages 15–16)
- **Item 8:** To identify shapes with a line of symmetry (see pages 17–18)
- **Item 9:** To identify patterns using color, size, or shape; to complete a pattern; to identify the shape that comes next in a pattern (see pages 19–22)

Test Objectives page 206

- **Item 1:** To use one-to-one correspondence to compare the objects in two sets without counting; to match objects from two groups to compare for more or less (see pages 25–28)
- **Items 2–5:** To review the numbers 1 to 5 (see pages 29–34, 39–42)
- **Item 6:** To write the numbers 1 to 5 (see pages 35–38, 41–42)

Chapter 3 Alternate Test

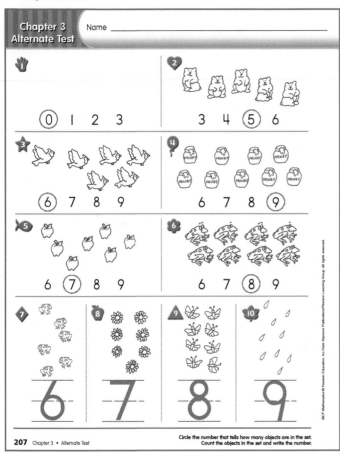

Circle the number that tells how many objects are in the set.
Count the objects in the set and write the number.

Chapter 4 Alternate Test

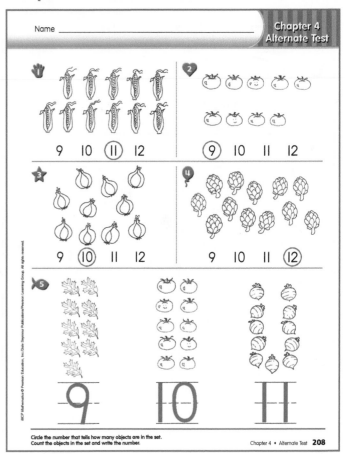

Circle the number that tells how many objects are in the set.
Count the objects in the set and write the number.

Test Objectives page 207

- **Items 1–6:** To review numbers 0 through 9 (see pages 45–50, 53–56, 61–64)
- **Items 7–10:** To write numbers 6 through 9 (see pages 51–52, 57–58, 61–62)

Test Objectives page 208

- **Items 1–4:** To count 0 to 12 objects and choose the correct number (see pages 67–72, 75–76)
- **Item 5:** To write 9, 10, and 11 in order (see pages 73–74, 79–80)

Chapter 5 Alternate Test

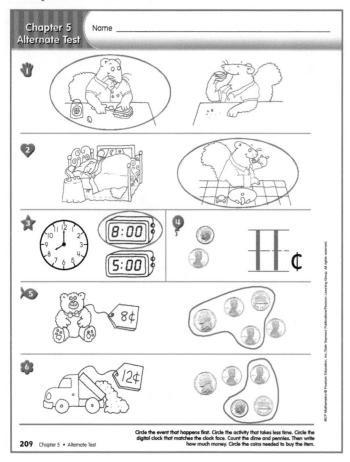

Circle the event that happens first. Circle the activity that takes less time. Circle the digital clock that matches the clock face. Count the dime and pennies. Then write how much money. Circle the coins needed to buy the item.

Chapter 6 Alternate Test

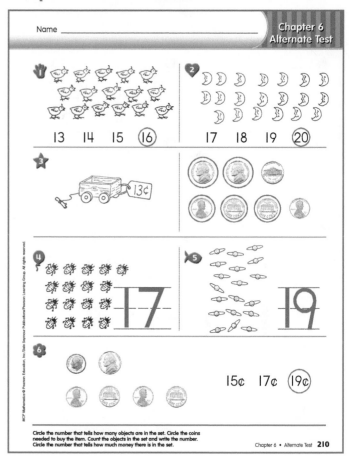

Circle the number that tells how many objects are in the set. Circle the coins needed to buy the item. Count the objects in the set and write the number. Circle the number that tells how much money there is in the set.

Test Objectives page 209

- **Item 1:** To arrange pictures in sequence (see pages 85–86)
- **Item 2:** To compare the duration of two activities; to identify which activity takes more time or less time (see pages 87–88)
- **Item 3:** To match time on a digital clock to time on an analog clock (see pages 91–92)
- **Item 4:** To count coins for amounts through 12¢ (see pages 93–98)
- **Items 5–6:** To choose coins to buy an item priced through 12¢ (see pages 96, 99–100)

Test Objectives page 210

- **Items 1–2, 4–5:** To count 13 through 20 objects and choose or write the correct number (see pages 105–108, 113–116)
- **Item 3:** To count combinations of pennies, nickels, and dimes through 16¢; to match items priced through 16¢ to the correct set of coins (see pages 111–112)
- **Item 6:** To circle the amount for combinations of pennies, nickels, and dimes through 20¢ (see pages 117–118)

Chapter 7 Alternate Test

Chapter 8 Alternate Test

Test Objectives page 211

- **Items 1–2:** To count and write the number for 10 or less; to circle a group having fewer or more objects than another group (see pages 125–126)
- **Item 3:** To use ordinal numbers first through fifth (see pages 127–128)
- **Item 4:** To use ordinal numbers sixth through tenth (see pages 129–130)
- **Items 5–10:** To count 0 through 31 objects; to write 11 through 31; to use groups of 10 to identify and write numbers through 31 (see pages 131–132, 135–138)

Test Objectives page 212

- **Item 1:** To identify figures with equal and unequal parts (see pages 143–144)
- **Item 2:** To identify one-half, one-fourth, and one-third of a figure (see pages 145–150)
- **Item 3:** To identify a container that holds more or less than another (see pages 153–154)
- **Item 4:** To identify an object that is heavier or lighter than another (see pages 157–158)
- **Item 5:** To measure lengths using nonstandard units (see pages 155–156)
- **Item 6:** To split groups with no more than 10 objects into 2 or 3 equal groups (see pages 151–152)

Chapter 9 Alternate Test

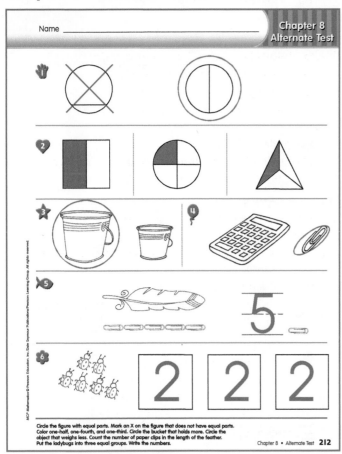

Test Objectives page 213

- **Items 1–2, 4–5:** To join 2 groups for sums through 9; to write an addition problem using both vertical and horizontal forms (**see pages 161–171, 177–178**)
- **Item 3:** To add pennies for sums through 6¢ (**see pages 173–176**)

Chapter 10 Alternate Test

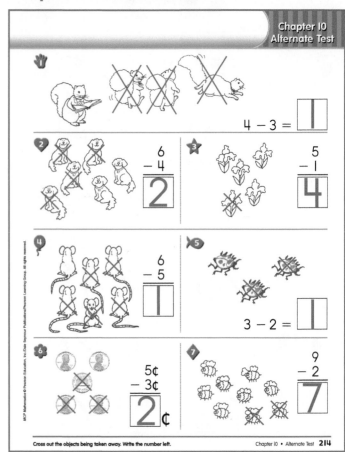

Test Objectives page 214

- **Items 1–5:** To understand take-away subtraction; to find differences with minuends through 6 (**see pages 183–194**)
- **Item 6:** To solve subtraction problems with minuends through 6¢ (**see pages 195–196**)
- **Item 7:** To subtract 1 or 2 from 7, 8, or 9 (**see pages 201–202**)

Concepts and Directions for the Reproducible Masters

CHAPTER ONE

"Little Spot's Surprise," page RM1 (*Use to open the chapter.*)
After you read the story, check that students understand each character's role in the story.

Position, page RM2 (*Use with Lessons 1-1 and 1-2.*)
Have students practice using position words.

Colors, page RM3 (*Use with Lesson 1-3.*)
Have students practice using color words and colors.

Patterns, page RM4 (*Use with Lessons 1-10 and 1-11.*)
Display patterns of various colors that students can copy in each line of squares. Then, discuss the next color in each pattern.

CHAPTER TWO

"Three Nice Mice," page RM5 (*Use to open the chapter.*)
After you read the story, check for students' understanding of position words.

As Many As, page RM6 (*Use with Lesson 2-1.*)
Show students one-to-one correspondence and two equal sets with the same number of objects.

Numbers, page RM7 (*Use with Lessons 2-6 through 2-9.*)
Have students practice writing the numbers 1 through 5.

Numbers 1 Through 5, page RM8 (*Use after Lesson 2-8.*)
Have students practice number recognition.

CHAPTER THREE

"Scampy's Party," page RM9 (*Use to open the chapter.*)
After you read the story, check that students have listened carefully for details in the story. Note that Scampy's plate has four cookies, whereas the other plates have three each.

Numbers 1 Through 9, page RM10 (*Use after Lesson 3-6.*)
Have students practice identifying numbers 1 through 9 by counting different groups in the picture.

CHAPTER FOUR

"The Nine-Bee Sneeze," page RM11 (*Use to open the chapter.*)
After you read the story, have students explain the action in the pictures using details from the story.

CHAPTER FIVE

"The Flea Race," page RM12 (*Use to open the chapter.*)
After you read the story, have students explain the action in the picture in relation to the story.

Telling Time, page RM13 (*Use with Lesson 5-3.*)
Have students practice placing the numbers on a clock face in the correct order and then show different times.

Digital Clocks, page RM14 (*Use with Lesson 5-4.*)
Have students practice properly reading analog and digital clocks.

CHAPTER SIX

"Time for 12-Nut Pie," page RM15 (*Use to open the chapter.*)
Have students practice counting through 12 and higher.

Numbers 0 Through 12, page RM16 (*Use with Lesson 6-1.*)
Have students practice ordering and recognizing numbers.

Money, page RM17 (*Use with Lesson 6-8.*)
As students find the total amount of money in each group, listen for how they are counting.

CHAPTER SEVEN

"Ronnie Raccoon's Petunia Patch," page RM18
(*Use to open the chapter.*)
After you read the story, have students draw the correct number of petunias to make 20.

Calendar, page RM19 (*Use with Lesson 7-6.*)
Have students practice writing the numbers 1 through 31 in order on a calendar.

CHAPTER EIGHT

"Big Mouth Bass," page RM20 (*Use to open the chapter.*)
After you read the story, have students illustrate the action of the story by matching by sizes.

One-Half, page RM21 (*Use with Lessons 8-1 and 8-2.*)
Discuss matching parts and the concept that two halves make one whole.

One-Fourth, page RM22 (*Use with Lesson 8-3.*)
Discuss the concept that four fourths make one whole.

One-Third, page RM23 (*Use with Lesson 8-4.*)
Discuss the concept that three thirds make one whole.

Fractions, page RM24 (*Use with Lessons 8-2 through 8-4.*)
Show the relationship of one whole strip to two halves, three thirds, and four fourths.

CHAPTER NINE

"Long Enough for School," page RM25 (*Use to open the chapter.*)
After you read the story, have students measure lengths with knots, a nonstandard unit of measurement.

Adding 1 More, page RM26 (*Use with Lessons 9-2 and 9-3.*)
Have students practice adding one more item.

CHAPTER TEN

"Silly Squirrel's Cake," page RM27 (*Use to open the chapter.*)
After you read the story, have students count sets of pennies and relate this action to the story.

Subtracting 1 or 2 More, page RM28
(*Use with Lessons 10-2 through 10-3.*)
Have students practice subtracting one or two more items.

Name _____

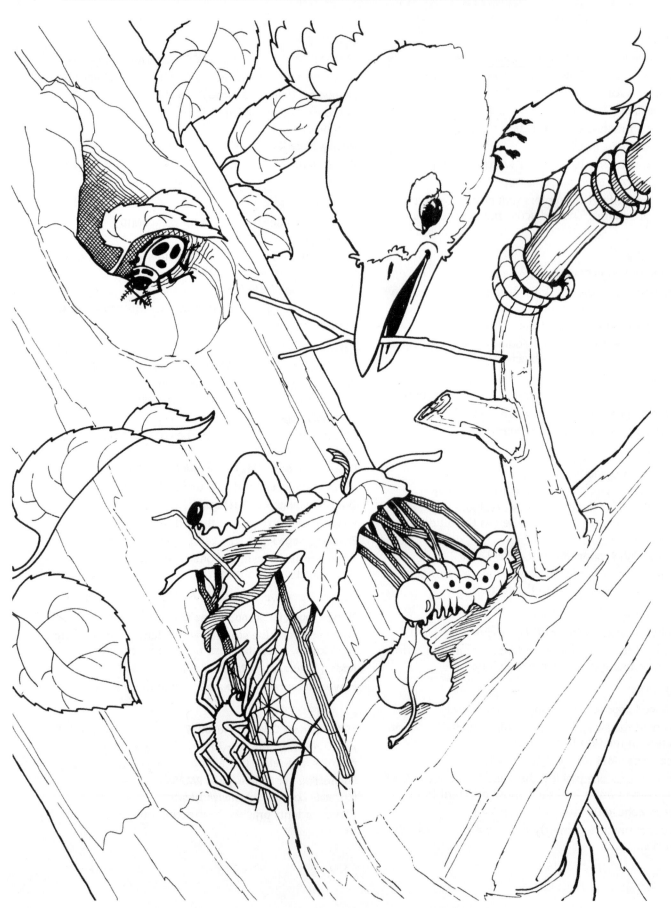

Name _____

Mark an X on each branch and circle each set of animals.
Describe the picture using position words.

Name _____

Color the picture with the appropriate colors.

Name _____

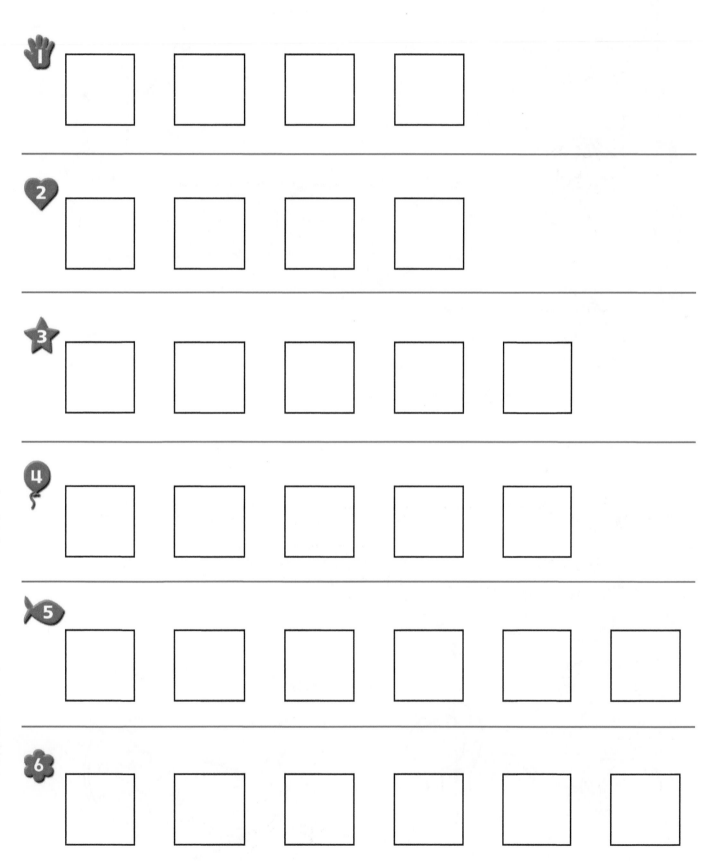

Name _____

Describe the picture using position words.

Name _____

Draw a line from the service person to the correct vehicle.

Practice writing the numbers.

Name _____

Find and color the hidden numbers I through 5.

Name _____

Cut out the five animals and paste them in position
around the table using clues from the story.

Name _____

Count by writing numbers on the members of each set of pigs, butterflies, and fence posts.

Name _____

**Count the number of bees in each picture.
Then describe what is happening in each picture.**

Name _____

Draw a line from flea 1 to flea 12 in order.

Name _____

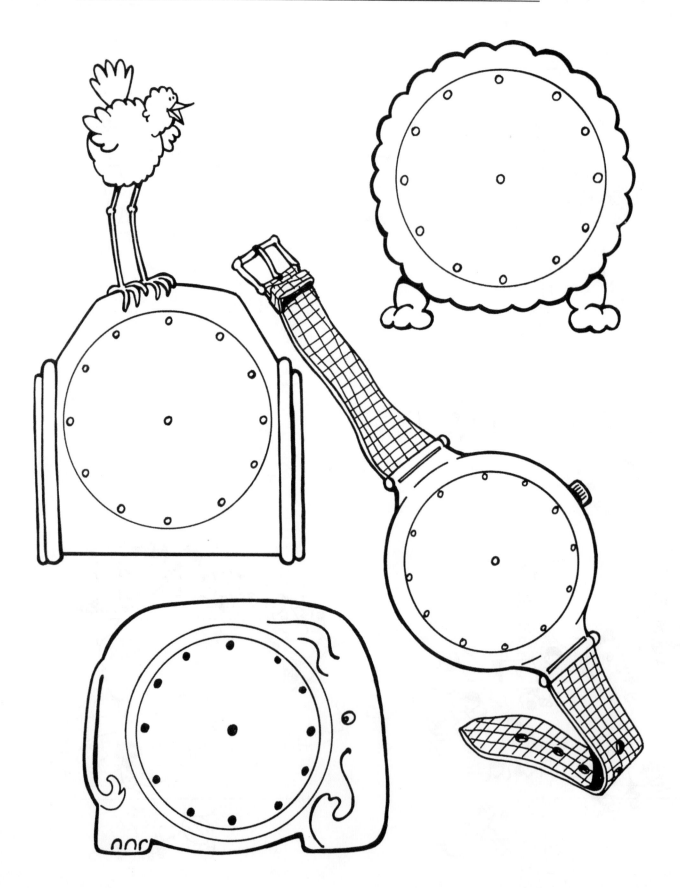

Write the numbers I to I2 on each clock face.

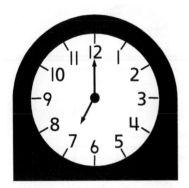

11:00

7:00

2:00

4:00

Draw a line from each clock face to the matching digital clock.

Name _____

**Count various objects in the picture
such as the nuts or the hearts on the napkin.**

Name _____

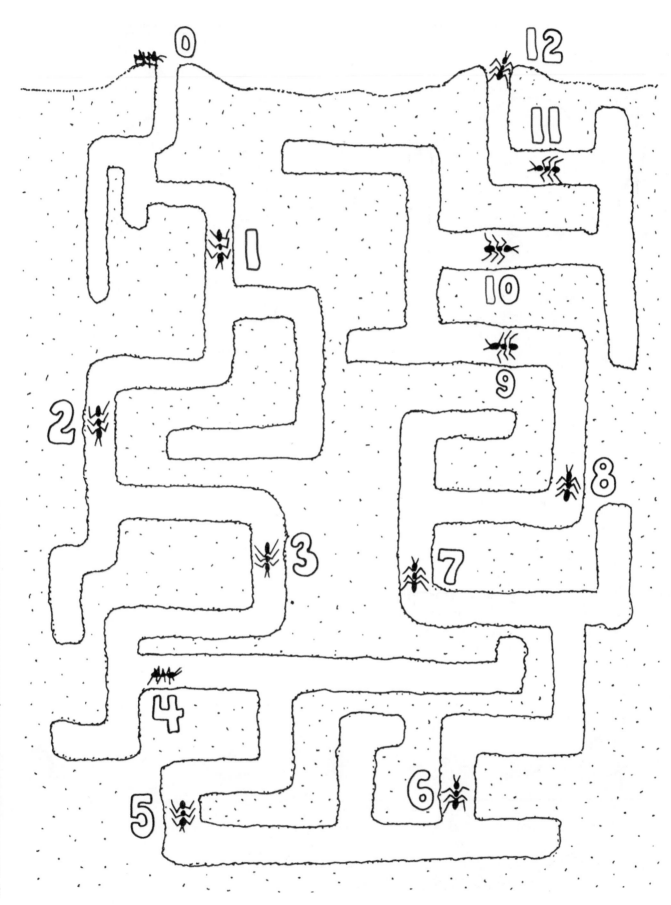

Draw a line through the maze from 0 to 12 in order.

Chapter 6 • Numbers 0 Through 12 **RM16**

10,   15,   16,   17,   18,   19¢

       ¢

      ¢

       ¢

       ¢

Count the money. Write the total amount.

Name _____

**Count the petunias around the stump. Draw and
color enough petunias to make 20 in all.**

Name _____

SUNDAY	MONDAY	TUESDAY	WEDNESDAY	THURSDAY	FRIDAY	SATURDAY

Write the numbers for the present month on the calendar, starting with 1.

Name _____

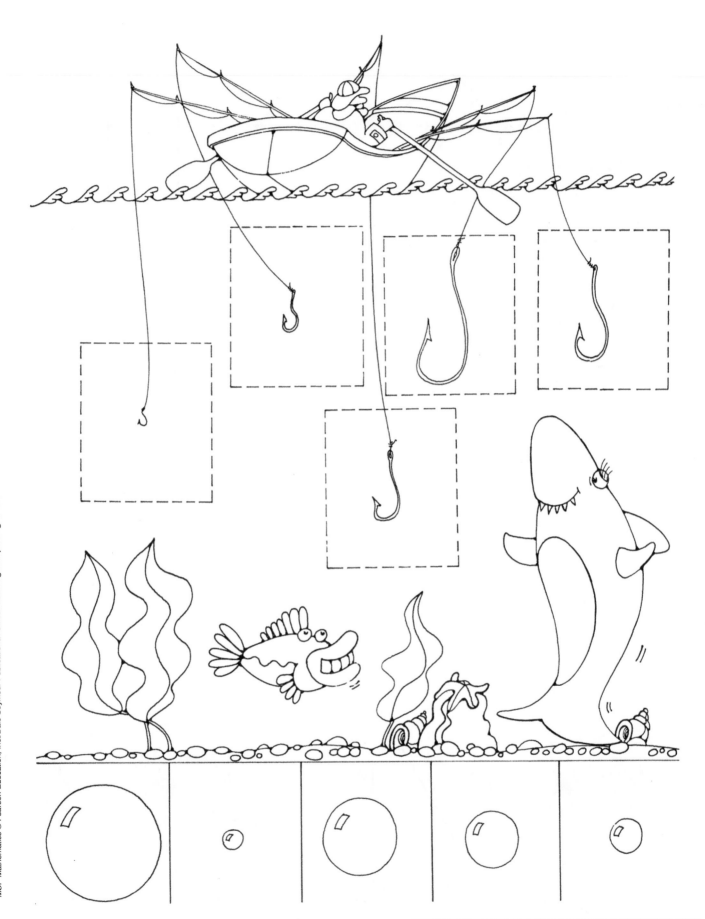

**Cut out the five bubbles and paste one on
each of the five hooks, matching by size.**

Chapter 8 • "Big Mouth Bass" **RM20**

Name _____

Color and cut out each half. Match two halves together to make a shape with two matching parts.

Name _____

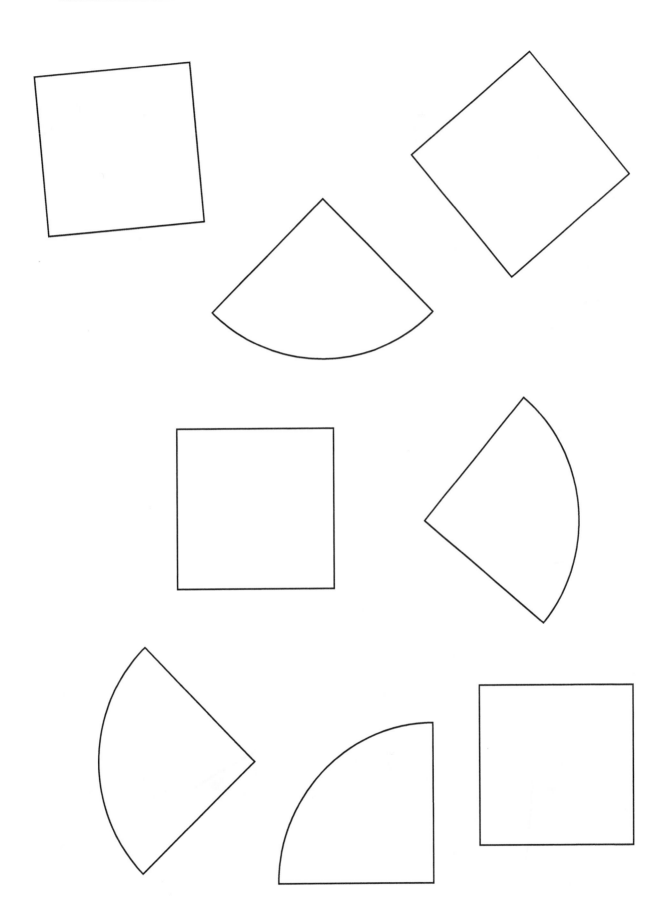

Color and cut out each fourth. Match four fourths together to make a shape with four matching parts.

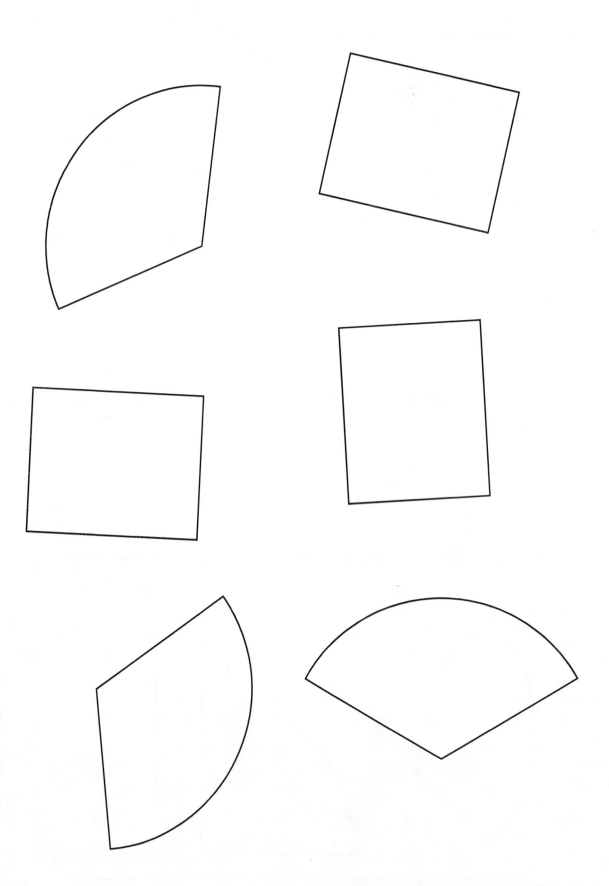

Color and cut out each third. Match three thirds together to make a shape with three matching parts.

Color each row of rectangles a different color. Then cut all rectangles apart.
Now reassemble them to make shapes with matching parts.

Name _____

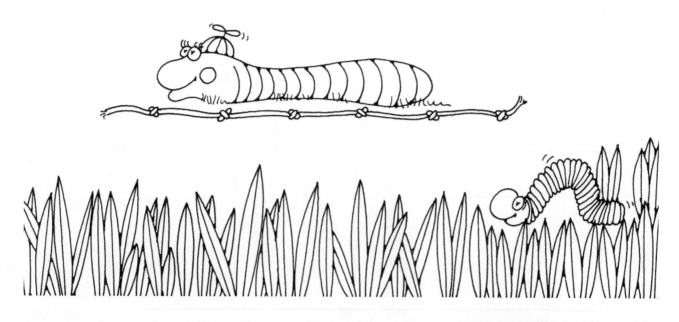

**Count the number of lengths between
knots to find the length of each worm.**

 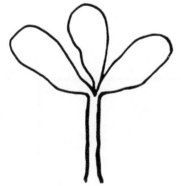

Draw the same number of petals plus one on the stem to the right.

Name _____

Cut out the sets of pennies and paste one set near each animal.

Name _____

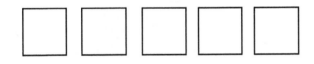

Draw a set of one or two fewer shapes than the number of shapes found in the set on the left.

Index

Addition
equal sign, 165–168, 171, 173, 177, 179–180
facts, 161–180, 182, 199–200, 204
horizontal form, 165–168, 171–173, 177, 179–180
money, 173–176
plus sign, 165–180, 182, 197–200, 204
sums through 6¢, 173–176
sums through 9, 161–180, 182
understanding, 161–168
vertical form, 169–170, 174–176, 178, 198–200

Algebra readiness, 19, 21, 25, 27, 81, 125, 137, 139, 163, 165, 167, 173, 175, 177, 185, 187, 197

Calendar, 133–134

Clocks
analog, 89–92
digital, 91–92

Colors, 5–6, 24, 89, 140, 182, 204

Comparison
as many as, 25–26
bigger/biggest, 7
colder, 160
fewer, 28
greater than, 125
hotter, 160
larger/largest, 8
less than, 126
longer/longest, 10
more, 27, 109–110
shorter/shortest, 9–10
smaller/smallest, 7–8
taller/tallest, 9

Counting
money through 20¢, 117–118
numbers from 1 to 100, 139
numbers by 2s, 5s, and 10s, 140

Fractions
equal parts, 143–150
one-fourth, 147–148
one-half, 145–146
one-third, 149–150

Geometry
flat surfaces, 16
matching parts, symmetry, 17–18
plane shapes, 11–12
sorting and classifying, 13–14
solid figures, 15–16

Graphs
bar, 122, 142
picture, 109–110

Measurement
capacity, 153–154
length, 7–10, 155–156
comparing, 7–10
nonstandard units of, 155–156
weight, 157–158

Money
addition, 173–176
counting, 93–100, 111–112, 117–118, 173–174, 195–196
dimes/nickels/pennies, 93–100
making change, 102
purchasing objects with, 96, 99–100, 112, 118, 175–176
subtracting, 195–196
writing, 93–95, 97–98

Number line, 82, 137

Numbers
before, after, or between, 81–82
on hundred board, 139–140
0 through 31, 29–42, 45–64, 66–82, 84, 103–120, 123–126, 131–140
in order, 44, 79–80, 119–120
writing, 35–38, 41–42, 51–52, 57–58, 62, 64, 68, 73–74, 77–78, 105–108, 113–116, 119

One-to-one correspondence, 25–28

Ordinal numbers, 127–130

Patterns
looking for, 19–20, 139–140
making, 21–22

Place value, 135–136

Position vocabulary, 1–4, 24, 85–86

Problem Solving
act it out, 99–100, 151–152
choose an operation, 197–198
draw a picture, 179–180
look for a pattern, 19–20, 139–140
make and use a picture graph, 109–110
try, check, and revise, 59–60

Sequencing, 81–82

Subtraction
facts, 183–202, 204
horizontal form, 187–188, 193–195, 197, 201
minus sign, 187–202, 204
money, 195–196
understanding, 183–190
vertical form, 189–192, 196, 198–200, 202
from 9 and less, 183–202, 204

Tally marks, 84

Temperature, 160

Time
clock faces, 89–92
comparing lengths of, 87–88
digital clocks, 91–92
hour hand, 89
minute hand, 89
month, 133–134
order of events, 85–86
week, 133–134
sequences, 85–86